D0078040

MICROBIOLOGY
PERSPECTIVES

MICROBIOLOGY PERSPECTIVES

A Photographic Survey
of the Microbial World

Second Edition

GEORGE WISTREICH
East Los Angeles College

PEARSON

Prentice
Hall

Upper Saddle River, NJ 07458

Library of Congress Cataloging-in-Publication Data

Wistreich, George A.
 Microbiology perspectives : a photographic survey of the microbial world /
George Wistreich. --2nd ed.
    p. ; cm.
  Includes index.
  ISBN-13: 978-0-13-239688-2
  ISBN-10: 0-13-239688-2
  1. Microbiology--Atlases. 1. Title
   [DNLM: 1. Microbiology--Atlases. 2. Microbiological Techniques--Atlases.
QW 17 W817m 2007]
  QR54. W57 2007
  579.022'2--dc 22

                        2006029121

Executive Editor: Gary Carlson	Creative Director: Juan R. López
Project Manager: Crissy Dudonis	Art Director: Maureen Eide
Editor-in-Chief, Science: Dan Kaveney	Cover Design: Daniel Conte
Executive Managing Editor: Kathleen Schiaparelli	Senior Managing Editor, Art Production and Management: Patricia Burns
Assistant Managing Editor: Beth Sweeten	Manager, Production Technologies: Mathew Haas
Production Editor: Pine Tree Composition	Managing Editor, Art Management: Abigail Bass
Manufacturing Manager: Alexis Heydt-Long	Art Production Editor: Eric A. Day
Manufacturing Buyer: Alan Fischer	Illustrations: Laserwords
Director of Creative Services: Paul Belfanti	Composition: Laserwords

Cover Images: **1st row**: GEORGE A. WISTREICH - The paraffin-plug technique. Left tube, uninoculated with a paraffin plug; center tube, turbidity (cloudiness) indicating growth; right tube, no growth under anaerobic conditions

   2nd row: *left*: ROBERT GALLO (Light microscopy of cytopathic effects. Eosinophilic nuclear inclusion body formation (arrow) (Purple); *right*: DR. TSU-YI CHUANG - Sporothrix schenckii. Microscopic view grown at 25 C stained with lacto-phenol cotton blue. Note the flower-like arrangement of the oval conidia. Original magnification, 400X

   3rd row: *left*: COURTESY OF DR. FRANZ-RAINER MATUSCHKA/FREE UNIVERSITY OF BERLIN OF BERLIN - A female deer tick lxodes dammini that transmits B. burgdorferi, the cause of Lyme disease; *right*: DR. HANNELE F. SOMER - P. intermedia on two different agar media. Colonies on a laked blood medium

   4th row: *left*: Tube a, uninoculated, Tube b, positive for motility (general turbidity in the medium). Tube c, positive for hydrogen sulfide production (blackening of the medium), positive for indole (red layer), and positive for motility (cloudy). Tube d, positive for indole, and motility; *right*: GEORGE A. WISTREICH - The results of a spore stain using carbolfuchsin as the primary stain and methylene bule as the secondary or coutestain. Spores stain red

© 2007, 1999 by Pearson Education, Inc.
Pearson Prentice Hall
Pearson Education, Inc.
Upper Saddle River, New Jersey 07458

All rights reserved. No part of this book may be
reproduced, in any form or by any means,
without permission in writing from the publisher.

Pearson Prentice Hall ™ is a trademark of Pearson Education, Inc.

Printed in the United States of America

10 9 8 7 6 5 4 3 2 1

ISBN 0-13-239688-2

Pearson Education Ltd., *London*
Pearson Education Australia Pty. Limited, *Sydney*
Pearson Education Singapore, Pte. Ltd
Pearson Education North Asia Ltd., *Hong Kong*
Pearson Education Canada, Ltd., *Toronto*
Pearson Education de Mexico, S. A. de C. V.
Pearson Education—Japan, *Tokyo*
Pearson Education Malaysia, Pte. Ltd.

To Renée, my wife, and to my sons Eddie and Phillip, whose love and support have been such an important part of my personal and professional life, and to my colleagues, former instructors, and students, who have provided the motivation and inspiration to devolop my microbiology perspectives.

Contents

SECTION 4   Bacteriology: A Survey of the Bacteria and Archeae Domains   48

SECTION 5   Mycology   110

SECTION **8**   Virology   **157**

Preface

Microbiology Perspectives: A Survey of the Microbial World provides a visual guide to many of the specific properties, activities, and procedures associated with microorganisms, and helminths (worms). It also discusses selected features of a number of infectious diseases. In short, this book is intended to present microorganisms to anyone who is interested and curious about the microbial world. Individuals, whether familiar or not familiar with this microscopic world of life, can gain some perspective as to the many types of microorganisms and their activities that exist in the world around them. For those not directly involved in industrial, laboratory, or medically related activities, this orientation and exposure provides a broad perspective of how microorganisms and their various activities influence everyday situations.

Microbiology Perspectives is not intended to replace currently available texts and manuals; rather, it is designed to show the reader several timely and relevant properties of microorganisms and the results of their actions, many of which either are not included or are not adequately presented in texts or manuals. Toward this end, a limited amount of explanatory material accompanies the many color photographs. This information also is included so that individual sections may serve as self-contained study guides and as aids in laboratory situations.

Perhaps more than most science specialties, microbiology depends heavily on color and visual representation of microbial reactions and activities. In the selection of illustrations for this publication, care was taken to include microorganisms and activities commonly encountered in the laboratory, as well as microorganism activities and situations that are not included in texts but are nevertheless significant to microbial or disease agent identification.

The color plates are used to illustrate the microbial colonial and microscopic features, selected biochemical characteristics of the most frequently encountered microorganisms, various disease states associated with microorganisms, and clinical and diagnostic features of selected infectious diseases. The illustrations demonstrate most effectively the way in which microorganisms, helminths, and arthropods appear in clinical laboratories and the various ways in which preliminary or definitive identifications are made. It should be noted, however, that no standard magnification or enlargement has been applied throughout. The sizes of organisms are not necessarily proportionate from one illustration to another. Magnifications and enlargements have been selected so as to provide the greatest clarity possible.

The second edition of *Microbiology Perspectives* expands the view of the microbial world with effective, updated explanations of basic laboratory procedures, increased coverage of laboratory procedures and associated results, the domains of microbal life, immunology, and infectious diseases. In addition, particular attention is given to the features of viruses and prions. The use of selected agents of bioterrorism also is discussed. Furthermore, the Appendix figures showing the specific body locations used by pathogens to cause infections and disease have been updated to incorporate such microorganisms as the SARS agent, Ebola and Marburg viruses, and the human metapneumovirus. The new illustrations incorporated throughout the second edition were selected to emphasize the variety of microorganisms and their activities, and thereby increase the microbiology perspective.

It is hoped that *Microbiology Perspectives* will be a useful learning tool and functional reference, not only for undergraduate students taking an introductory biological science or microbiology course but also for other students taking more advanced programs in the biological sciences, medical technology, medicine, biotechnology, dentistry, and veterinary medicine.

Acknowledgments

As with any book, numerous people have been involved and have helped in its production, and I would like to take this opportunity to acknowledge them. I am particularly grateful to my fellow microbiologists and investigators from around the world who generously contributed their marvelous photographs of clinical states, microorganisms, and related subjects; to Jann Flores from Hardy Diagnostics, Ms. Linda Puyear from Remel, Inc., Ms. Jessica Paciolla from Becton, Dickinson, and Company, and Ms. Brenda Gurza from the Centers for Disease Control and Prevention for providing selected illustrations; to Patty Donovan and John Shannon for their expertise in guiding the various aspects of publication; and to Crissy Dudonis and Gary Carlson for their support and interest in the development of true microbiological perspectives.

Pronunciation Guide

In *Microbiology Perspectives* you will find phonetic pronunciations for many terms and specific microorganisms. Taking time to sound out new terms and to say them aloud once or twice will help you master one of the tasks in microbiology-related activities—learning its specialized vocabulary. The following key explains the system used for the pronunciations.

1. The strongest accented syllable appears in capital letters: e.g., microbe (MĪ-krōb) and microscope (MĪ-krō-skōp). A syllable that has a secondary accent is followed by a single prime ('): microaerophilic (mi-krō-Â-er-o-fil-Ī'k) and micrococcus (mi-krō-KOK-us).

2. Vowels pronounced with long sounds are indicated by a line above the vowel and are pronounced as in the following words.
   ā as in māke
   ē as in bē
   ī as in ivy

   ō as in pōle
   ū as in ūnit

3. Vowels not marked for long sounds are pronounced with the short sound, as in the following words.
   a as in above
   e as in bet
   i as in sip
   o as in not
   u as in bud

4. Other phonetic symbols are used to indicate the following sounds.
   soo as in sue
   kyoo as cute
   oy as in oil

A list of most of the microorganisms helminths and arthropods described in *Microbiology Perspectives* and their phonetic pronunciations can be found on the front and back inside covers.

About the Author

George Wistreich is Professor of Microbiology and former Chair of the Department of Life Sciences, East Los angeles College, Monterey Park, California. He has taught for over 30 years in the various areas of microbiology, general biology, and electron microscopy. He received his B.A. in Bacteriology and his M.S. in Infectious Diseases from the University of California, Los Angeles. After receiving his Ph.D., he completed a National Institutes of Health postdoctoral fellowship in the Department of Medical Microbiology and Immunology at the School of Medicine, University of Southern California. He has also completed several programs, seminars, workshops, and courses in molecular biology, genetic engineering, development in infectious diseases, and curriculum design. Dr. Wistreich belongs to a number of scientific and professional organizations; in addition, he has been elected as Fellow to the American Academy of Microbiology, the Linnean Society of London, the American Institute of Chemists, and the Royal Society of Health. He has served on the Board of Education, American Society of Microbiology for 12 years; nine of these years were in the capacity of chair of the Precollege Education Committee. Dr. Wistreich also develops continuing education units for healthcare professionals.

Among the awards he has received during his years of teaching are the Chancellor's Award for Student Success and the Outstanding Educator Award from Chicanos for Creative Medicine. Dr. Wistreich is the author or co-author of over 20 scientific articles and over 60 laboratory manuals, workbooks, and textbooks.

Introduction

It is characteristic of Science and Progress that they continually open new fields to our vision.

—Louis Pasteur

Microbiology as a science includes the study of bacteria (**bacteriology**), fungi (**mycology**), viruses (**virology**), protozoa (**protozoology**), and certain algae (**algology** or **phycology**). The study of worms, or helminths (**helminthology**), is frequently also incorporated into microbiology courses because of the ability of some of these forms of life to cause disease. In addition, the study of the immune system (**immunology**), which includes approaches to disease prevention and diagnosis, is considered a subdivision of microbiology.

Microbiology had its beginnings with the discovery of microscopic forms of life by Anton van Leeuwenhoek in 1685. The major developments in microbiology did not begin, however, until, in the 1870s, Louis Pasteur and others (Figure 1) conclusively demonstrated that microorganisms are living entities (organisms) capable of self-multiplication and are not spontaneously generated from nonliving organic matter. This section provides an introduction to the world of microorganisms.

A Word about Microscopes

Observing cells, the basic units of life, in action is made difficult by the fact that they are, in general, very small and transparent to visible light and thus are invisible to the unaided eye. The approaches used to overcome these limitations include **microscopy** and the application of dyes that stain different cellular parts. Staining techniques used in microbiology are discussed in a later section.

Since the early 1950s, the technical advances in microscopy and associated techniques have steadily provided scientists with new approaches to probe more deeply into the structure, organization, and functions of microorganisms and other forms of life (Figures 2 and 3).

The objectives of microscopy include (1) magnifying the observed image, (2) maximizing the resolving power (the ability to see detail), and (3) distinguishing the various elements in the material being viewed. The third objective involves providing **contrast** and is best achieved with the use of stains.

Historically, a significant amount of information about cells has been provided by light microscopy. Light microscopes can magnify about 2,000 times and show details as fine as 0.25 μm (micrometer). A micrometer is one-millionth of a meter (about 1/25,000th of an inch). With electron microscopy (a more specialized system for examining materials), a concentrated beam of electrons (negatively charged particles) can be used to examine cellular features. Two general types of electron microscopes are currently in use. The **scanning electron microscope (SEM)** adds depth to a microscopic view (Figure 2). The **transmission electron microscope (TEM)** provides flat, two-dimensional views (Figure 3A). A combined version of these two instruments, the **scanning transmission electron microscope (STEM)**, is also used in microbiology and related scientific areas.

A number of specific TEM techniques are of value in the study of microorganisms, their activities, and other types of biological specimens. These include ultrathin sectioning or slicing (Figure 3A), shadow casting (Figure 3B), negative staining (Figure 3C), and freeze-etching (Figure 3D). Table 1 lists these techniques together with brief explanations and uses of the procedures involved. A photograph of a microscopic image is called a **photomicrograph**. Most of the following sections contain numerous examples of a variety of micrographs.

Newer microscopes have been developed and also are used to make valuable contributions to understanding microorganisms and various aspects of their activities. Among these newer instruments is the **scanning probe microscope (SPM)**, which is capable of generating three-dimensional surface views of cells and molecular structures that reveal detail (resolving

Figure 1
Photograph, taken in Paris in 1894, showing several outstanding French physicians and microbiologists of the time. Front row, left to right: A. Calmette, L. Martine, E. Roux, L. Pasteur, E. Nocard, H. Pottevin, and F. Mesnill, Back row, left to right: E. Viala, Rebound, L. Merieux, A. Fernbach, R. Chaillou, A. Borrel, H. Marnier, A. Marie, A. Veillon, and E. Fernback. (Courtesy of the Pasteur Institute.)

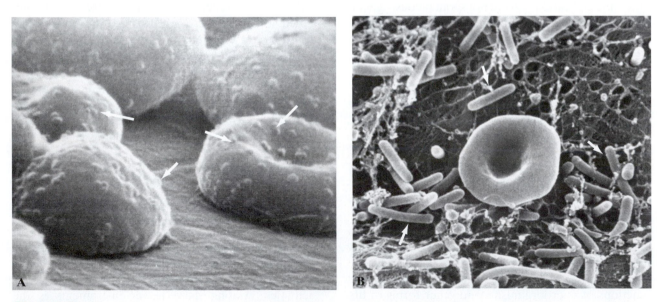

Figure 2
Size relationships and surface features of microorganisms as shown by scanning electron microscopy. (A) Influenza viruses (arrows) attached to the surfaces of red blood cells. (B) Bacteria (arrows) surrounding a single much larger human red blood cell. Refer to Figure 6 for actual sizes.

power) at a nanometer level. A nanometer is one-billionth of a millimeter. Two examples of scanning probe microscopes are the **scanning tunneling microscope (STM)** and the **atomic force microscope (AFM)**. Selected features of the different types of microscopes are summarized in Table 2.

A Brief View of Classification

There is a great unity as well diversity in the biologic world. Moreover, a number of similarities also exist. As more and more forms of life were discovered, the need for a system with which to catalog them became crucial for various purposes, such as communication, showing relationships, and research studies. Such a system provides the means with which to identify previously unknown forms of life (organisms) and then group them into taxonomic categories or taxonomic ranks (**taxa**) with others having similar characteristics. In 1969, Robert H. Wittaker proposed the five-kingdom system, based on the cellular form of organization found in animals, plants, protozoa, algae, and bacteria. Thus were created the five biological categories called kingdoms: **Prokaryotae** or **Monera** (the bacteria), **Protista** (protozoa and certain algae), **Fungi**, **Plantae**, and **Animalia**.

Table 1 Specimen Preparation Techniques for Electron Microscopy Viewing

Preparation Technique	Brief Description	Uses (Include)
Thin sectioning (slicing)	Used for specimens too thick for direct viewing; before sectioning, specimens must be treated in some manner to preserve a specific structure or structures; procedures for biological materials include chemical fixation, dehydration, and embedding in plastic.	study of cells and tissues, structural arrangements, internal organization and development cycles
Shadow-casting	Depositing a thin film (atoms) of a heavy metal, such as platinum, uranium, or alloys of gold and pallidium, in a vacuum	increases contrast, provides bases for making measurements of cells and/or surfaces
Surface replicas	Depositing a thin layer of a low molecular weight material, such as carbon, on a specimen in a vacuum; followed by floating the coated specimen on a water surface and then transferring it to a strong acid or alkali solution capable of dissolving the biological material and leaving a surface impression or replica of it.	provides views of surface of details of microorganisms and/or their structures, such as spores
Electron Staining	Electron stains are solutions containing heavy metal elements, such as phosphotungstic acid; this technique has two forms: positive and negative staining; solutions are applied directly to specimens, and no specialized equipment is required.	often used to determine numbers of microorganisms, identify specific structures, and diagnose various diseases
Freeze-fracture	Technique does not require fixation, plastic embedding, sectioning; unfixed (non-preserved) specimens can be studied directly in the electron microscope; specimens are quick-frozen, in water, mechanically split or cleaved, and treated to expose their interior structures; exposed surfaces are shadow-cast to provide contrast, and a replica is made as described above	used to show the features of inner and external surfaces of membranes, and to reveal materials passing through membranes

Table 2 Comparisons of Microscopes

Microscope	Type of Illumination	Magnification	Types of Specimen; Preparation Techniques; Main Use(s)
Light (Bright-field)	Visible light	1,000–2,000	Living, dead; stained, unstained; observation of stained and unstained specimens, determination of microbe numbers, and/or special cell parts
Darkfield	Visible light	1,000–2,000	Living, dead, unstained; observe cells difficult to stain by other means, detection of cellular movements and molecular activity
Phase-contrast	Visible light	1,500+	Living, generally unstained specimens; observation of internal cell parts; rapid detection of microorganisms and/or their parts in various specimens
Fluorescent	Ultraviolet light	1,500+	Living; stained; Immunofluorescence (fluorescent-antibody technique)[a]
Confocal	Laser	1,500+	Living, dead, stained; obtain three-dimensional views of microbes and other cell types invarious specimens
Transmission electron microscope (TEM)	Electrons	100,000–1 million X	Dead; stained, coated with thin layers of heavy metals, and ultrathin sectioning (slicing); observe internal and submicroscopic features of various cell types, viruses, and in some cases molecules, as well as cellular processes
Scanning electron microscope (SEM)	Electrons	100,000–500,000 X	Dead, coated with thin layers of heavy metals; observe the surface features of prokaryotic and eukaryotic cells and viruses, as well as selected cellular processes
Scanning transmission microscope (STEM)	Electrons	100,000–500,000 + X	Similar to those for TEM and SEM; combination of uses listed TEM and SEM
Scanning tunnel microscope (STM)	Electrons from specimen	25–10,000,000	No special preparation procedures; provides means for quantitative analysis of surface features of various biological and nonbiological specimens, determination of atomic composition
Atomic force microscope (AFM)	Electrons	25–10,000,000	No special preparation; can be used with light, fluorescence, and other microscope techniques; allows three-dimensional imaging and measurement of biological and nonbiological structures and molecules from micrometer to atomic levels

[a]This diagnostic technique uses **antibodies** (naturally formed defense molecules) produced by humans and many other animals in response to foreign substances known as **antigens**. Such antibody molecules are coated (tagged) with fluorescent dye molecules and are used to detect microorganisms and/or their parts (antigens) and other foreign substances.

Eukaryotes and Prokaryotes

With the development of various types of microscopy and molecular biological techniques, more and more significant differences among members of the biologic world were uncovered. Among these differences was the existence of three different types of cellular organization, namely two types of prokaryotic cells and one type of eukaryotic cell.

The term *karyotic* refers to the nucleus, an envelope-enclosed, or membrane-bound, cellular structure containing genetic material; the prefix *eu* means "true" in Greek. Thus eukaryotic cells are typically recognized by the presence of an envelope-enclosed nucleus (Figure 4). In addition, eukaryotic cells are larger than prokaryotic cells and contain a variety of other **organelles** (specialized cellular structures

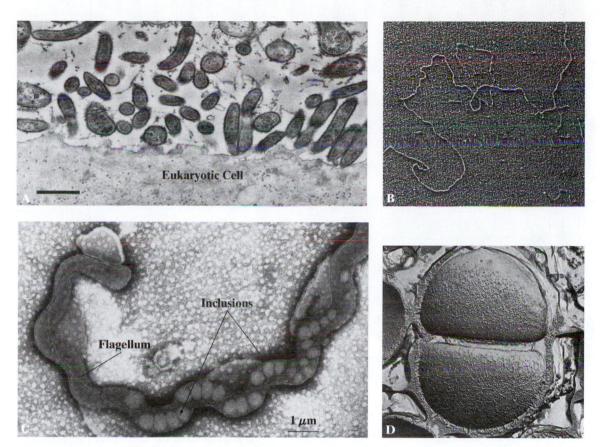

Figure 3

Examples of transmission electron microscope techniques. (A) An ultrathin stained, section of the layers of bacteria, forming a **biofilm**, on an eukaryotic cell surface. The bar marker was originally equal to 1 μm. (Courtesy of J. M. de Boer and R. H. F. Plantema, reproduced from the *Can. J Microbiology* **34**:757–766, 1988.) (B) A shadow-cast preparation showing a portion of a viral chromosome. (Courtesy of H. Bujard.) (C) A negatively stained preparation of a spiral bacterium showing a number of inclusions within the cell and a faintly appearing flagellum. The bar marker was originally equal to 1 μm. (Courtesy K. Hovind-Hougen.) (D) A freeze-fracture view of the external surfaces of the plasma membranes of two bacteria originally magnified 56,000 X.

enclosed by membranes). Organelles perform specific functions, such as the formation of organic molecules and the **production of energy**. Table 3 summarizes the typical components of eukaryotis and prokaryotic cells.

Prokaryotic cells do not have a well-defined nucleus, and lack other membrane-enclosed or -bound organelles (Figure 5). Their genetic material is found in a general cytoplasmic area known as the **nucleoid**.

Prokaryotes vary in size and can be as small as 0.2 micrometers (μm) in diameter and 0.5 to 3 μm in length. While prokaryotic cells are generally much smaller than eukaryotic cells measuring 0.1 μm in diameter and 0.5 to 3 μm in length (Figure 6), there are exceptions. One such exception is the giant bacterium *Epulopiscium fishelsoni* (EP-ū- l̄o-pis-ē-um, FISH-el-son-ē), which can be up to 80 μm in diameter and 500 to 600 μm in length (Figure 7). Interestingly, it can be seen without the aid of a microscope. This peculiar microorganism inhabits the intestinal tracts of surgeonfish species from both the Great Barrier Reef in Australia and the Red Sea in the Middle East.

▲ The Three Domains

Along with the discovery of the existence of three different types of cellular organization, a comparison of the nucleic acid sequences in the ribosomes from these cells showed that there are three distinctly different cell groups: the **eukaryotes** and two different prokaryotes, the *eubacteria* (true bacteria) and the *archaea*. This finding formed the basis of the 1978 proposal by Carl R. Woese to raise the three cell types to a taxonomic level above kingdom to one known as the **domain**. According to this scheme, animals, plants, fungi, and protists belong to the **Eurkarya** domain. While disease-causing, non-disease-causing bacteria, and photoautotrophic bacteria such as the cyanobacteria form the **Bacteria** domain. The **Archaea** domain consists purely of bacteria that lack a typical cell wall, live in extreme environments, and perform unusual metabolic activities. Figure 8 shows this arrangement and the evolutionary relationship of the three domains.

Table 3 Summary of Typical Eukaryotic and Prokaryotic Cell Components and a Comparison with Viruses

Cell Part and/or Related Structures	Functions	Eukaryotic				Prokaryotic, Bacteria, Cyanobacteria, etc.	Viruses
		Animal	Plant	Protist	Fungus		
Cell wall	Protection, structural support	None	X	X[a]	X	X	None
Plasma membrane	Control of substances moving into and out of cell	X	X	X	X	X	None
Nucleus	Physical separation and organization of DNA	X	X	X	X	None[b]	None[c]
DNA	Encoding of hereditary information	X	X	X	X	X	Either DNA or RNA, never both
RNA	Transcription translation of DNA messages into specific proteins	X	X	X	X	X	Either DNA or RNA, never both
Nucleolus	Assembly of ribosomal subunits	X	X	X	X	None	None
Ribosome	Protein synthesis	X	X	X	X	X	None
Endoplasmic reticulum (ER)	Initial modification of many newly forming proteins	X	X	X	X	None	None
Golgi body	Final modification of proteins and lipids, sorting and packaging them for use inside cell or for export; lipid synthesis	X	X	X	X	None	None
Lysosome	Intracellular digestion	X	X	X	X	None	None
Mitochondrion	(ATP) formation	X	X	X	X	None	None
Photosynthetic pigments	Light-energy conversion	None	X	X	None	X	None
Chloroplast	Photosynthesis, some starch storage	None	X	X	None	None[d]	None
Central vacuole	Increasing cell surface area, storage	None	X	None	X	None	None
Cytoskeleton	Cell shape, internal organization, basis of cell motion	X	X	X	X	None	None
9+2 Flagellum	Movement	X	X	X	X	None[e]	None
9+2 Cilium	Movement	X	X	X	X	None	None
Pilus	Attachment or transfer of genetic material	None	None	None	None	X	None
Spore	Reproduction	None	None	X[f]	X	X[g]	None

[a]Found only with algae.

[b]Not a typical nucleus; part is known as a nucleoid.

[c]Viruses contain a nucleic acid core.

[d]Photosynthesizing prokaryotes have structures called chromatophores.

[e]Flagella not showing the 9+2 arrangement are found among prokaryotes.

[f]Found among some algae.

[g]Found among certain genera only; not used for reproduction.

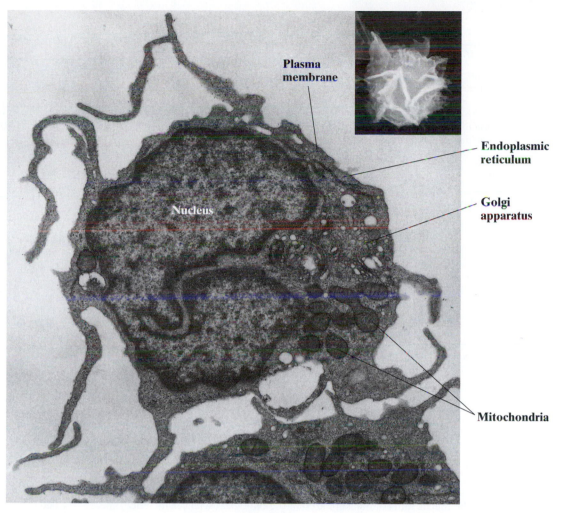

Figure 4

A transmission electron micrograph of an ultrathin section (slice) of a human white blood cell. A number of typical organelles (cell parts) can be seen. These include: an endoplasmic reticulum, Golgi apparatus, mitochondria, a lobed (2-part) nucleus, and a plasma membrane. The insert at the upper right shows a scanning electron micrograph of a similar cell. (Both micrographs are courtesy of Dr. Hilary Christensen.)

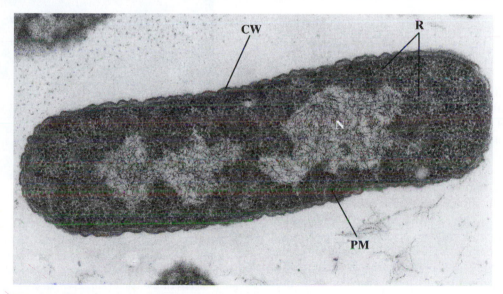

Figure 5

A transmission electron micrograph of a prokaryotic cell. A number of typical parts are shown. CW, cell wall; N, nucleoid; R, ribosomes; PM, plasma (cell) membrane.

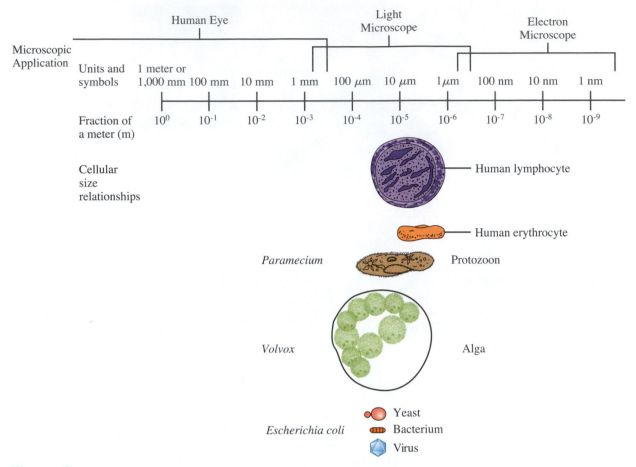

Figure 6

The size relationships of selected microorganisms and types of cells and the respective magnification ranges of specific microscopes. The metric units of measure used in the biological sciences include: 1 meter = 1,000 millimeters (mm); 1 mm = 1,000 micrometers (μm); 1 μm = 1,000 nanometers (nm). A prion is not shown.

The respective members of the three domains differ in other properties in addition to their cellular organization. Other differences include sensitivity to antibiotics, transfer ribonucleic acid molecules, and membrane lipid structure.

All organisms need energy to perform their various activities and life processes, such as reproduction, growth, metabolism, responding to and adjusting to environmental factors, movement, and storage and /or transport of molecules. Some forms of life capture photosynthetic energy from sunlight and store it in molecules such as sugars and oils. These forms of life, which include plants, some prokaryotes, and algae (protists), are called **autotrophs**, meaning "self-feeders." Organisms incapable of carrying out photosynthesis must obtain their energy prepackaged in the molecules manufactured by other forms of life. These organisms are referred to as **heterotrophs**, meaning "other feeders."

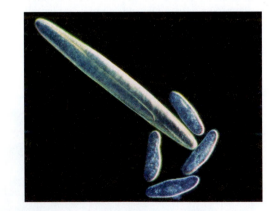

Figure 7

A bright-field micrograph showing *Epulopiscium* (ep-ū-LO-pē-sē-um), one of the largest bacterial species known. Four protozoa, *Paramecium* (par-a-MES-ē-um) species are also present. *Epulopiscium* measures of 0.5 mm in length and can be seen with the naked eye. One million cells of the bacterium *Escherichia coli* could fit into one *Epulopiscium*. (Courtesy of Esther R. Angert and Norman R. Pace, Indiana University.)

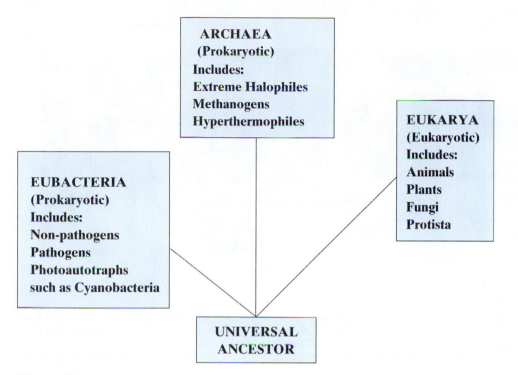

Figure 8
The three domains of life originally proposed by Carl Woese. This phylogenetic tree is based on comparisons of ribosome RNA sequences.

A large number of prokaryotic microorganisms, and protozoa, all fungi, and animals are heterotrophs. **Pathogens** (disease-causing organisms) also are heterotrophic.

Viruses

Viruses are given a position independent of prokaryotic and eukaryotic organisms because of several significant differences. Clearly, viruses are not cells (Figure 9). They have no membranes of their own and none of the other structures found among prokaryotes or eukaryotes. In addition, viral nucleic acid is contained within a nucleic acid core. Viruses cannot be seen with an ordinary light microscope. Special instruments, electron microscopes, must be used to observe these submicroscopic forms. In addition, viruses cannot move or grow, make their own source of energy, or replicate (reproduce) outside of a living cell. Viruses lack the enzymatic machinery to totally replicate themselves and to perform essential metabolic processes. These submicroscopic forms are totally dependent on living cells for their survival. They are obligate intercellular parasites. Section 8, "Virology," discusses viruses more fully.

Prions

In 1982, the American neurobiologist and Nobel Prize recipient Dr. Stanley Pruisner proposed that an infectious protein was the cause of a neurological disease in sheep called scrapie. Because of its chemical composition and infectious nature, Prusiner coined the term prion (**pr**, *proteinaceous*, **i**, *infectious*, **on**, particle) to distinguish the new form of disease agent from all others. Prions are the cause of several diseases known as the transmissible spongiform encephalopathies (**TSEs**). Prion infection results in numerous holes (sponge-like lesions) in various parts of the central nervous system, hence the term *spongiform*. The TSEs are a group of rapidly progressive, invariably fatal, neurodegenerative diseases that affect both humans and lower animals. They include the conditions commonly known as mad cow disease and Creutzfeld-Jakob disease (CJD). Cows with TSE developed the condition as a result of being fed animal feed containing the scrapie-infected remains of dead sheep.

Prions occur in two different forms, namely the *normal cellular prion protein* (PrP^C) and the *disease-causing (scrapie) form* (PrP^{Sc}). The polypeptide chains of both forms are identical in composition but

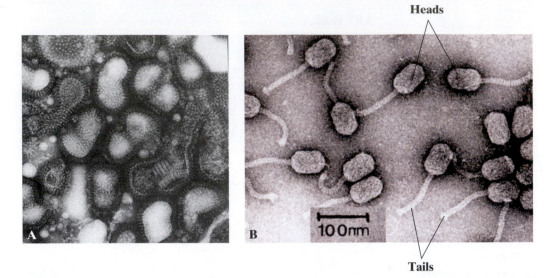

Figure 9

Transmission electron micrographs of two different viruses both of which clearly lack the structures found in eukaryotic and prokaryotic cells. (A) Influenza virus particles. The coiled nucleic acid component (nucleic acid core) of some individual viruses can be seen (195,000X). (B) Bacterial viruses. Note the presence of heads and tails.

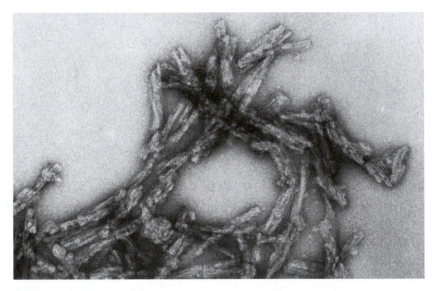

Figure 10

A transmission electron micrograph of a large number of scrapie prions. Note the complete absence of a cellular organization.

differ in their three-dimensional arrangements, referred to as *conformations*. Figure 10 shows a transmission electron micrograph of scrapie prions.

Normal cellular prion protein (**PrP^C**) can be found in the neurons and glia of the brain and spinal cord, as well as in several peripheral tissues and in leukocytes. The protein is localized on cell surfaces. Its normal physiologic function, however, currently is unknown. Additional features of prions are presented in Section 8.

One enlarges science in two ways; by adding new facts and by simplifying what already exists.

—Claude Bernard

For most infectious diseases, microbiological isolation and identification techniques are the fastest and most specific approaches to the determination of the causative agent. However, there are also a number of clinical conditions for which a reliable method of locating the site of infection would be extremely useful to diagnosis and determining the effectiveness of treatment. Such conditions include bacterial and fungal infections involving the entire body (systemic infections), situations in which no specific diagnosis is made, and situations in which fever of an unknown origin exists.

In the last 25 years, techniques for the diagnosis and treatment of a wide range of diseases and previously unsuspected disorders have advanced as a consequence of the improved capability and availability of medical imaging. **Radiography** ($\overline{RA}$-d$\overline{e}$-$\overline{o}$'-graf-$\overline{e}$), or the conventional x-ray (Figure 11), is well known for its diagnostic value and has been in use since the late 1940s. This imaging technique involves the passage of a single barrage of x-rays through the body to produce a two-dimensional image of the internal parts of the body.

Newer techniques, which include **computerized tomography (CT) scanning**; **magnetic resonance imaging**, or MRI; **ultrasonography**; and **scintigraphic techniques**, provide sharply focused objective information helpful in precisely locating sites of infections, abnormalities, and inflammation. These techniques also contribute to the understanding of body functions and disease processes. Brief explanations of the techniques are provided here because several of them are associated with the detection or diagnosis of infectious diseases.

Computerized Tomography (CT) Scanning

After the injection of material to provide contrast of body structures, an x-ray beam is moved in an arc around the body. A series of cross-sectional images (slices) of an individual's body is produced by a scanning device that is processed by a computer and displayed on a video monitor. Three-dimensional views of body parts are constructed by stacking a series of images taken at different levels through an organ, one on top of another. These CT scans are widely used for observing evidence of disease (such as lesions) or changes in an organ (Figure 12).

Magnetic Resonance Imaging (MRI)

Magnetic resonance imaging uses radio-frequency radiation in the presence of a carefully controlled magnetic field. It measures the response of positive subatomic particles (protons) to a pulse of radio waves while they are subjected to a strong magnetic field. The response produces high-quality cross-sectional images of the body (Figure 13). Magnetic resonance imaging is valuable in providing images of the brain, heart, large blood vessels, and soft body tissues, and in finding certain disease agents. The injection of contrast material is not required.

Ultrasonography

Ultrasonography (ul-tra-son-OG-ra-f$\overline{e}$) is a technique that uses high-frequency sound waves that bounce off tissues and are recorded by a scanning device as it is passed over the body. Signals from the scanner are transmitted to form an image called a **sonogram** (SO-no-gram) on a video monitor (Figure 14). The technique is used to detect cardiovascular disorders, abnormal masses in certain body organs, and multiple pregnancies and to find gallstones and related disorders.

Scintigraphic Techniques

Scintigraphy involves the injection of a chemical to which some radioactive material has been attached.

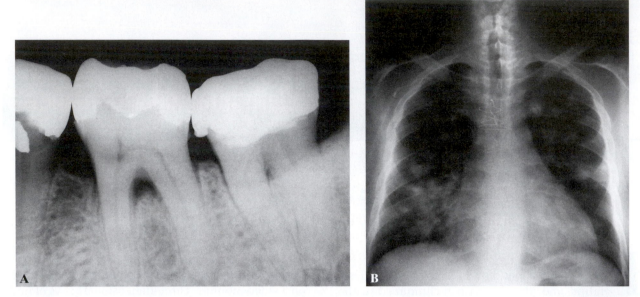

Figure 11

Two examples of x-rays or radiographs. (A) Results of a dental x-ray examination showing crowns (top) on molars and the beginning of gum disease (dense areas). (B) A chest x-ray showing extensive disease in the lungs. This imaging technique is used as a major screening approach for the detection of a variety of respiratory diseases and abnormalities.

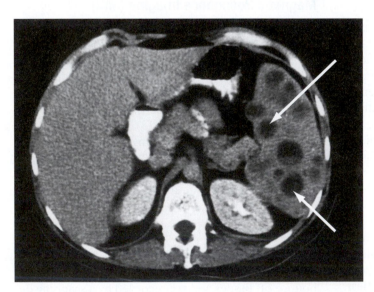

Figure 12

A CT scan showing the liver injury (dark circular areas) caused by a microbial infection (arrows).

As this radiopharmaceutical (rā-dē-ō-far-ma-SOO-ti-kal) circulates, radiation detectors determine its uptake or distribution within the body. The detector is a chemical substance that scintillates (sin-ti-LĀTES), or gives off energy in the form of a flash of light. The energy is picked up by a camera, and the information is processed by a computer to produce an image (Figure 15). Scintigraphic techniques have a major role in the detection of various types of cancer and in locating sites of infection and inflammation.

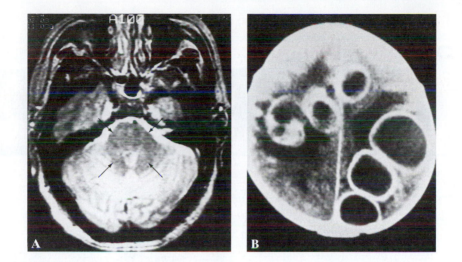

Figure 13
Two examples of magnetic resonance imaging (MRI). (A) The damaging effects of neurosyphilis (arrows). (From M. Tuite, L. Ketonen, K. Kieburtz, and B. Handy. *Amer. J. Neuroradiology* 14 (1993):257–263.) (B) The extensive destruction (large holes) caused by a microbial infection of the brain.

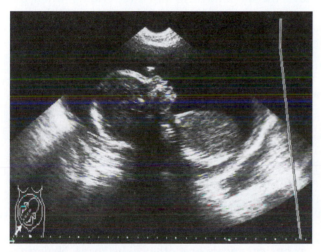

Figure 14
A sonogram of a human fetus. The general outline of the body is evident.

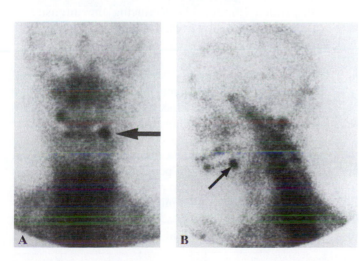

Figure 15
Scintigram showing the signal produced by an increased uptake of radioactive material and the location of a dental bacterial infection. (A) frontal view. (B) A lateral (side) view. (From K. Siminoski. *Clin. Inf. Dis.* 16 (1993):550–554.)

Bacteriology — Laboratory Techniques

. . . it is of the highest importance, therefore not to have useless facts elbowing out the useful ones.

—*Sir Arthur Conan Doyle*

Bacteria, fungi, protozoa, algae, and viruses constitute a group of biological forms that differ in a variety of properties. However, they resemble one another in their small size and relative simplicity of structure and organization.

The first portions of this section present various general laboratory approaches, techniques, and materials used not only to study bacteria, but also to identify them. This presentation is followed in Section 4 by a limited survey of several of the better-known, cultured, and interesting species of bacteria. A number of the distinctive properties of bacteria commonly studied in the laboratory as well as those of pathogens and representative diseases also are considered.

▲ Bacterial Morphology, Morphological Arrangements, and Staining Techniques

The first application of the microscope to the study of microorganisms occurred only in the late 1600s, when a highly curious Dutch merchant, Anton van Leeuwenhoek, applied his talents to lens grinding and the construction of simple single-lens microscopes. His clear, accurate descriptions of the "little animalcules" he saw in specimens taken from teeth, the throat, pond water, rain barrels, and other sources defined the major microscopic types of microorganisms known to this day.

The objectives of microscope usage today are not appreciably different from those of van Leeuwenhoek. These include magnifying the image, maximizing the detail **(resolution)** of cells and their parts, and achieving sufficient contrast with which to distinguish microorganisms and cellular parts from the background of a microscopic viewing area.

The oil-immersion microscope lens is an especially important tool for studying microorganisms. To obtain the best possible results, the objective is immersed in a medium cedar oil or a similar substance that has approximately the same index of refraction as glass. Consequently light rays entering the objective will not be distorted. In addition, oils have the advantage of not evaporating when exposed to air for long periods of time. If the oil-immersion objective is used without immersion oil, some light rays between the specimen and the objective will be lost. The image observed is usually fuzzy, the finer details cannot be seen (Figure 16).

The type of specimen preparation used is determined by the condition of the specimen (living or dead), the purpose of the viewing or examination, and the type of microscope involved.

Living Preparations versus Simple Staining

The direct examination of live microorganisms can be extremely useful in determining size and shape relationships, movement (motility), and reactions to various chemicals and other substances. Nevertheless, live and unstained organisms are difficult to see and to distinguish from the fluid in which they are suspended (Figure 17). Reducing the intensity of the light source, however, can increase contrast, and thereby one can see the outline and arrangement of cells with less difficulty.

Dark-field techniques also can be used to see cell outlines and arrangements by increasing resolving power (the ability to detect detail). Dark-field microscopes are equipped either with special condensers or with an opaque disk placed below an existing condenser to block the direct illumination of specimens. Only thin beams of light penetrate the visual field, so that when a moving object passes through one of these beams it is brilliantly illuminated against a dark background. Dark-field microscopy is generally limited to living microorganisms that are in a narrow size range,

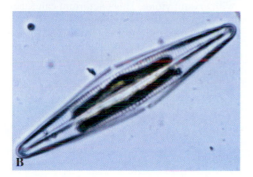

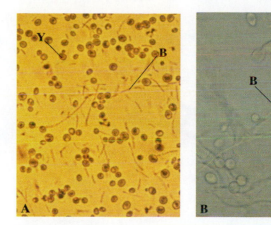

Figure 17
A preparation of living bacteria (**B**) and yeast (**Y**). (A) A low-power view (100X). (B) A higher magnification (450X).

Figure 16
A comparison of an image of a diatom (alga) when viewed with the oil-immersion objective. (A) The view without oil. Air was present between the objective and the specimen. (B) The result with the placement of immersion oil between the objective and the specimen. A clearer image and greater detail can be seen.

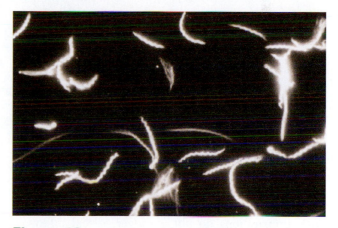

Figure 18
A photomicrograph showing the dark-field image produced with bacteria. Note the bright appearance of several corkscrew shaped spirochetes. Some cells appear quite clear against the dark background (field). Others, because of their movements, are blurred.

difficult to see in a bright-field microscope, or are not stained well by standard techniques. These techniques produce a black background against which objects appear bright (Figure 18).

From a practical standpoint, contrast is increased by using dyes, which are either natural or artificial organic compounds. Such dyes are used in the laboratory for the direct staining of specimens to make cells or their parts more visible than they would be unstained (Figure 19). The surfaces of microorganisms are negatively charged. Therefore, positively charged (basic) dyes or their components will be naturally attracted to those negatively charged surfaces. Dyes commonly used include crystal violet, safranin, carbol fuchsin, and methylene blue.

Before they are stained, bacteria are usually suspended in water or some other liquid on a clean microscope slide and are then spread in a thin, even film. The film is allowed to dry in air, and the organisms are "fixed" (attached) to the slide by gentle heating. The preparation is known as a **fixed smear** and is ready for staining.

Simple staining procedures involve the application of a single stain to the fixed smear. The time required for staining varies with the dye being used. After the staining, the smear is briefly rinsed to remove excess stain. The slide is then dried and examined with the appropriate microscope.

Simple stained smears are used to detect the morphology (shape), cellular arrangement, and relative size

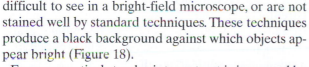

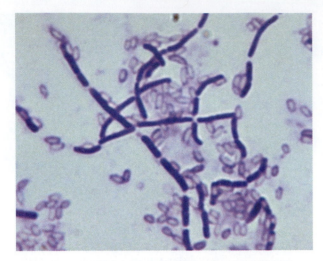

Figure 19

The bacterial rod, *Bacillus subtilis,* simple stained with crystal violet. This bacterium can form heat-resistant structures known as spores. In simple stained preparations such as this one, spores do not stain but appear as clear zones within or outside the bacterial cells in which they were formed.

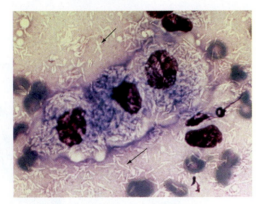

Figure 20

A negative stained preparation showing tissue cells covered and surrounded by large numbers of unstained or colorless bacteria (arrow).

of bacteria. They can also be used to detect the presence of certain structures, such as spores (Figure 19). Such preparations do not, however, reveal internal structural details.

Negative staining is an example of a particular technique generally using a single acidic dye, such as acid fuchsin, eosin, or nigrosin. Most bacterial surfaces repel the negatively charged molecules of such dyes, so the dye does not stain the individual bacterial cells, but outlines them against a dark background (Figure 20). The unstained bacteria appear as colorless forms highlighted against a dark or colored background. The terms "negative images" or "ghost cells" are used to describe the unstained cells. Negative staining is of value in showing the general shapes, sizes, and a limited number of bacterial structures, such as capsules. The technique is also used in the examination of certain tissues. Cell distortions are minimal because heat-fixing of preparations is unnecessary.

Bacterial Morphology and, Morphological Arrangements

The general shapes of bacteria—**rod, coccus** (spherical), and **spiral**—were described by Anton van Leeuwenhoek in the late 1600s. A new morphological type of prokaryote, the square, was described in 1980 by A. E. Walsby. These cells appear as flat, rectangular boxes with perfectly straight edges. In addition to these differences in cellular shape, definite patterns in the number and arrangement of cells are known to exist among different bacterial species (see Figure 24). In the case of spherical, or coccus, forms, five patterns can be found (Figure 21). These are pairs of cells **(diplococci),** chains of four or more cells **(streptococci),** four cells in a square arrangement **(tetrads),** irregular groups of cells resembling grape clusters **(staphylococci),** and eight cells grouped into a cuboidal packet **(sarcinae).**

Rod-shaped bacteria (Figure 22), also referred to as **bacilli** (singular, *bacillus*), can be found occasionally in pairs **(diplobacilli)** and in chains **(streptobacilli).** However, these morphological patterns of rod-shaped forms are not as constant as in the case of the cocci and therefore should not necessarily be considered characteristic for particular types of bacteria (species). In certain cases, because of the limitations of a microscope, it is difficult to distinguish the shape of extremely small rods, which appear to be almost cocci. Such cocci-like organisms

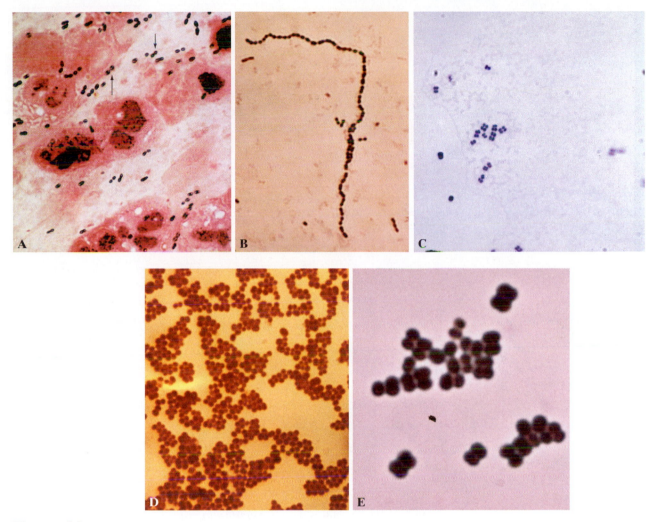

Figure 21
The morphological arrangement (pattern) of cocci. (A) Diplococcus (two cells). Note arrows. (B) Streptococcus (chains). (C) Tetrads (four cells) forming a square. (D) Staphylococcus (grapelike clusters of cells). (E) Sarcinae (cuboidal packets of eight of cells). Note the thickened appearance of the cocci (arrows) exhibiting this pattern.

are referred to as **cocco-bacilli** or **coccoid** in shape (Figure 22C).

Two groups of coiled or spiral-shaped bacteria are known. One group, the spirochetes (Figure 23A), consists of flexible, waving forms with several coils. The second group, the spirilla (Figure 23B), are rigid bacteria possessing one or several curves. Spiral forms that are short and do not form complete coils are called vibrios (Figure 23C).

Figure 24 summarizes the characteristic cell shapes and morphological arrangements found among bacteria (see page 5).

Variations in the general shapes and sizes of bacteria are frequently seen and can be explained in terms of environmental factors. **Pleomorphism** is the term used to denote these modifications when they occur

under favorable conditions. Under unfavorable conditions, these variations are called **involution forms**.

Differential Staining Techniques

As indicated earlier, for most cases of infectious diseases microbial isolation and identification techniques offer the fastest and most specific determinations of causative agents. On the other hand, finding disease agents in the course of preparing cells, tissue sections (slices), or both often adds important information and in some situations is crucial to a timely diagnosis. Several examples of microorganisms in tissues are shown in later sections.

New dimensions in techniques for staining and identifying microorganisms have been provided by

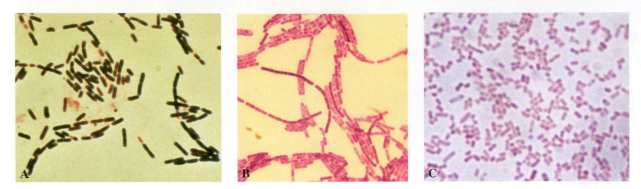

Figure 22
Patterns of rod-shaped bacteria. (A) Diplobacillus (single cells also are present).
(B) Streptobacillus (chains). (C) The coccobacillus (coccoid form).

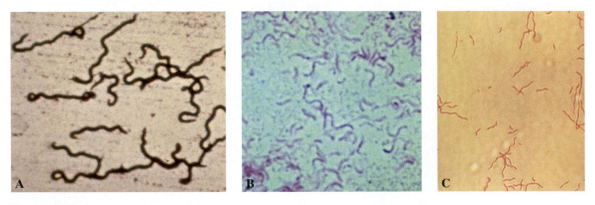

Figure 23
Patterns of spiral and related forms of bacteria. (A) Spirochetes. (B) Spirilla.
(C) Vibrios.

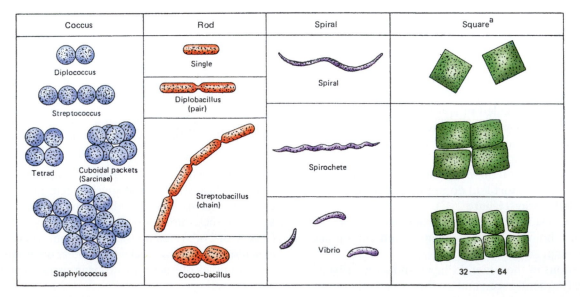

Figure 24
A summary of bacterial cell shapes and arrangements.

biotechnological advances. Some of these are briefly discussed in this section.

Differential staining procedures distinguish structures within a cell or distinguish one type of cell from another. In the laboratory, all stains and related chemicals or processes can functionally be divided into one of four categories: **primary stain, mordant, decolorizer,** and **secondary stain,** or **counterstain.**

A primary stain is generally a basic (alkaline), positively charged dye. Examples include crystal violet, safranin, carbol fuchsin, and methylene blue. Mordants usually are chemicals that fix (firmly attach) the primary stain to the target microorganisms. A common example is Gram's iodine used in the Gram-staining procedure (Table 4). A decolorizer is a chemical that removes unattached stain from target microorganisms. After a decolorizer has been applied,

the target microorganisms are the color of the primary stain, and nontarget organisms and the surrounding background should be colorless. Acetone-alcohol, used in the Gram stain, and acid-alcohol (a combination of 3% hydrochloric acid and alcohol), used in the acid-fast staining procedure, are examples of common decolorizers. Most, but not all differential staining procedures require the application of a counterstain to color and show the presence of nontarget organisms (Table 4). Safranin used in the Gram-staining procedure and methylene blue in the acid-fast procedure are counterstains.

Differential stains are widely used for the detection and identification of bacteria and other microorganisms. Table 4 summarizes the features of differential and related staining procedures, and Figures 25 through 33 show the results of these procedures.

Differential Staining Techniques

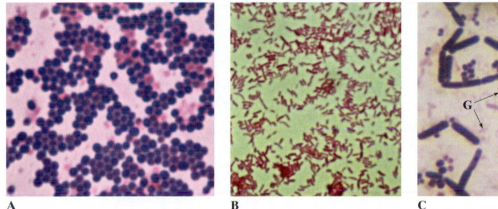
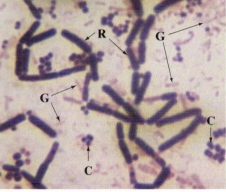

A **B** **C**

Figure 25
Gram reactions. (A) Gram-positive cocci. (B) Gram-negative rods. (C) A mixed smear showing Gram-positive cocci (C) and rods (R) and Gram-negative rods (G).

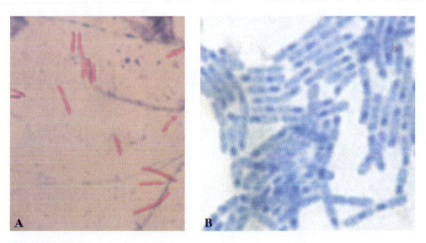

A **B**

Figure 26
The acid-fast reaction. (A) Red, acid-fast cells. (B) A non-acid-fast reaction showing blue cells.

Table 4 Summary of Differential and Related Staining Reactions

Staining Procedure	Description	Reaction(s) and/or Cellular Appearance	Application(s) and/or Examples
Gram stain	*Primary stain*: crystal violet *Mordant*: Gram's iodine *Decolorizing agent*: acetone-alcohol or 95% ethanol *Counterstain*: safranin	Gram-positive: purple cells. Gram-negative: pink cells.	Divides bacteria into one of two groups: Gram-positives: *Staphylococcus, Bacillus.* Gram-negatives: *Escherichia, Pseudomonas, Salmonella,* etc.
Acid-fast stain	*Primary stain*: carbol fuchsin *Decolorizing agent*: acid-alcohol *Counterstain*: methylene blue	Non-acid fast: blue cells. Acid-fast: red cells.	Divides bacteria into one of two groups: Acid-fast: *Mycobacterium, Nocardia.* Non-acid fast: most other bacteria.
Spore stain (Schaeffer-Fulton)	*Primary stain*: malachite green *Counterstain*: safranin *Decolorizing agent*: none	Spores present: green oval structures inside and outside of red cells. Spores absent: red cells without any green oval structures.	Separates spore-formers from non-spore-formers.
Spore stain	*Primary stain*: carbolfuchsin *Counterstain*: methylene blue *Decolorizing agent*: none	Spores present: red oval structures inside and outside of blue cells. Spores absent: blue cells without any red oval structures.	Spores: *Bacillus, Clostridium.* Non-spore formers: most other bacteria.
Capsule stain	Culture is mixed with 1 drop of India ink. After air-drying, safranin is applied.	Clear (halo-like) area surrounding pink cells.[a] No clear area surrounding pink cells.	Demonstrates the presence of capsule-formers. Capsule present: *Klebsiella, Streptococcus pneumoniae.* Capsule absent: *Escherichia, Mycobacterium.*
Flagella stain	A combination of mordant and stain is applied to the smear. The mordant acts to deposit more stain onto the flagella.	Flagella present: long wavy structures extending from cells in various patterns. Flagella absent: no wavy structures extending from cells.	Shows presence and arrangement of flagella. Flagella are found with several organisms, including *Salmonella, Escherichia, and Bacillus.*
Fluorescent-antibody staining	Specific fluorescent-dye-tagged (attached) antibody is applied to smear containing unknown bacteria or other types of cells.	Cells fluoresce: the color(s) exhibited by cells are determined by the fluorescent dye molecules used; colors can include red, green, yellow, and orange. Cells do not fluoresce in negative tests.	Identifies specific organisms; can be used for diagnosis of diseases. Only microorganisms and other cells containing specific antigens fluoresce.

[a]The capsule itself does not stain.

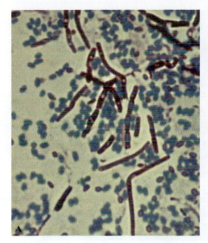
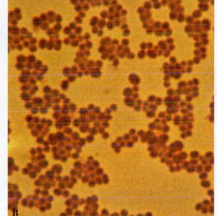

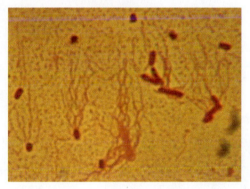

Figure 27
The Schaeffer-Fulton spore stain.
(A) Green spores inside and outside
of non-spore forming cells. (B) A non-
spore-former.

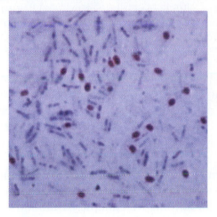

Figure 29
Flagella at the ends of the bacterium
Salmonella typhi, the cause of typhoid fever.

Figure 28
The results of a spore stain using carbol fuchsin as the primary stain and
methylene blue as the secondary or
counterstain. Spores stain red.

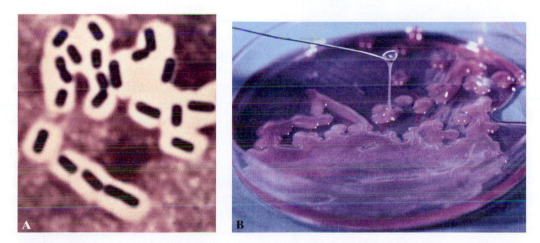

Figure 30
The capsule stain. (A) Cells are surrounded by clear areas formed by capsules. (B) In the
laboratory, the presence of a capsule-producer usually is indicated when colonies exhibit a
stringy consistency when touched with an inoculating loop.

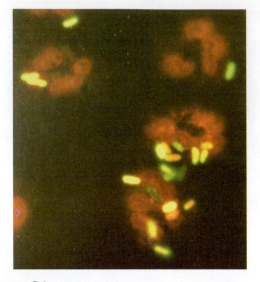

Figure 31
Immunofluorescence (fluorescent-antibody staining). Specific antibody molecules to which fluorescent dye molecules are attached can differentiate among cells. Bacteria appear as greenish yellow cells.

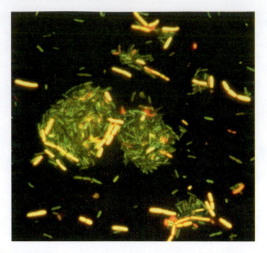

Figure 33
A variation of the Gram stain using LIVE Bac Light. Gram stains with living Gram-positive *Bacillus cereus* (yellow-orange) and Gram-negative *Pseudomonas aeruginosa* (green). (Courtesy of Molecular Probes, Inc.)

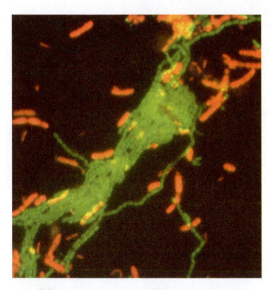

Figure 32
The use of fluorescent dye–labeled nucleic acid probes to detect and identify bacteria. (From M. Wagner, R. Amann, H. Lemmer, and K.-H., Schleifer. *Appl. Environ. Microbiol.* 59(1993):1520–1525.)

Variations of differential staining procedures include the applications of fluorescence microscopy. Certain dyes, called **fluors** or **fluorochromes,** have the property of becoming excited (raised to a higher energy level) after absorbing ultraviolet (UV) light (light of short wavelength). As the excited molecules return to their normal state, they release the excess energy in the form of visible light or a longer wavelength than that which first excited them. This property of becoming self-luminous is called **fluorescence.** Fluorescing objects appear brightly lit against a dark background (Figures 31–33), with the color depending on the fluorochrome being used.

The principal use of fluorescence microscopy is a diagnostic technique called **immunofluorescence,** or the **fluorescent-antibody technique. Antibodies** are natural defense protein molecules produced by humans and many other animals in response to substances recognized by their respective immune systems as foreign. These foreign substances or cells are known as **antigens.** Two general types of fluorescent-antibody techniques are in use. Direct fluorescent-antibody techniques are used to detect antigens in specimens, whereas indirect fluorescent methods are used to show the presence of antibodies in a variety of specimens, such as blood or other body fluids.

Fluorescent antibodies for a particular antigen are obtained when a laboratory animal is injected with a specific antigen, such as a bacterium. The animal then begins forming specific antibodies against the injected antigen. After a sufficient amount of antibodies have been produced, they are removed from the blood of the animal and chemically combined with fluorochrome molecules. The resulting fluorescent antibodies can then be used to detect and identify the same antigen that was injected into the laboratory animal, on cells, tissues, or other types of clinical specimens. Fluorescent antibodies are applied to specimens on slides containing an unknown organism. If the unknown organism is the same bacterium that was injected into the animal, the fluorescent antibodies bind to the surface of the bacteria, causing them to fluoresce, or glow.

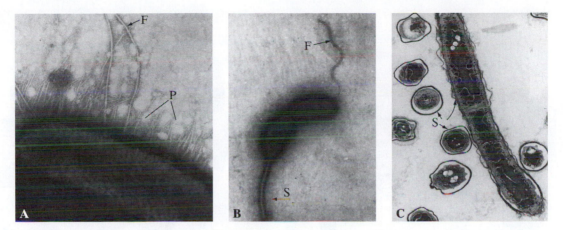

Figure 34

Transmission electron micrographs of selected bacterial specialized parts. (A) The submicroscopic pili (P). These structures are much thinner than the flagella (F) also shown. (B) A bacterium bearing a stalk (arrow). (C) The specialized covering known as a sheath (S). This structure enables bacteria to attach to surfaces and also provides protection against predators.

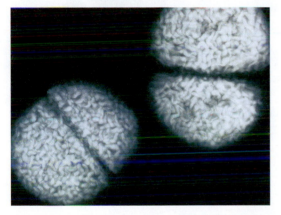

Figure 35

A transmission electron micrograph showing the intricate surface of negatively stained bacteria.

Modern technology has provided alternatives to fluorescent antibodies for identifying bacteria. These include the attachment of fluorochromes to nucleic acid components to detect highly specific ribonucleic acid parts of bacterial ribosomes (Figure 32).

An alternative technique to the Gram stain has been developed that can be applied only to living bacteria (Figure 33). The **LIVE Bac Light** procedure uses a mixture of nucleic acid–fluorescent dye stains that can differentiate between living gram-positive and gram-negative organisms. When stained with the unique nucleic acid–fluorescent dye combinations, gram-positive cells fluoresce yellow-orange, and gram negative organisms fluoresce bright green.

Selected Specialized Structures

Several prokaryotes have complex and/or special structures. These include parts used for attachment, such as *pili* and *stalks* (Figures 34A and B), and *sheaths*, which enable bacteria to attach to surfaces and provide protection against attacks by predators (Figure 34C). Certain pili are also used for the transfer of deoxyribonucleic acid (DNA).

Some bacterial species also have unusual surfaces (Figure 35), which together with the structures mentioned are not readily visible with the use of the usual laboratory staining techniques. Electron microscopy and associated staining methods are required.

▲ Basic Cultivation Techniques

The cultivation of microorganisms requires the use of nutrient preparations called **culture media** (singular, *medium*). Natural media such as milk, vegetable slices, and certain meat infusions contain soluble organic and inorganic substances that are the necessary factors for growth. However, the exact chemical compositions of natural media are unknown and quite variable. On the other hand, in the case of **chemically defined**, or **synthetic**, **media**, the kinds and exact amounts of all ingredients are known. Media of this type can be duplicated to specifications and used to study the effects of specific compounds on microbial growth.

There are three general forms of culture media: **solid**, **semisolid**, and **broth** (liquid). Each type of medium has particular properties that make it more or less suitable for certain growth situations (Figure 36). Most solid and semisolid media contain **agar** as a solidifying agent. It is a complex polysaccharide extracted commercially from certain species of red marine algae, such as *Gelidium*, *Gracilaria*, and *Rhodophyta*.

Figure 36
An assortment of different media that may be used in bacterial detection, isolation, and identification. Plate, broth, and agar slant media are shown. (Courtesy of Becton Dickinson Microbiology Systems.)

The introduction of agar as a solidifying is credited to Fannie Eilshemius Hesse (Figure 37), an American, born in Jersey City, New Jersey. This contribution, which was first used extensively by Robert Koch, made it possible to prepare solid, transparent, and sterile media that would keep their consistency at all temperatures at which bacteria could grow.

Isolation Techniques

Certain procedures have become indispensable to bacteriologists. Among them are the standard **pour plate** (Figure 38) and **streak plate** (Figure 39) **techniques.** These methods can be effective in both the detection and the determination of the numbers and kinds of different microorganisms present in specimens. Microorganisms such as bacteria grow as visible accumulations of identical cells forming **colonies** on the surfaces of agar media contained in Petri dishes. These dishes are constructed to allow air but not dust to pass through.

The pour plate technique consists of: (1) cooling a melted agar-containing medium (1.5% agar) to

Figure 37
Fannie Eilshemius Hesse (1850–1934), a pioneer woman of microbiology. Her practical application of agar-agar into bacteriological media was an essential contribution to modern microbiology. (Courtesy of Dr. Dieter H. M. Groschel, University of Health Sciences Center, University of Virginia.)

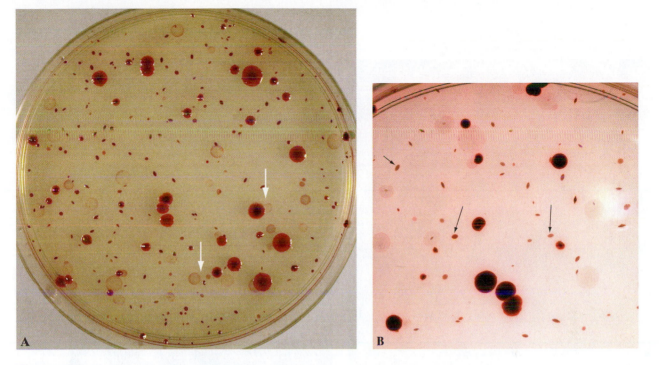

Figure 38
The pour plate technique. (A) A plate of *Micrococcus luteus* and *Serratia marcescens*. Note the distribution of differently pigmented colonies. (B) Higher magnification of this preparation showing the various shapes of bacterial colonies (arrows).

approximately 42° to 45°C, and (2) inoculating the medium with a specimen just before pouring it into a sterile Petri dish. Thus, bacteria are distributed throughout the agar and trapped in position as the medium hardens. Although the solidified medium restricts bacterial movement from one area to another, it is of a soft enough consistency to permit growth. Growth occurs both on the surface and within the inoculated medium. Unfortunately, there are several disadvantages to this technique, including the following: (1) colonies of several species may present a similar appearance in the agar environment; (2) certain species of bacteria may not grow in this environment; and (3) difficulty may be encountered in removing (picking) colonies for further study. Figure 38 shows the features of the pour plate.

The streak plate procedure is another example of a dilution technique. It was originally developed by two bacteriologists, Friederich Löffler and Georg Gaffky, in the laboratory of Robert Koch. The preparation of a streak plate involves the spreading of a single loopful of material (**inoculum**) containing microorganisms over the surface of an agar medium that has been allowed to solidify. Numerous streaks are made to allow for a greater separation of the organisms in the inoculum. Figure 39 shows the results obtained with the streak plate technique.

After inoculation by any method, plates are incubated at desired temperatures. Water condensation may form in Petri dishes as a consequence of the high concentration of water in agar media. Prepared plates are incubated in an inverted position to prevent water from forming on media surfaces and causing bacterial colonies to run together.

Agar Plate, Broth, and Agar Slant Characteristics

The pigmentation, size, shape, and overall general appearance of bacterial colonies growing in or on agar plates (Figure 40) can serve as identifying features when viewed under both reflected and transmitted light. Pigmentation and several other cultural characteristics are influenced by incubation temperature. Colonies may appear to be transparent, opaque, or translucent. Some species produce a distinguishing fluorescent pigment that is evident when cultures are examined under ultraviolet light. Other colonies exhibit specific surface properties and may appear as dry (powdery), contoured, rough, smooth (glistening), or rugose (wrinkled), or they may have concentric rings or ridges. Some organisms also produce characteristic odors. Such odors may be sweet, foul, soil-like, or musty.

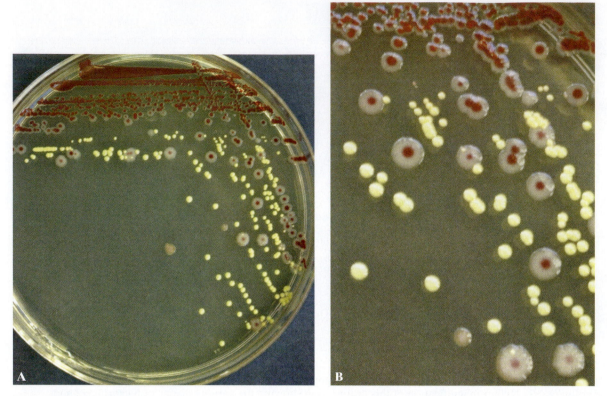

Figure 39

The streak plate technique. (A) A plate with a mixed culture containing *Micrococcus luteus* and *Serratia marcescens*. (B) A close-up of the agar surface. Note the well-separated colonies.

Agar Plate, Agar Slant, and Broth Characteristics

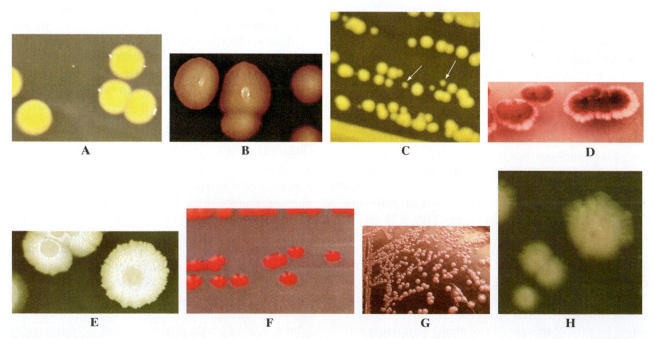

Figure 40

Examples of bacterial colony characteristics. (A) Circular, entire, smooth, opaque, yellow-pigmented colonies. (B) Irregular, undulate, umbonate, opaque, cream-colored colonies. (C) Punctiform (less than 1 mm in diameter), smooth, entire, opaque colonies among larger circular ones. (D) Unusual red-and-white-pigmented colonies that also exhibit irregular, lobate, and opaque properties. (E) Irregular concentric, undulate, opaque, somewhat raised colonies. (F) Red-pigmented, circular, smooth, entire, opaque, convex colonies. (G) Filamentous, opaque, colonies. (H) Irregular, opaque, umbonate colonies.

In addition to the agar plates, solid medium preparations can be made into slants. Such preparations are made by pouring a melted agar medium into test tubes and allowing the material to solidify at an angle. Agar slants are valuable in studying growth characteristics and in maintaining pure cultures. The patterns of growth on agar slants are important characteristics used in bacterial identification (Figure 41). Liquid or broth media generally are contained in test tubes closed by nonabsorbent plugs or special closure caps to prevent the entering of unwanted microorganisms. After the inoculation of broth media, bacteria may exhibit a particular form of growth. These include clouding of the medium **(turbidity),** accumulations of cells at the tube bottom **(sediment),** and the formation of a thin surface film **(pellicle).** Broth media can also be used to perform various biochemical tests. Several of these are presented later. Figures 41 and 42 show selected properties of agar slant and broth cultures, respectively.

Anaerobic Cultivation

Anaerobic bacteria are unable to grow on or near the surfaces of semisolid or solid media in air at

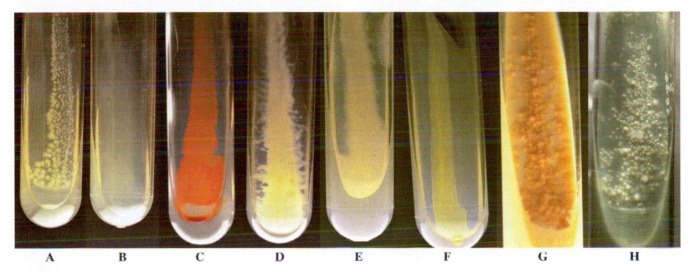

Figure 41
Selected agar slant patterns. (A) Beaded form of growth. (B) Spreading growth. (C) Flat, red, and filiform. (D) Arborescent (branched). (E) Rhizoid growth. (F) Filiform, yellow growth. (G) Orange growth of *Mycobacterium phlei*. (H) Nonpigmented growth of a *Mycobacterium* species. Both species of *Mycobacterium* are growing on a form of Löwenstein-Jensen medium.

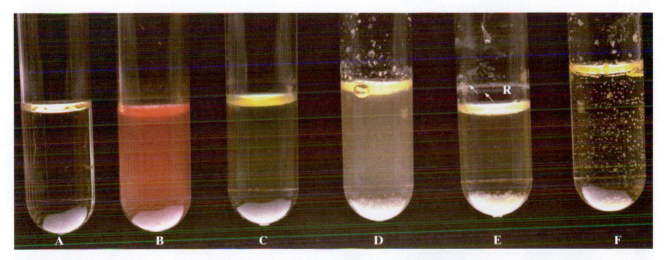

Figure 42
Selected broth culture patterns. (A) Clear, uninoculated broth. (B) Turbid (cloudy) growth and red pigmentation. (C) Turbid and green pigmentation in the upper portion of the broth. (D) Turbid growth. (E) Slightly turbid and ring formation (1 ml of the broth was removed to demonstrate the presence of the ring). (F) Somewhat clear broth with flocculent growth.

atmospheric pressure. The early bacteriologists believed that successful cultivation of strict anaerobes (organisms that can survive and grow in environments completely without free oxygen) could be achieved only through methods that excluded all free oxygen. Providing anaerobic conditions requires only the removal of free oxygen from the immediate environment of the microorganisms or simply the maintenance of a low oxidation-reduction (redox) potential, or E_h, in the media. The E_h is a measure of the tendency of a preparation to be oxidized or reduced. In routinely used laboratory media, oxygen is primarily responsible for the increasing E_h. Various redox dyes can be used to estimate the E_h of a medium or culture. Useful dyes are reversibly oxidized or reduced and are colored in the oxidized state and colorless in the reduced state (see Figure 45).

Procedures for cultivating and identifying anaerobic bacteria are much like those for aerobes. The difference lies in the incubation atmosphere. Reducing agents, which are nontoxic, are added to most anaerobic media to depress and maintain the redox potential at low levels. Examples of such agents are sodium thioglycollate and cysteine hydrochloride.

Various types of containers are used to provide an anaerobic environment for organisms. Among the most widely used are the Brewer anaerobic jar with a heat-activated catalyst and the GasPak Anaerobic System (Figure 43). With the Brewer apparatus,

hydrogen or a mixture of gases is introduced into the anaerobic jar after it has been sealed. Electrical heat activation of a platinum catalyst present in the lid creates anaerobic conditions.

The GasPak jar differs from all other anaerobic jars in that it has no external connections and utilizes a room-temperature catalyst system, thus rendering electrical connections or other means of heating the catalyst unnecessary. This anaerobic jar is specifically designed to be used with a disposable hydrogen and carbon dioxine–generating system. When water is added to a GasPak envelope, hydrogen gas is released. This gas, in turn, reacts with oxygen in the presence of a catalyst to produce anaerobic conditions. Carbon dioxide is also produced in quantities sufficient to support the growth of anaerobes that require it. A disposable anaerobic indicator containing methylene blue is used to determine whether an anaerobic condition exists within the system.

Procedures that do not involve complicated pieces of equipment or elaborate techniques are also commonly used for the cultivation of some anaerobic, **facultative** organisms (which can grow with or without oxygen) and **microaerophilic** organisms (which prefer low oxygen tension). These include the paraffin-plug technique (Figure 44) and the use of media containing a reducing compound, such as sodium thioglycollate (Figure 45) to reduce the oxygen content. In the paraffin-plug technique,

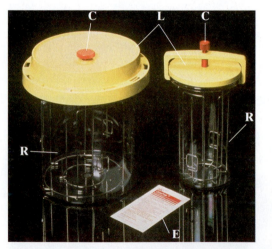

Figure 43
A commerical anaerobic cultivation device. The components of the GasPak Anaerobic System. GasPak anaerobic jar lid (L), clamp screw (C), charged catalyst reaction chamber (R), and a disposable hydrogen + GasPak carbon dioxide–generator envelope (E). The GasPak disposable anaerobic indicator is not shown.

Figure 44
The paraffin-plug technique. (Left tube) Uninoculated with a paraffin plug. (Center tube) Turbidity (cloudiness) indicating growth. (Right tube) Growth under anaerobic conditions.

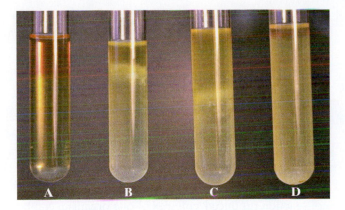

Figure 45
Representative growth characteristics of bacteria in sodium thioglycollate broth. (A) Uninoculated. Note the presence of the upper pink zone (for aerobes) and the lower yellow zone (for anaerobes). (B) Shows the growth of a typical anaerobic organism. (C, D) Show facultative (adjustable) organisms.

the tube of medium is heated for several minutes, cooled quickly, and then inoculated. A layer of melted paraffin approximately $\frac{1}{4}$ inch thick is poured onto the top of the medium. The culture is then ready for incubation.

Thioglycollate media are used in the cultivation of anaerobic, microaerophilic, and aerobic bacteria and for the detection of bacteria in normally sterile materials. The ingredients of these media support the growth of a wide variety of microorganisms having a broad range of growth requirements. The presence of sodium thioglycollate lowers the oxidation-reduction potential of the media, while resazurin serves as the oxidation-reduction (OR) indicator and indicates the status of oxidation.

General or all-purpose media can be used to study the effects of a variety of physical and chemical substances and factors on the properties of bacteria. The results of such procedures can have a number of practical applications in the control of various types of microorganisms.

The physical factors include heavy metals, and ultraviolet light (UVL). These factors can cause a variety of easily recognized changes in organisms. For example, heavy metals, such as copper, mercury, and silver, are particularly reactive with proteins and are capable of interfering with the metabolism of microorganisms.

Ultraviolet light is well known as an agent that can bring about changes at the genetic level. For example, UVL exposure of the pigment-controlling genes of some bacteria can cause affected cells to lose the ability to form a characteristic pigment. The effect of UVL exposure brings about alterations in the nucleotide base sequences of deoxyribonucleic acid molecules and can result in permanent changes in an organism's genetic properties known as **mutations.** The mutated organism also can change in physical appearance (Figure 46A).

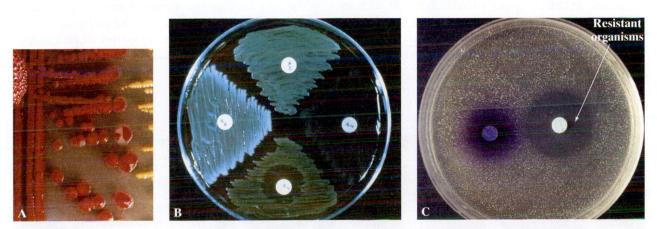

Figure 46
The effects of physical and chemical agents using general or all-purpose media. (A) The effect of ultraviolet light exposure on pigment production. A sectored colony of *Serratia marcescens*. Such colonies are caused by a mutation in a pigment-producing gene. (B) The relative effects of different antibiotic concentrations. The greater the inhibitory effects of an antibiotic, the larger the clear zone surrounding the specific disk containing it. (C) The relative bacteriostatic effects of crystal violet (purple) and malachite green. Note the appearance of resistant organisms.

Bacteriological media also can be used to test the effectiveness of a number of chemicals, such as antibiotics and dyes. The term *antibiotic* originally referred to chemicals produced by one microorganism that could kill or inhibit the growth of other microbial forms. Today, the term is applied to a variety of antimicrobial drugs, including totally synthesized laboratory products and chemically modified (semisynthetic) forms of natural antibiotics. Several methods are in use to determine the smallest amount of an antibiotic to inhibit the growth of a microorganism. The resulting value is known as the *minimal inhibitory concentration* (MIC). The agar-diffusion method and its variations are in common use for this purpose. In the basic procedure, commercially available filter paper disks, each containing a defined concentration of specific antibiotics, are placed on the surface of a suitable medium that has been heavily inoculated with a test organism, and incubated. During incubation the antibiotics in the different disks diffuse into the agar, and their respective effectiveness is subsequently indicated by the extent to which bacterial growth is prevented. Clear areas surrounding the individual disks show the extent of growth inhibition, or *bacteriostatic* effect (Figure 46B). A common application of the agar-disk method is known as the Kirby-Bauer test, named after W. M. Kirby and A. W. Bauer, the developers of the procedure in the l960s.

At times all-purpose media are not appropriate for all situations, especially when an organism present in low numbers in a specimen or in a mixed culture is to be isolated. The challenge here can often be met by the incorporation of a chemical agent into a medium, thus making it a *selective* preparation. Selective agents inhibit the growth of one or more unwanted organisms in a specimen without preventing the growth of the wanted organisms. Selective agents include chemicals such as acids, and certain types of dyes such as crystal violet and malachite green. The effectiveness of such substances can be be determined using general media (Figure 46C).

▲ Biochemical Activities of Microorganisms

Differential, Selective, and Selective and Differential Media

A large number of media have been developed to aid in the isolation, differentiation, and identification of microorganisms such as bacteria and fungi. Differential media, which do not prevent the growth of organisms, contain one or more substances **(substrates)** that can be enzymatically attacked. On the basis of the type of reaction produced, colonies of one organism can be distinguished from others growing on the same plate. Blood agar (which contains whole blood) often is used to isolate and distinguish organisms that can enzymatically attack hemoglobin differently (Figure 47).

A selective medium is defined as one that permits the growth of certain organisms while preventing or retarding the growth of others. Selection, in general, can be carried out through (1) control of ingredients of the medium, (2) alteration of atmospheric components, or (3) adjustment of incubation temperature.

Several media in use today incorporate both selective and differential substances (Table 5). One example of such preparations is Bacto Brilliant Green agar (Figure 48). This medium is a highly selective preparation used for the isolation of *Salmonella* species other than *S. typhi,* the causative agent of typhoid fever, from stools or other specimens suspected of containing the organisms. The growth of other bacteria is almost completely inhibited by the presence of brilliant green dye. This medium also contains lactose and sucrose as substrates for enzymatic action. Typical *Salmonella* form lightly pink-white opaque colonies surrounded by brilliant red areas. The few lactose- or sucrose-fermenting organisms that can grow on the medium are easily differentiated from *Salmonella* by the formation of yellow-green colonies surrounded by intense yellow-green zones.

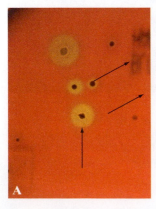
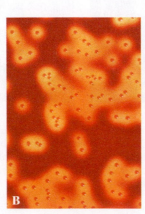
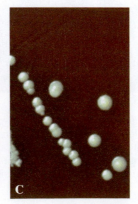

Figure 47
Representative hemolytic reactions on the differential medium blood agar. (A) Alpha hemolysis. Note the greenish discoloration of the medium surrounding the bacterial colonies. (B) Beta hemolysis. Here, clear zones surround the individual bacterial colonies. (C) Gamma hemolysis, or nonhemolytic reactions. Discolorations, or clear zones, are not present.

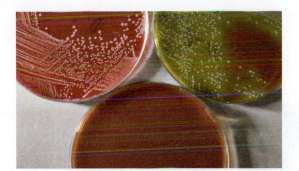

Figure 48
Bacto Brilliant Green agar, a useful medium for the isolation and identification of *Salmonella* species. Typical non-lactose-fermenting *Salmonella* colonies appear as slightly pink-white opaque colonies surrounded by a brilliant red medium. Colonies of lactose-fermenting organisms form yellow-green colonies surrounded by intense yellow-green zones on the medium. An uninoculated plate is shown as a control.

A medium frequently used for the isolation of *Staphylococcus* species is mannitol-salt agar. This preparation favors the growth of bacteria that can tolerate high sodium chloride concentrations, which in this case is 7.5%. Organisms unable to tolerate the high salt concentration either exhibit poor growth or do not grow at all. The carbohydrate substrate mannitol and the pH indicator phenol red are also included in the medium. Species that grow on the medium and are able to ferment mannitol with the production of acid result in the appearance of yellow zones surrounding their colonies. The presence of pink zones or no change in the medium constitutes a negative reaction (Figure 52). Table 6 summarizes a few selective and differential media, their substrates, and associated reactions and Figures 48 through 53 show some examples. The

Table 5 Selective and Differential Media Properties

Medium	Selective Agent(s)	Organisms Encouraged to Grow
Brilliant green agar	Brilliant green	Gram-negative rods[a]
Eosin-methylene blue agar	Eosin Y, methylene blue	Gram-negative rods
Hektoen enteric agar	Bile salts	Gram-negative rods
MacConkey agar	Bile salts, crystal violet	Gram-negative rods
Mannitol-salt agar	Sodium chloride	*Staphylococcus aureus*

[a]This medium is not used for the isolation of *Salmonella typhi*.

Table 6 Summary of Reactions Associated with Selected Differential (D) and Selective Differential (SD) Media Used for Isolation and/or Identification

Medium	Substrate(s)	Type of Medium	Reaction and Descriptions
Blood agar	Hemoglobin	D	1. Alpha hemolysis (green zones around colonies) 2. Beta hemolysis (clear zones around colonies) 3. Gamma hemolysis (no zone around colonies)
Brilliant green agar	Lactose, sucrose	SD	1. Lactose-fermenter (yellow-green colonies) 2. Non-lactose fermenter (pink to white colonies surrounded by brilliant red zones)
Eosin-methylene blue agar	Lactose, sucrose	SD	1. Lactose-fermenter (dark purple colonies or colonies with dark centers and transparent colorless borders) 2. Non-lactose or non-sucrose fermenters (colorless colonies)
Hektoen enteric agar	Lactose, sucrose, salicin, and amino acids containing sulfur	SD	1. Lactose-fermenter (salmon-pink colonies) 2. Non-lactose-fermenters (green, most colonies) 3. Salicin-fermenters (pink zones around colonies) 4. Non-salicin-fermenters (no change) 5. H_2S producers (colonies with black centers)
MacConkey agar	Lactose	SD	1. Lactose-fermenter (pink-red colonies surrounded by pink zones due to precipitated bile) 2. Non-lactose fermenter (colorless and translucent colonies)
Mannitol-salt agar	Mannitol	SD	1. Mannitol-fermenter (colonies surrounded by yellow zones) 2. Non-mannitol fermenter (small colonies with no color yellow change)

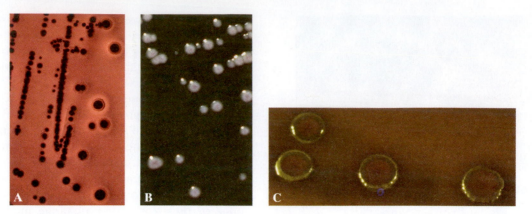

Figure 49
Eosin-methylene blue agar reactions. (A) Colonies of a lactose-fermenter. (B) Colonies of a non-lactose-fermenter. (C) A metallic sheen can form on the colony surfaces of certain lactose-fermenters.

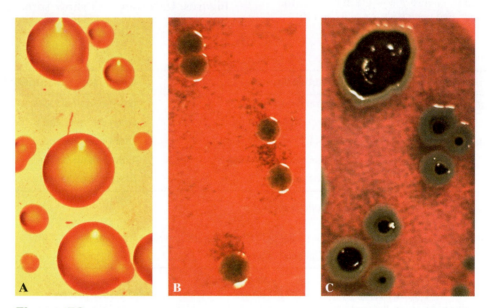

Figure 50
Hektoen enteric agar reactions. Three major types of reactions are shown. (A) Rapid lactose-fermenters (yellow to salmon-pink colonies). (B) Non-lactose-fermenters (green colonies). (C) Hydrogen sulfide (H_2S) producers (black-centered colonies). *Escherichia coli* is a rapid lactose fermenter; *Proteus vulgaris* and related species are well-known H_2S producers.

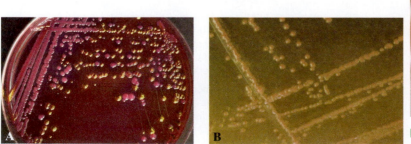

Figure 51
MacConkey agar reactions. (A) Colonies of a lactose-fermenter.
(B) Colonies of a non-lactose-fermenter.

Figure 52
Growth of *Staphylococcus aureus* on mannitol-salt agar. The presence of yellow zones around the bacterial growth indicates acid formation from mannitol.

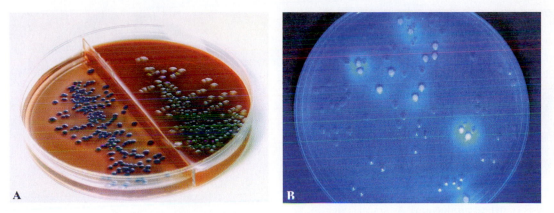

Figure 53

Chromogenic media. (A) The BluEcoli urine biplate. *Escherichia coli* colonies form as blue colonies on this chromogenic media (left side). The growth of this bacterium on blood agar is shown on the right side. (Courtesy of Hardy Diagnostics www.Hardy Diagnostics.com). (B) An example of a chromogenic medium with the fluorogenic substrate methlyumbelliferyl-beta-D-glucuronide also known simply as MUG. Colonies producing a positive reaction exhibit a bright blue fluorescence when exposed to ultraviolet light.

distinctive reactions produced by pathogenic bacteria on other types of media are described later.

Chromogenic Media

A new group of preparations known as *chromogenic* (color-producing) media have been developed to permit the detection, making colony counts, and presumptive identification of a number of bacterial species. Such media incorporate chromogenic and/or fluorogenic (fluorescence-emitting) substrates. These substrates are initially colorless and, when cleaved (enzymatically attacked), are oxidized producing a characteristic color (Figure 53A) or fluorescence (Figure 53B). The detection of a specific enzyme or enzymes serves as a means to identify specific bacterial species or groups of organisms.

Extracellular Degradation of Polysaccharides, Proteins, Lipids, and DNA

Metabolism is the sum total of the biochemical reactions required to maintain adequate nutritional levels and functional cellular activities. These reactions include the degradative, or breakdown, process called **catabolism** and the synthetic process known as **anabolism.** The two processes occur simultaneously and complement each other to provide for the essential needs of the cell. All metabolic reactions occur as a series of steps leading from one compound to another and are the bases of **metabolic pathways.** Such biochemical reactions are of value in bacterial identification.

Several microorganisms form and then release enzymes into their environment, such as a culture medium. Some of these microbial extracellular enzymes can degrade, or break down, large molecules in the environment surrounding the cells. Most microbial extracellular enzymes are referred to as

hydrolytic. These enzymes are so named because they break large molecules into small ones and are generally classified according to the large types of molecules they can degrade. Such enzymes include **esterases,** which decompose fats and lipids, **glycosidases,** which break apart polysaccharides, and **proteinases,** which break down proteins.

Most of the carbohydrates available to microorganisms are in the form of polysaccharides. Two common examples are cellulose and starch, both of which are composed of smaller units **(polymers)** of glucose. The basic differences between the chemical and physical characteristics of these substances depend upon the structural arrangement of their glucose units. If this were not the case, then cellulase, the enzyme that breaks down cellulose into simple sugar units, would also degrade starch. The enzyme responsible for hydrolysis of starch is called amylase. Cellulase and amylase are examples of extracellular enzymes **(exoenzymes);** they can be secreted through the cell wall in order to degrade complex substances into units that can readily enter the cell. An example of this type of metabolic pattern is the action of amylase on starch (Figure 54) yielding maltose (a disaccharide composed of two glucose molecules).

The degradative action of an organism on an intact protein is analogous to such action on carbohydrates. In the case of starch, the enzyme amylase degrades the polysaccharide into units that can readily be absorbed by the cell. With a protein, such as casein (milk protein), the enzyme caseinase accomplishes much the same result in yielding polypeptides (Figure 55).

Many microorganisms are able to degrade fats and oils (in the process called **lipolysis**) and thus obtain acetate for carbohydrate metabolism and amino acid synthesis. The presence of a lipase in the enzymatic functions of a microorganism can be considered a potential indication of its invasiveness because

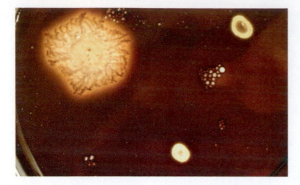

Figure 54
Starch hydrolysis (amylase production). A typical starch hydrolytic reaction as produced by the bacterium *Bacillus subtilis*. The complete absence of starch (hydrolysis) upon the addition of an iodine reagent is indicated by the yellow background surrounding the bacterial colonies. A negative reaction is indicated by a dark purple background.

Figure 56
The demonstration of lipid hydrolysis using Bacto-spirit blue agar containing Bacto-lipase reagent. The lipolytic activity of a bacterial species is recognized by the deep blue color or clearing that develops in the medium surrounding the test organism. A comparable color change is not observed with nonlipolytic microorganisms.

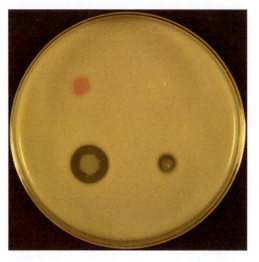

Figure 55
Caseinase activity. Casein digestion can be seen as clear zones around bacterial growths. One clearly negative result is also shown.

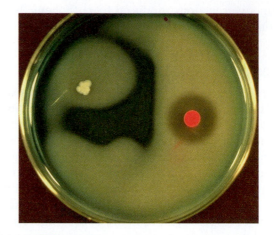

Figure 57
DNAse reactions. The presence of the enzyme is indicated by a clear zone around the bacterial growth. The cloudy area (negative reaction) is caused by the addition of hydrochloric acid, which precipitates the DNA.

animal cell membranes are largely composed of lipid. This degradative ability is yet another characteristic of microorganisms that can be useful in classification. The lipolytic activity of microorganisms appears in the area of lipolysis (Figure 56). Spirit blue agar can be used to show lipolytic activity. The medium contains a mixture of olive oil and spirit blue dye which gives the preparation a sky blue appearance. Lipase-producing bacteria growing on the medium hydrolyze the olive oil substrate causing the formation of clear areas or halos surrounding their colonies (Figure 56).

The production of deoxyribonuclease (DNAse), another extracellular enzyme, is useful for the isolation and differentiation of several bacterial species. DNAse agar is used to demonstrate the breakdown of DNA. Flooding inoculated plates after incubation with 0.1 N hydrochloric acid will show clear areas around the colonies that degraded the DNA in a medium (Figure 57). The hydrolytic breakdown of DNA in most cases produces fragments of two to four nucleotides in length.

Because microbial species vary in the kinds of extracellular enzymes they possess, demonstrating the presence or absence of a particular enzyme is of value in the identification of unknown organisms. Table 7 summarizes the features of tests used to demonstrate the presence of extracellular enzymes with specific organic macromolecules of importance in metabolism.

Carbohydrate Metabolism: An Introduction to Intracellular Metabolism

Carbohydrates are the prime sources of energy and carbon for the synthesis of cellular substance

Table 7 Summary of Extracellular Enzymes

Test	Substrate	Reagent(s) and Incubation Times	Positive Reaction	Negative Reaction
Casein degradation	Casein (milk protein)	No reagent; 24–48 hours	Clear zones around bacterial colonies	No clear zones around colonies
DNAse production	DNA	0.1 *N* hydrochloric acid (HCl); 24–48 hours	Clear zones around colonies after the addition of HCl	Cloudy zones around colonies after addition of HCl
Lipid hydrolysis	Lipid	Spirit blue dye; 24–48 hours	Clear and/or dark areas surrounding growth	No clear or dark area around colonies
Starch hydrolysis	Starch	Lugol's or Gram's iodine; 24–48 hours	Yellow zones around colonies after the addition of iodine	Purple or dark zones around colonies after the addition of iodine

and carbon skeletons. Simple sugars such as glucose and galactose have been described as the initiators of metabolic reactions. Most of the carbohydrates are available to microorganisms in the form of polysaccharides. Intracellular enzymes are used by cells to further the metabolic breakdown of carbohydrates produced by extracellular enzymes or to synthesize more complex cellular molecules. Two examples of intracellular enzymes are maltase and lactase. The former acts upon maltose to yield two molecules of glucose; the latter decomposes the carbohydrate lactose and yields one molecule of glucose and one of galactose.

The ability of an organism to attack and break down various carbohydrates can be determined easily by the use of a suitable nutrient medium containing the carbohydrate and a pH (acid-base) indicator. A pH indicator commonly used is phenol red, which is yellow at pH 6.9 (acid) and red at pH 8.5 (alkaline).

Gas production in a carbohydrate medium may be detected through the use of an inverted (Durham) tube that serves to trap any gas formed. The formation of acid and gas is an indication that the carbohydrate is enzymatically attacked (Figure 58).

Oxidative and fermentative production of acid may often be distinguished in a carbohydrate medium made semisolid by the addition of 0.3% agar. The preparation also contains a pH indicator. In this determination, two tubes of medium are inoculated by stabbing (inserting the inoculating tool down to the bottom of the tube), and the medium in one tube is layered with sterile mineral oil or similar material to exclude oxygen. Fermentative organisms produce acid throughout the medium in both tubes, whereas oxidative organisms produce acid only in the tube without mineral oil (Figure 59). Strict anaerobes produce acid only in the tube with mineral oil. Facultative anaerobes, organisms that can metabolize under

Carbohydrate, Nitrogen, and Other Metabolic Reactions

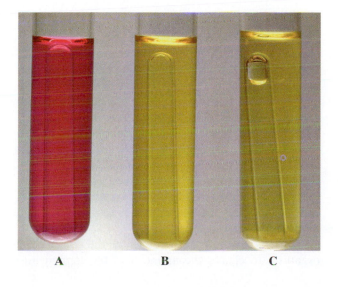

Figure 58
Carbohydrate fermentation employing Durham fermentation tubes. The indicator used is phenol red
(A) Uninoculated. (B) Acid production. (C) Acid and gas production. Note the collection of gas in the inverted vial.

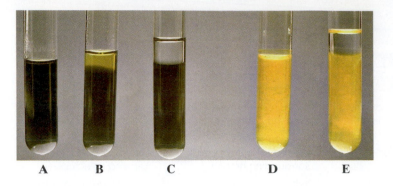

A   B   C   D   E

Figure 59
Carbohydrate oxidation or fermentation using tubes of glucose O/F agar. The indicator used is bromothymol blue. (A) Uninoculated. (B) Shows glucose oxidation. (C) Shows no glucose oxidation or fermentation. (D, E) Show glucose oxidation and fermentation, respectively. Note that the fermentation reaction is detected specifically in the mineral oil–covered medium.

Table 8   Oxidation/Fermentation Reactions

Type of Metabolism	Aerobic Conditions (No Mineral Oil Layer)	Anaerobic Conditions (Mineral Oil Layer)
Oxidative	Acid (yellow)	Alkaline (green)
Fermentative	Acid (yellow)	Acid (yellow)
Nonsaccharolytic	Alkaline (green)	Alkaline (green)

aerobic conditions, produce acid in both tubes. Table 8 summarizes the results obtainable with a glucose oxidative/fermentative (O/F) agar medium.

Other carbohydrate tests used to differentiate bacterial species are the methyl red (MR) and Voges-Proskauer (VP) reactions. In the methyl red test, organisms first metabolize glucose aerobically, thus exhausting all of the available oxygen by means of respiratory metabolism. This is then followed by one of two types of glucose fermentation, **mixed-acid** or **butylene glycol.** The type of fermentation and the end products formed are of value in species identification. Enteric bacterial species that perform the mixed-acid fermentation of glucose excrete large quantities of **acetic, formic, lactic,** and **succinic acids** and **ethanol.** These excreted metabolic products lower pH significantly to approximately 4.2, which is detectable by the methyl red indicator (Figure 60).

The Voges-Proskauer (VP) test is used to detect **acetoin** (a-SET-ō-in), also known as **acetylmethyl carbinol**. This is an intermediate compound produced by organisms carrying out the butylene glycol type of glucose fermentation. A positive reaction is indicated by the formation of a pink or cherry-red color upon the addition of alpha-naphthol and potassium hydroxide (KOH) plus creatine solutions (Figure 61).

Figure 60
Methyl red test: Left tube, negative. Right tube, positive.

Nitrogen Metabolism

Many different nitrogen-containing compounds are involved in the metabolism of a living cell. Proteins are involved in enzymatic and structural activities; purines and pyrimidines of nucleic acids are concerned with genetic mechanisms and certain syntheses; and inorganic compounds act as electron donors or acceptors.

Several differential media and tests are routinely used to detect enzymatic reactions involving nitrogen-containing compounds. These reactions include protein and protein-like compound degradation (gelatin, tryptophan, and phenylalanine), urease activity, and nitrate reduction (Table 9).

The indole test is used to determine the ability of an organism to cleave the amino acid tryptophan into

acetaldehyde

alpha-D-glucose ⟶ pyruvic acid ↘

                                 CO_2 ↗diacetyl

alpha-acetolactate ↗ acetoin ⟨

                                     ↘2,3-butanediol

Table 9   Summary of Selected Carbohydrate and Protein Intracellular Metabolic Reactions

Test	Substrate(s)	Reagents and Incubation Times	Positive Reactions	Negative Reactions
Carbohydrate fermentation (Durham fermentation system)	Glucose, lactose, etc.	Phenol red indicator; 24–48 hours	Acid (yellow color); gas (gas bubble in inverted Durham tube)	Alkaline (red color) no gas (no bubble in Durham tube)
Methyl red	Glucose	Methyl red indicator; 48 hours	Acid production (red color)	No acid production (any color other than red)
Voges-Proskauer	Glucose	Alpha-naphthol and KOH plus creatine; 48 hours	Acetoin produced (red color)	No acetoin produced (any color other than red)
Indole	Tryptophan	Kovac's reagent; 24–48 hours	Indole production (dark red) color in layer on top of broth	Negative indole production (any color other than dark red)
Gelatin hydrolysis	Gelatin	None; refrigeration after incubation is used to test for liquefaction; 48 hours	Gelatin medium is liquid after 30 minutes of refrigeration	Gelatin medium is solid after 30 minutes of refrigeration
Phenylalanine deaminase	Phenylalanine	Ferric chloride; 24–48 hours	Formation of green or brown color upon ferric chloride addition	No color change
Urease	Urea	Phenol red; 24–48 hours	Red color formation	Color other than red
Nitrite reduction	Nitrate	Sulfanilic acid and dimethyl-alphanaphthyl-amine; 24–48 hours	Red color after addition of reagents	No red color

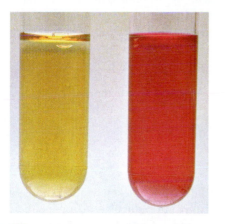

Figure 61
The Voges-Proskauer test. Left tube, negative. Right tube, positive.

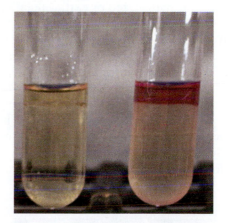

Figure 62
The Indole test. Left tube, negative. Right tube, positive.

indole, ammonia, and **pyruvic acid** (Figure 62). During the metabolic activities of microorganisms, amino acids enter cells and are subsequently degraded by specific intracellular enzymes, the **decarboxylases** and **deaminases**. The former remove the carboxyl group (—COOH) from an amino acid, and the latter remove the amino group (—NH) from an amino acid. If the organism being studied has the enzyme **tryptophanase**, the amino group will be removed and indole will be formed from tryptophan. Indole is easily

detected by the addition of Kovac's reagent and the appearance of a dark cherry-red layer (Figure 62).

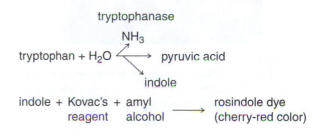

Carbohydrate, Nitrogen, and Other Metabolic Reactions

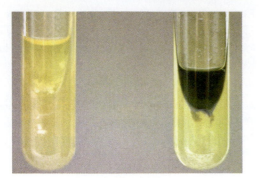

Figure 63

Phenylalanine agar (PA). This medium is used to detect the presence of phenylalanine deaminase. The formation of a green or dark brown color in the medium upon the addition of ferric chloride is a positive reaction.

Detecting the deamination of phenylalanine requires the use of phenylalanine agar. After incubation, ferric chloride is added directly to a growing culture. The development of a green or brown color indicates the presence of phenylpyruvic acid, a product of phenylalanine deamination (Figure 63).

Gelatin, a protein formed from collagen (an animal protein), can be hydrolyzed by extracellular enzymes (proteinases) produced by several microbial species. As microorganisms hydrolyze this protein, it changes from a solid to a liquid, thus its value as a solidifying agent is destroyed. Several tests for proteases are based on such gelatin liquefaction (Figure 64).

Many bacteria are capable of synthesizing acids from by-products of carbohydrate and lipid metabolism when provided with ammonia as a nitrogen source. Some of these organisms are able to split the compound urea, a major organic waste product of

Figure 64

Gelatin hydrolysis. A positive result is indicated by a liquefaction of the gelatin agar (right tube). A negative reaction is indicated by the persistent solid form of the medium (left tube).

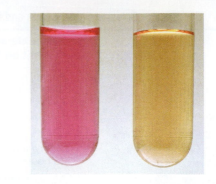

Figure 65

Urease activity. A positive reaction is indicated by the red color of the medium. A negative reaction is represented by an orange color. *Proteus* species generally produce positive results. The indicator in this medium is phenol red.

animal metabolism, into ammonia and carbon dioxide (Figure 65). The medium becomes highly alkaline (basic) owing to ammonia production. The product, carbon dioxide, is incorporated into carbohydrate and nitrogen metabolism through a variety of important reactions.

A major aspect of metabolism is the generation of transport electrons, yielding the energy required for metabolism. One facet of this electron transport system is the ability to reduce nitrate to nitrite as shown in the following reaction. Some microorganisms may go beyond nitrite to nitrogen gas.

$$NO_3^- + 2H^+ + 2e^- \longrightarrow NO_2^- + H_2O$$

Nitrate   Nitrite
Ion   Ion

$$\xrightarrow{7H^+ + 7e^-} N_2O \xrightarrow{2H^+ + 2e^-} N_2$$

              Nitrous   Nitrogen
              oxide   gas

Nitrate Reduction. The enzyme responsible for the reaction is **nitrate reductase**. After incubation, a test for the presence of **nitrate** is performed by the addition of **sulfanilic acid** first and then **demethyl-alpha-naphthylamine.** The appearance of a red color is a positive test. If a red color does not occur, an additional test is necessary to determine whether nitrate was not reduced to nitrite (the absence of nitrate reductase) or whether nitrate was converted to gaseous nitrogen by means of **denitrification** (the presence of nitrite reductase). What actually occurred is determined by adding a small amount of powdered zinc to the culture. The zinc will reduce nitrate to nitrite, producing a positive (red) reaction. This reaction will show that the organism did not possess the necessary enzyme. A negative result at this point will confirm the presence of nitrate reduction

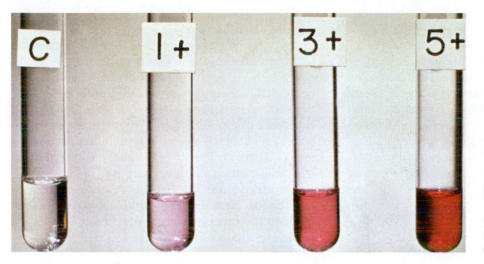

Figure 66
Nitrate reduction. A control tube is included with the positive nitrate reduction reactions to emphasize the range of color changes that can occur.

(Figure 66). The presence of gaseous nitrogen can be demonstrated with the use of an inverted, medium-filled vial and appears as a bubble.

Table 9 summarizes the tests and reactions of nitrogen metabolism presented here.

Oxygen Utilization: Oxidase and Catalase Activities

Cytochrome *c* and catalase are two enzymes involved in the use of oxygen by aerobically respiring bacteria. Cytochromes are enzymes that have nonprotein, iron-containing portions called **heme** and are tightly bound in prokaryotic plasma membranes.

The oxidase test is used to detect the presence of **cytochrome c.** The test uses the reagent tetramethyl-p-phenylenediamine, which can give its electrons (electron donor) to the oxidized form of cytochrome *c*. The resulting oxidized form of this reagent forms a dark blue, violet, or purple color (Figure 67), whereas the reduced form is colorless. Commercially available reagent-containing swabs such as **Oxyswab** and disposable slides such as the **DrySlide Oxidase** make testing for cytochrome *c* relatively simple (Figure 67b-c).

The oxidase test is of value in detecting the presence of cytochrome *c*–containing gram-negative cocci belonging to the genera *Moraxella (Branhamella)* and *Neisseria*. In addition, the test can distinguish between

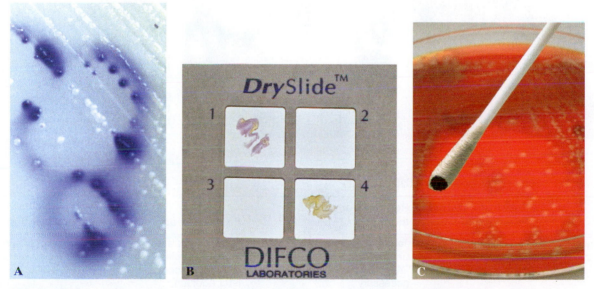

Figure 67
The oxidase test. (A) Colonies demonstrating a positive reaction are blue and turn darker later. Negative reactions are indicated by colorless colonies. (B) Results on a DrySlide. (Courtesy of Difco Laboratories, Detroit, MI.) (C) A positive Oxyswab test. The reagent is contained in the swab material. (Courtesy of Remel, Lenexa, KS.)

oxidase-negative, gram-negative enteric bacterial rods and gram-negative rods belonging to the genera *Pseudomonas* and *Aeromonas*.

Toxic forms of oxygen such as the superoxide radical (O_2) are found in all liquid environments that contain dissolved oxygen. These chemicals are toxic to living cells, so microorganisms in aerobic environments must be able to detoxify them. Such microbes produce an enzyme, **superoxide dismutase** (SD), which adds protons to the superoxide radical to form hydrogen peroxide, as shown in the following equation:

$$2O_2 \xrightarrow{2e^-} 2O_2 \xrightarrow{4H^+} 2H_2O_2$$

Superoxide   Hydrogen peroxide

Because hydrogen peroxide also is toxic, microorganisms also must produce the enzyme **catalase,** which is capable of decomposing the chemical into water and molecular oxygen:

$$2H_2O_2 \xrightarrow{O_2} 2H_2O$$

Catalase

Catalase is produced by all actively growing aerobic microbes. Because strict anaerobes lack the ability to use oxygen in their respiration, they also lack catalase.

The test for catalase is simple to perform. All it requires is the addition of a few drops of 3% hydrogen peroxide directly to a young broth culture (Figure 68A), or to colonies on an agar surface (Figure 68B), or to a clump of cells on a glass slide. The vigorous evolution of oxygen bubbles is a positive result. Cultures growing on blood agar are not tested because the agar medium itself may produce a positive reaction.

The catalase test is very useful in differentiating bacteria that have similar morphological features but differ in their metabolic activities. Table 10 summarizes the features of the oxidase and catalase tests.

Motility Media

Bacterial motility can be shown with the aid of several types of motility media. The composition of these preparations is such that they offer no more resistance to movement during incubation than would a broth culture. Motility Medium S, which contains 2, 3, 5-triphenyltetrazolium chloride (TTC),

Oxygen Utlization, Motility, and IMViC Reactions

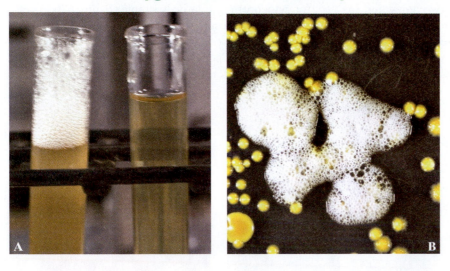

Figure 68
The catalase reaction. Catalase is an enzyme that catalyzes the breakdown of hydrogen peroxide (H_2O_2), thereby releasing oxygen gas. Formation of a white froth if a few drops of 3% H_2O_2 are added to a microbial colony or to a broth culture is a positive reaction. (A) A positive reaction in a broth culture. (B) The agar plate reaction.

Table 10   Summary of Oxidase and Catalase Tests

Test	Substrate, Detects Presence of	Reagents and Incubation Time	Positive Reaction	Negative Reaction
Oxidase (cytochrome c)	No substrate; enzyme cytochrome c	Tetramethyl-phenylenediamine; depending on procedure, 24–48 hours may be necessary	Formation of a deep violet or purple color within 60 seconds after addition of reagent	No major color change after addition of reagent
Catalase	No substrate; enzyme catalase	Hydrogen peroxide; cultures 18–24 hours old	Formation of visible bubbles after addition of reagent	No visible bubbles formed after addition of reagent

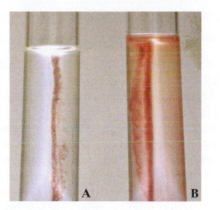

Figure 69
Motility medium. This preparation, to which a dye called 2,3,5-triphenyltetrazolium chloride is added, can be used to detect motile organisms. (A) In a nonmotile culture, growth appears only along the line of inoculation. (B) In a motile culture, growth spreads from the line of inoculation, making most of the medium turbid.

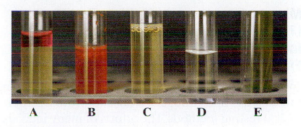

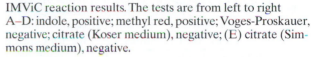

Figure 70
IMViC reaction results. The tests are from left to right A–D: indole, positive; methyl red, positive; Voges-Proskauer, negative; citrate (Koser medium), negative; (E) citrate (Simmons medium), negative.

is used. Motility can be recognized by the presence of a diffuse red growth away from the line of inoculation, or stabline (Figure 69). Nonmotile organisms grow only along the stabline.

Differential Test Patterns: The IMViC Test

The IMViC pattern of reactions reflects the enzymatic makeup or biochemical fingerprint of the microorganism being studied. For example, *Escherichia coli,* a normal inhabitant of the intestinal tracts of humans and lower animals, resembles *Enterobacter aerogenes,* a bacterial species that is widely distributed in nature, especially on plants and plant products. Both organisms are gram-negative, and they are similar in morphological and cultural characteristics. These bacteria can be differentiated by means of the IMViC set of tests (Figure 70). Each letter in the series (except *i*) refers to a separate procedure. *I* stands for the indole test; *M,* the methyl red test; *Vi,* the Voges-Proskauer reactions; and *C,* the citrate

test. The small *i* after the *V* is used to make pronunciation easier. The indole, methyl red, and Voges-Proskauer tests were described earlier.

The remaining test in the IMViC series, the **citrate test,** determines the presence of enzymes that enable citrate to enter the cell, which can then use the citrate as the sole source of carbon for its metabolism and growth. Two media can be used for the citrate test: **Koser citrate** and **Simmons citrate.** The reaction in Koser citrate medium (a clear colorless liquid) is positive if the preparation becomes cloudy after incubation (Figure 71A). A light inoculum must be used with Koser citrate. Simmons citrate agar (Figure 71B) contains the pH indicator bromthymol blue, which is green under acidic conditions and dark blue when the medium becomes alkaline. Organisms utilizing citrate produce an alkaline reaction, as indicated by a Prussian blue color. Carbon dioxide released during the reaction causes the color change in the medium.

$$\text{citrate} \xrightarrow{\text{citrase}} \text{oxaloacetic acid} + \text{acetic acid}$$

$$\text{acetic acid} + \text{formic acid} \longleftarrow \text{pyruvic acid} \xrightarrow{} CO_2$$

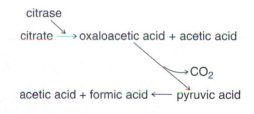

Figure 71
The citrate test. (A) Koser's citrate. Left tube, negative; right tube, positive. (B) Simmons citrate. Left tube, negative; right tube, positive.

Microbial Reactions in Multiple-Test Media: Litmus Milk; Triple Sugar Iron Agar; Sulfide, Indole, Motility Medium, and Four-in-One Entero-Screen 4

Investigators have long been interested in reducing the expense and drudgery associated with microbiological testing methods. The efforts of many individuals have resulted in the development of at least two general categories of improved biochemical testing procedures and materials: (1) several test substrates combined in one or two tubes that are inoculated and incubated in the conventional manner, and (2) separate biochemical tests contained in miniaturized, multi-compartment devices that are inoculated by other than conventional methods but are incubated according to standard practices.

The first category includes several widely accepted media, such as **litmus milk** (Figures 72 and 73), **triple sugar iron agar** (Figure 74), and **sulfide, indole, motility medium** (Figure 74).

Very few media, after inoculation and incubation, can yield as much information on microbial reactions as litmus milk. The medium includes materials that most bacteria require for growth and, specifically, the substrates lactose and casein. Litmus serves as an indicator of the organism's acid or alkali production and oxidation and reduction activities. Characteristic reactions observed with litmus milk are as follows (Figures 72 and 73).

1. Acid and acid curd formation
2. Alkaline conditions
3. Rennin curd production
4. Peptonization
5. Litmus reduction
6. Gas formation

Production of acid by an organism is a function of its ability to utilize lactose, one of the most abundant sugars in milk. Acid production is demonstrated when a litmus medium changes from blue to pink. If the organism is able to produce a considerable amount of acid, an insoluble complex of calcium and casein may be formed, resulting in a curd. An organism that cannot attack lactose might well use the milk proteins as a source of nitrogen and carbon, and an alkaline reaction might result. This reaction is indicated by an intensification of the blue color in the litmus medium. If the amount of acid or alkali produced is low, no apparent change in the medium may be observed. However, a rennin curd may develop even with low acid or alkali production if the organism has enzymes that can produce the insoluble calcium-casein complex. This type of curd usually retracts and yields a grayish liquid known as **whey.** Acid or rennin curds are quite palatable dairy products known as cottage cheese. Either type of curd can be digested by proteinases, a process known as **peptonization** (or **solubilization**). Peptonization is usually characterized by a reduction in curd size and the formation of a brownish liquid. This reaction can be observed easily when it occurs along with the curd. Reduction of the litmus by oxidation-reduction activities of the bacteria can be shown by loss of the litmus color. The reaction is quite apparent when a pink acid curd begins to turn white. The change develops at the bottom of the tube and moves upward. Another readily apparent characteristic is gas production, which can be observed as holes or tears in curds or as separated curd strands.

Multiple-Test Media Reactions and Miniaturized and Rapid Microbiological Systems

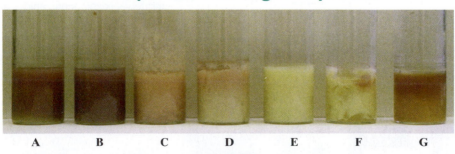

Figure 72
Selected litmus milk reactions. (A) Uninoculated blue medium. (B) An alkaline reaction (dark blue to purple), indicating the inoculated organism's use of the milk proteins as a source of carbon and nitrogen. (C) An acid result (pink), demonstrating the production of a considerable amount of acid and lactose fermentation. (D) Acid production at the surface (pink) and the beginning of litmus reduction (white portion of the medium). The reaction is caused by the oxidation-reduction activities of the inoculated bacteria. This reaction begins at the bottom of the tube and spreads upward. (E) Litmus reduction. (F) Acid clot formation with extreme gas production causing stormy fermentation. (G) Casein curd formation and peptonization (proteolysis). With peptonization, a reduction in curd size occurs and a brownish, purple, or clear supernatant, called whey, is formed.

Triple sugar iron agar is a characteristic preparation used to differentiate Gram-negative enteric organisms by their ability to ferment dextrose (glucose), lactose, or saccharose (sucrose) and to reduce sulfites to sulfides. The medium is dispensed as agar slants. It is inoculated by the insertion of an inoculating needle into the bottom, or butt region, of the tube, and, as the needle is withdrawn from this area, the slant surface is streaked. Several types of reactions can be recognized (Figure 74 and Table 11).

When a culture ferments only dextrose, the concentration of which is one-tenth of lactose or sucrose, the bottom portion of the medium will turn yellow from the acid produced. Because its concentration is low, glucose is quickly utilized. In the event that the organism in the culture cannot ferment either of the other two carbohydrates in the medium, growth in the butt stops. However, the organisms on the slant continue to utilize, as a source of energy, the peptone in the medium or the intermediate products of dextrose fermentation, or both. Use of these intermediate products results in their reduction, and the use of peptone results in the eventual secretion of ammonia into the medium. The ammonia will neutralize the intermediate products, causing the slant to turn red.

The entire medium will turn yellow and remain so if the culture can utilize lactose, sucrose, or both, in addition to dextrose. Apparently, the organisms in such cases will not exhaust the available fermentable carbohydrates. The peptone, or the intermediate end-products, are not limited. Gas production can be detected when holes in the agar are formed or the medium is broken into several fragments.

Hydrogen sulfide (H_2S) production by a culture results in the blackening of the medium. This color is caused by the production of H_2S from an ingredient of the medium, sodium thiosulfate, which, when combined with another component of the medium, ferrous ammonium sulfate, results in the formation of the black insoluble compound ferrous sulfide FeS.

$$H_2S + FeSO_4 \rightarrow H_2SO_4 + FeS\downarrow$$

Sulfide, indole, motility (SIM) medium can be used to determine hydrogen sulfide production, indole production, and motility. The presence of H_2S is indicated by a blackening along the line of inoculation in the medium. Quite often, the dark color diffuses when a motile organism is involved. The presence of indole is detected by the use of Kovac's reagent. Motility is recognized by the appearance of a diffuse growth or turbidity spreading from the inoculation site (Figure 75).

The **Four in One EnteroScreen 4** is a one-tube system designed for the screening of clinical stool

Table 11 Summary of Selected Multiple Test System Biochemical Reactions

Test	Substrate(s) and Incubation Times	Reagent(s)	Positive Reaction	Negative Reaction
Litmus milk	Lactose and/or casein; 24–48 hours	Litmus indicator	Acid (pink) Alkaline (light to dark blue) Litmus reduction (white) Curd: milk changes to solid state with fluid (whey) Proteolysis (decrease in milk turbidity and eventual formation of clear brownish or purple fluid)	No change (light blue)
Triple Sugar Iron Agar	Dextrose, lactose, sucrose; 24–48 hours	Phenol red	Acid from dextrose (yellow bottom) Acid from lactose and sucrose (yellow slant)	Alkaline, no change (red bottom) Alkaline (red slant)
	Sodium thiosulfate; 24–48 hours	Ferrous ammonium sulfate	H_2S positve (blackening of medium)	H_2S negative (no blackening of medium)
Sulfide, indole, motility medium	Sodium thiosulfate; 24–48 hours Tryptone; 24–48 hours None; 18–24 hours	Peptonized iron Kovac's reagent None	H_2S positive (blackening of medium) Indole positive (red layer) Motility (turbidity throughout medium)	H_2S negative (no blackening of medium) Indole negative (no red layer) No motility (growth only on line of inoculation)

Figure 73

Coagulation and peptonization. Note the solidification of the milk protein and the clear liquid, whey.

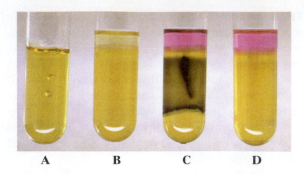

Figure 75

Bacto-SIM medium reactions. (A) Uninoculated. (B) Positive for motility (general turbidity in the medium). (C) Positive for hydrogen sulfide production (blackening of the medium), for indole (red layer), and positive for motility (cloudy). (D) Positive for indole, and for motility.

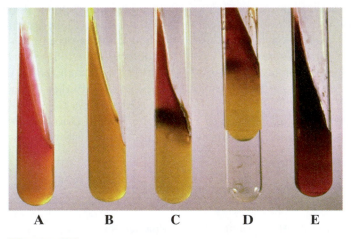

Figure 74

Selected triple sugar iron agar medium reactions. Alkaline reactions (Alk) are indicated by a red coloration; acid (A) production by a yellowing of the medium; gas formation by the presence of air pockets in the preparation; and the presence of H_2S by a blackening of the medium. The reactions shown are as follows:

	Tube A	Tube B	Tube C	Tube D	Tube E
Slant	Alk	Acid	Alk	Alk	Alk
Butt	A	A	A	A	Alk
Gas	Absent	Absent	Absent	Present	Absent
H_2S	Absent	Absent	Present	Present	Absent

Note: The dark color in tube E is caused by the purple pigment of the organism. Acid in glucose (bottom) must be produced before H_2S can occur.

cultures for the presence of *Salmonella* and *Shigella* species. The medium consists of specific substrates arranged in layers. These substrates are: lysine with ingredients for hydrogen sulfide production and to detect a deaminase reaction; lysine to detect a decarboxylase reaction without ingredients for hydrogen sulfide production; and urea agar to show the presence of urease. A sterile petrolatum layer separates the urea agar from the upper components of the medium. No additional reagents are needed to interpret the results after incubation of an inoculated system (Figure 76).

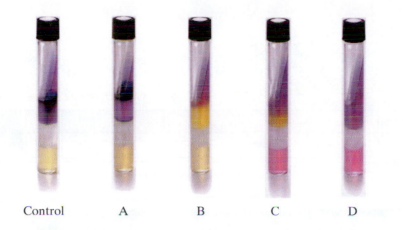

| | Control | A | B | C | D |

Figure 76

The Four-in-One EnteroScreen 4, an accurate one-tube system for the screening of clinical specimens for *Salmonella* and *Shigella* species. Positive reactions for the substrates in the medium are: lysine deamination, yellow color in the top layer; H_2S formation, blackening of the medium in the top layer; a positive lysine decarboxylase reaction, yellow color in the layer above the petroleum layer; and a positive urease test, a pink color in the bottom layer. Note the presence of the sterile petroleum white layer just above the urease test agar. The reactions shown in the representative tubes are shown below. No reactions (NR) are indicated for the control tube. (Courtesy of Hardy Diagnostics, www.HardyDiagnostics.com)

Test	Control	Tube A	Tube B	Tube C	Tube D
Lysine deamination	NR	–	–	–	–
H_2S	NR	+	–	–	–
Lysine decarboxylase	NR	–	+	+	–
Urease	NR	–	–	+	+

Miniaturized and Rapid Microbiological Systems

Improved biochemical procedures include several commercially available miniaturized multiple test systems and devices designed to facilitate the identification of bacterial cultures. Complete instructions together with accurate identification keys based on established biochemical reactions are provided by the respective manufacturers. Two systems—the **Enterotube II** (Figure 77) and the **API 20E** (Figure 78)—consist of manufacturer-selected combinations of substrates that not only are ready for use but require minimal storage space and are stable at either room or refrigerator temperature for significant time periods. The carefully selected tests in each case also form the bases for computer-developed identification systems. These two systems emphasize the importance placed on the need for rapid analysis of test results for the identification of clinically isolated bacteria.

The Enterotube II, which is used for the identification of lactose-fermenters, incorporates conventional media into a single, ready-to-use, multicompartment tube with an enclosed inoculating rod (Figure 77). The unique inoculating arrangement permits simultaneous inoculation of all compartments and the performance of 15 biochemical tests. The Enterotube II

is used by first removing the end caps of the system to expose a long inoculating rod. The rod is used to pick an isolated colony from a selective and differential agar, or related plate medium (Figures 48–53). The rod is then drawn through a series of substrate-containing agar compartments in the plastic tube of the system. The end caps are then replaced, and the system is incubated. After incubation, appropriate reagents are introduced into the chambers requiring them. The color changes resulting from the biochemical activity of unknown bacterial cultures used as inocula are interpreted and used in their identification.

The API system employs a plastic strip holding 20 miniature compartments, or cupules. Each cupule contains a dehydrated substrate for a different test (Figure 78). At least 22 standard biochemical tests can be performed, including *o*-nitrophenyl-β-D-galactosidase (ONPG), arginine dihydrolase, lysine and ornithine decarboxylases, citrate utilization, hydrogen sulfide (H_2S) production, urease, tryptophan deaminase, indole production, acetoin production, gelatinase, and the fermentation of glucose, mannitol, inositol, rhamnose, sucrose, melibiose, amygdaline, and arabinose.

The dehydrated substrates are inoculated with a bacterial suspension and subsequently incubated according to a procedure described by the manufacturer.

A

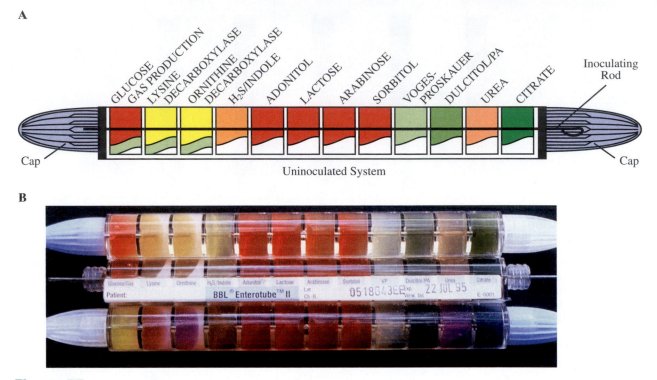

Uninoculated System

B

Figure 77

The rapid and easy-to-use Enterotube II System. Specific substrates are contained in individual compartments which can be inoculated by pulling a special inoculation needle through each of the compartments. With this system 14 standard biochemical tests can be performed. (A) A diagram of an inoculated system before incubation. Test symbols; D = dextrose; Gas = gas production from dextrose; Lys = lysine; Orn = ornithine; H$_2$S = hydrogen sulfide production; I = indole; Lac = lactose; Phenyl = phenylalanine; Dul = dulcitol; U = urease; Cit = citrate. (B) An actual inoculated system before incubation (Top system). A system with the end caps removed and exposing the inoculating needle (Middle system). The bottom system after incubation showing several positive reactions for the presence of specific enzymes: Glucose, acid, no gas; Lysine, acid; Ornithine, acid; H2S +,/ indole −; Urea +; Citrate +. (The explanation of the individual test symbols is listed above.)

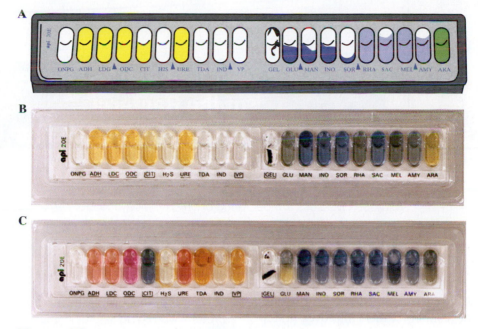

Figure 78

The rapid and ready-to-use API 20 Enteric (E) System. Specific substrates are contained in individual compartments known as *cupules*. With this system 22 standard biochemical tests can be performed. (A) A diagram of an inoculated system before incubation. Test symbols; ONPG = α-glactosidase; ADH = arginine dihydrolase; LDC = lysine decarboxylase; ODC = ornithine decarboxylase; CIT = citrate; H₂S = hydrogen sulfide production; URE = urease; TDA = tryptophan deaminase; IND = indole; VP = acetoin; GEL = gelatin; GLU = glucose; MAN = mannitol; INO = inositol; SOR = sorbitol; RHA = rhamnose; SAC = sucrose; MEL = melibiose; AMY = amygdalin; ARA = arabinose; OXI = oxidase; NO₃ = nitrate reduction. (B) An actual inoculated system before incubation. (C) The system after incubation showing several positive reactions for the presence of specific enzymes; ODC, pink; CIT, blue; URE, pink; TDA, orange; and GLU, yellow. (The explanation of the individual test symbols is listed above.)

Bacteriology: A Survey of the Bacteria and Archeae Domains

Microbes possess a wider range of physiological and biochemical potentialities than do all other organisms combined. Microbes represent forms of life that can persist in nature because they fill particular ecological niches.

— *C.B. Van Niel*

The classification of bacteria changes and expands as our understanding of them improves. In view of the fact that thousands of different bacterial species are known, it would not be possible to consider all of them in the limited survey presented here. However, consideration will be given to a significant number of representatives of both cultured bacterial species and other bacterial species known to exist but not yet cultured. Diverse nonpathogens as well as pathogens and representative disease states are included in the survey. The presentation of the bacteria here follows the classification scheme currently accepted by most microbiologists and contained in *Bergey's Manual of Systematic Bacteriology* and in *The Prokaryotes*.

▲ THE BACTERIA

The domain *Bacteria* consists of 16 phyla, and collectively contains thousands of different species. A substantial number of species from selected phyla are presented. Some of the representatives selected are well-known cultured species, while others, although not yet cultured in the laboratory, have distinctive phenotypic properties. Descriptions of diseases associated with specific pathogens also are included in this survey.

▲ Phylum 1: Proteobacteria

The phylum Proteobacteria is the largest of all the phyla in the domain and contains the broadest range of extremely physiologically diverse bacteria. Physiologically, these microorganisms can be either (1) *phototrophic* (obtain energy from the light), (2) *chemolithotrophic* (obtain energy from the oxidation of inorganic compounds, or (3) *chemoorganotrophic* (obtain energy from the oxidation of organic compounds). The name of the phylum is derived from the mythological Greek god Proteus, who could assume numerous shapes.

The Proteobacteria contains five distinctive subgroups, each of which consists of several genera. The subgroups are designated by the Greek letters *alpha* (α), *beta* (β), *gamma* (γ), *delta,* (δ), and *epsilon* (ϵ). These clusters make up most of the Gram-negative chemoorganotropic bacteria in the domain, and are believed to have arisen from a photosynthetic ancestor. Selected representatives from each of the five clusters are presented.

▲ The Alpha Proteobacteria

This subgroup contains a number of agriculturally important species, as well as several plant and human pathogens. The alpha proteobacteria are known for their ability to grow at very low levels of nutrients. In addition, several species have unusual shapes and exhibit protrusions known as stalks and buds.

▲ Aerobic/Microaerophilic Gram-Negative Coccobacilli and Stalked Bacteria

Alcaligenes

Alcaligenes faecalis
A common inhabitant of the intestinal tract, *Alcaligenes faecalis* can also be found in water and in soil. Strains of this organism have been recovered from pussy ear discharges, blood, urine, and sputum specimens and associated with certain hospital-acquired (**nosocomial**) infections.

Morphology and Cultural Properties. *Alcaligenes faecalis* is a Gram-negative short rod, or almost a coccobacillus, 0.5–1.0 μm wide and 0.5–2.6 μm long (Figure 79). The organism is motile and obligately aerobic. The growth temperatures range from 20° to 37°C. Colonies on nutrient agar are nonpigmented, glistening, and convex (Figure 80A). *Alcaligenes faecalis* is non-hemolytic (Figure 80B) and grows well on MacConkey agar (Figure 80C). The organism is oxidase and catalase positive.

Pathology. *Alcaligenes faecalis* occasionally causes opportunistic as well as nosocomial infections.

Brucella

Brucellosis, also known as undulent fever and Malta fever, is an acute or chronic recurring disease spread from lower animals to humans. The disease brucellosis exists worldwide, especially in countries of the Mediterranean basin, the Arabian gulf, the Indian subcontinent, and parts of Mexico and Central and South America. There are four species of *Brucella* that are pathogenic for humans, and these have a limited number of preferred natural host animals: *B. abortus* (cattle), *B. melitensis* (goats and sheep), *B. suis* (swine), and *B. canis* (dogs). The disease is an important cause of abortion and sterility, and it decreases milk production in farm animals.

Transmission. Common routes of infection include inoculation through cuts and abrasions in the skin or through the conjuctival sac of the eyes, inhalation of infectious aerosols, and ingestion into the gastrointestinal tract.

Morphology and Cultural Properties. *Brucella* species are Gram-negative cocco-bacilli, 0.5 μm wide and 0.3–0.9 μm long, occurring singly and in pairs (Figure 81). The organisms are nonmotile and may be encapsulated.

Brucella species require an aerobic environment and enriched media, such as blood agar. Smooth pinpoint colonies appear on primary isolation after 2 to 3 days of incubation. Differentiation among *Brucella* species can be on the basis of responses to dye-impregnated disks and biochemical tests. These organisms are usually catalase and oxidase positive.

Pathology and Clinical Features. In humans, the reticuloendothelial system is the main target of infection. If not treated adequately with antibiotics, the infection results in the formation of small tumors that serve as sites for bacterial reproduction. Eventually, bacteria are released periodically into the circulatory system. Such recurrent episodes are mainly responsible for the recurring chills and fever of the clinical illness.

Brucellosis begins with feelings of general discomfort, chills, and fever for 7 to 21 days after infection. Drenching sweats in the late afternoon or evening are common and may continue for several weeks or even 1 to 2 years. Other findings include weight loss, body aches, loss of appetite and enlargement of lymph nodes, spleen, and liver.

Diagnosis. Definitive diagnosis requires isolation of *Brucella* from blood or biopsy specimens. Serological tests also are used.

Caulobacter

Caulobacter species
Frequently found in aquatic environments attached to particulate matter, plant material, or other microorganisms. Caulobacters are nonpathogenic.

Morphology and Cultural Properties. *Caulobacter* (kō-lō-BAK-ter) species are Gram-negative rod-shaped, or fusiform (spindle-shaped) cells, 0.4–0.6 μm wide and 1.0–2 μm long (Figure 82). These bacteria belong to a large and heterogenous group of Proteobacteria that are noted for having various kinds of cytoplasmic extrusions: *stalks, hyphae,* or *appendages.* These extrusions are called **prosthecae** (singular, **prostheca**), contain cytoplasm, are bounded by the organism's cell wall, and are smaller than the mature cell. Cell division among the *Caulobacter* and members of another stalked bacterial genus, *Prosthecobacter,* involves the formation of a new daughter cell with the parent or mother cell keeping its identity after the completion of the cell division process. This unequal binary fission or splitting into two nonequivalent cells is in sharp contrast to the binary fission process resulting in the formation of two equivalent cells that generally is typical of other bacteria. One end of the cells bears a single flagellum, while the other bears the stalk (Figure 82). A holdfast device is formed at the end of the stalk for attachment purposes. Caulobacters are strict aerobes and chemoorganotrophic.

Rhizobium

Rhizobium species
These bacteria are found in plant root nodules, where they normally convert atmospheric nitrogen (N_2) into ammonia (NH_3), for use by the plant host (Figure 83A). This process, known as **nitrogen fixation**, is restricted to certain prokaryotes. Some bacteria, called *symbiotes,* fix nitrogen only in association with certain plants. *Rhizobium* (ri-ZŌ-bē-um) species specifically infect the roots of legumes, such as beans, peanuts, and peas, and contribute to the formation of nodules (Figure 83B).

Aerobic/Microaerophilic Gram-Negative Coccobacilli, Rods, and Stalked Bacteria

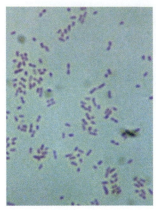

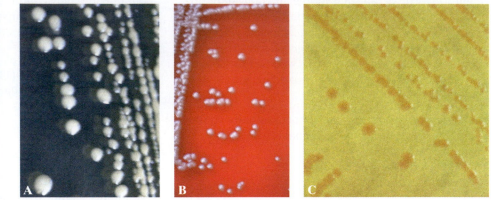

Figure 79
Gram-negative rods of
Alcaligenes faecalis
(1,000×).

Figure 80
(A) *Alcaligenes faecalis* on nutrient agar. (B) *Alcaligenes faecalis* on blood agar.
(C) Nonfermenting colonies of *Alcaligenes faecalis* on MacConkey agar.

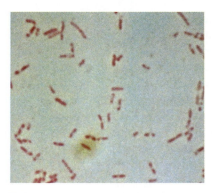

Figure 81
Gram-negative cocco-bacilli of *Brucella melitensis.*

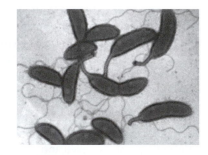

Figure 82
The stalked bacterium *Caulobacter*. This bacterium is
noted for its unequal division process. (Courtesy of S.
Koyasu, Tokyo Metropolitan Institute of Medical
Sciences.)

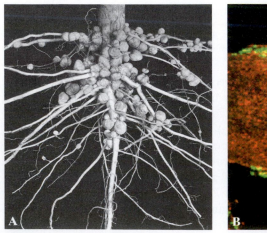

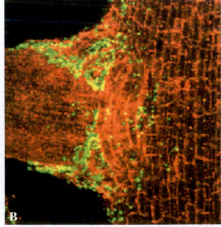

Figure 83
Symbiotic nitrogen fixation.
(A) Nodules that house symbiotic
nitrogen-fixing bacteria on the root
of a soybean plant. (Courtesy of
USDA.) (B) Confocal laser-
scanning micrograph showing a
root junction where fluorescent
tagged (green) *Rhizobium* cells
gain entrance into a plant host.
Rhizobia enter the interior of the
root by passing between surface
displaced cells. (From F. Chi,
S-H. Shen, H-P. Cheng, Y-X. Jing,
Y. G. Yanni, and F.B. Dazzo. *Appl.
and Environ. Microbio.* 71 (2005):
7271–7278.)

Morphology and Cultural Properties. *Rhizobium* species are Gram-negative rods, 0.5–0.9 × 1.2–3.0 μm. The bacteria move by means of a single polar flagellum or by two to six flagella distributed around individual cells (*peritrichous*). Rhizobia are aerobic and chemoorganotrophic. Optimum growth temperatures range from 25° to 30°C. The optimun pH is 6–7. Colonies are circular, convex, semitranslucent, raised, and slimy. Yeast–mannitol–mineral-salts medium can be used for cultivation. Cellulose and starch are not utilized.

▲ Rickettsia Ehrlichia, and Bartonella

Rickettsia

Rickettsias are a diverse collection of obligate intracellular bacteria that grow and reproduce within eukaryotic host cells (Figure 84). They are the causative agents of a variety of zoonotic diseases called rickettsioses and include the genera *Coxiella*, *Ehrlichia*, *Orienta*, and *Rickettsia* (Figure 85).

Historically, members of the genus *Rickettsia* have been divided into the spotted fever group (SFG), the typhus group, and the scrub typhus group. Examples of species included in each of these groups are listed in Table 12. The associated diseases and the means of transmission also are indicated.

Transmission. *Rickettsia* and *Orienta* species are spread by the bite of infected ticks (Figure 86B) or mites or by the feces of an infected louse (Figure 87B) or flea.

Morphology and Cultural Properties. The rickettsiae are mainly Gram-negative, often pleomorphic, small rods. Cells measure 0.25 μm in width and 0.5–1.25 μm in length (Figures 86A and 87B). Most rickettsiae can be isolated or cultured in embryonated eggs and cell cultures.

Pathology and Clinical Features. The severity of the different rickettsioses varies considerably. For example, rickettsialpox is a relatively mild disease that has never proved fatal (Figure 85), whereas Rocky Mountain spotted fever and epidemic typhus fever are serious, life-threatening illnesses with high fatality rates in untreated individuals.

Diagnosis. The rickettsioses are difficult to diagnose both clinically and in the laboratory. Laboratory diagnostic techniques include animal or cell culture inoculation with specimens and serological methods such as fluorescence-antibody tests (Figures 86A, 87A and 88), enzyme immunoassay, or the older Weil-Felix test. The last-named technique is used to show the presence of antibodies that agglutinate specific antigens of the bacterium *Proteus mirabilis*.

Coxiella

Coxiella burnetii is the causative agent of Q fever. The disease is a zoonosis that is widespread in a variety of animals, including cats, cattle, goats, and sheep. Organisms are found in their infected mammary glands, milk, urine, feces, and placentas. *Coxiella burnetii* is considered to be one of the most infectious of bacteria and among the most stable in the environment.

Transmission. Infection in humans occurs through inhaling *C. burnetii* in contaminated aerosols. Outbreaks of Q fever are also associated with domestic animals, especially infected newborns.

Table 12 Human Rickettsioses

Group	Pathogen	Disease	Vector
Spotted fever	*Rickettsia rickettsii* (Figure 86A)	Rocky Mountain spotted fever	*Dermacentor* species (wood ticks; Figure 86B)
	R. akari	Rickettsialpox	*Allodermanyssus sanguineus* (mite)
Typhus	*R. prowazekii* (Figure 87A)	Epidemic typhus fever	*Pediculus humanus corporis* (human body louse, Figure 87B) louse feces or bite
	R. typhi	Murine typhus fever	*Xenopsylla cheopis* (rat flea; Figure 155B) flea feces or bite
Scrub typhus	*Orienta (Rickettsia) tsutsugamuchi*	Scrub typhus fever	*Leptotrombidium akamushi* (mite)
Q fever	*Coxiella burnetii*	Q-fever	Inhalation of pathogen

Rickettsia, Ehrlichia, And Bartonella

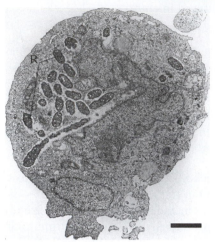

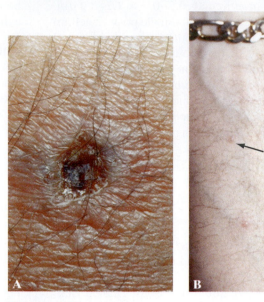

Figure 84

Transmission electron micrograph of a tissue culture cell infected with *Rickettsia prowazeki* (**R**), the cause of epidemic typhus fever. A nucleus, plasma (cell) membrane, and mitochondria of the host eukaryotic cell are shown. (Courtesy of Drs. David J. Silverman and Charles L. Wisseman, Jr., Department of Microbiology, University of Maryland School of Medicine.)

Figure 85

Clinical features of rickettsial infections. (A) A characteristic, painless lesion that can be found at the site of a tick bite. (B) The red, slightly elevated rash that commonly occurs. (From C.A. Kemper, A.P. Spivak, and S.C. Deresinki. *CID* 15(1992):591–94.)

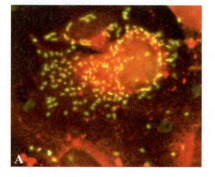

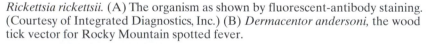

Figure 86

Rickettsia rickettsii. (A) The organism as shown by fluorescent-antibody staining. (Courtesy of Integrated Diagnostics, Inc.) (B) *Dermacentor andersoni,* the wood tick vector for Rocky Mountain spotted fever.

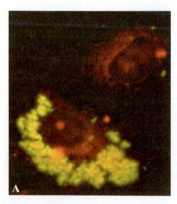

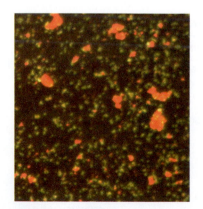

Figure 87

Rickettsia prowazekii. (A) The causative agent as shown by fluorescent-antibody staining. (Courtesy of Integrated Diagnostics, Inc.) (B) *Pediculus humanus corporis,* the body louse.

Figure 88

Fluorescent-antibody showing the presence of *Coxiella burnetii* (small, yellow-green cells). (Courtesy of Integrated Diagnostic, Inc.)

Morphology and Cultural Properties. *Coxiella burnettii* is an obligate intracellular, Gram-variable coccobacillus. It is quite pleomorphic and measures 0.25 μm in width and 0.5–1.25 μm in length (Figure 88).

Coxiella burnetii can be cultured under laboratory conditions in embryonated chicken eggs or cell cultures.

Pathology and Clinical Features. After infection with *C. burnetii,* a variety of clinical syndromes may result. Some patients are entirely asymptomatic, but most exhibit a mild flu-like illness. The major obvious clinical features of Q fever can be divided into two groups: acute and chronic. Acute Q fever is characterized by fever, general discomfort, headache, pneumonia, and hepatitis. The relative nonspecific nature of these signs and symptoms leads to difficulties in diagnosis.

The development of chronic Q fever complicates the infectious process, which ultimately leads to valvular endocarditis (the most common result), hepatitis, and vertebral osteomyelitis.

Diagnosis. Although *C. burnetii* can be isolated by inoculation of embryonated eggs or by cell culture, diagnosis is generally made by serological methods demonstrating antibodies to specific proteins. Direct light or electron microscopic demonstrations of *C. burnetii* in cardiac valves by means of immunofluorescence or immunohistochemistry also are used.

Ehrlichia

Ehrlichia chaffenis (Figure 89) and *E. sennetsu* are two species known to cause human ehrlichiosis. Other species cause several important veterinary diseases. Blood-forming cells are the major sites of invasion by ehrlichiae.

Transmission. Ehrlichiosis is transmitted by ticks.

Morphology and Cultural Properties. The genus *Ehrlichia* consists of small, obligate intracellular bacteria that average 0.5 μm–1.5 μm in length. The ultrastructural morphology is highly variable and ranges from cocci or cocco-bacilli to unusual shapes and forms. Ultrastructural studies have revealed a Gram-negative-type cell wall. Specialized laboratories can be used for cell cultures, isolations, and cultivations.

Ehrlichia form **morulae**, large solid masses of cells. When stained, they appear as round or ovoid purple bodies measuring 2–5 μm in diameter and surrounded by a single membrane (Figure 89A). *Ehrlichia* also can cause cellular damage (cytopathic) that appears as clear areas (plaques) in cell cultures (Figure 90B).

Pathology and Clinical Features. Human ehrlichiosis ranges from asymptomatic illness or mild fever with spontaneous recovery to a severe disease with hepatitis and kidney failure. The disease has an incubation of about 2 weeks in adults. Its onset is usually abrupt, and individuals experience fever, chills, muscle pain, and headache.

Diagnosis. Microscopic examination of Giemsa-stained blood smears may be of value in some cases. However, fluorescent-antibody tests using *E. chaffenis* as an antigen are most commonly the method of choice. An increase in antibody levels and the polymerase chain reaction also are of value.

Bartonella

Bartonella (formerly *Rochalimaea*) species are now associated with several clinical syndromes, including trench fever, bacillary angiomatosis (Figure 90), bacilliary peliosis hepatitis (Figure 91A), and cat scratch fever disease. These disease states are mainly found in, but are not limited to, adults infected with human immunodeficiency virus (HIV) and transplant patients.

Morphology and Cultural Properties. *Bartonella* species are nonmotile, Gram-negative, slightly curved rods 0.5–0.6 μm wide and 1.0–2.0 μm long (Figure 91B). Because these organisms can be cultivated in or on cell-free enriched media, they do not fit the obligate intracellular property of rickettsias (Figures 86A and 87A).

Bartonella species are catalase, oxidase, and urease negative. They also are nonreactive in carbohydrate test media.

Diagnosis. *Bartonella* can be detected with several techniques. These include staining tissue biopsy specimens, culturing on media such as chocolate agar (Figures 92 and 93) or trypticase soy, or brain-heart infusion agar supplemented with 5% sheep blood and incubated with CO_2, and commercially available serological tests.

Bartonella henselae

B. henselae is known to cause persistent bacteremia, bacillary angiomatosis, and cat scratch fever disease. Contact with cats and lowered resistance, as in the case of HIV infection, are known risk factors.

Pathology and Clinical Features. Bacillary angiomatosis (BA) can present in several ways. The commonest form is an enlarged, red elevated skin area that sometimes resembles a cranberry. This lesion is generally surrounded by numerous additional growths that can penetrate deeper into the skin. Bacillary angiomatosis has been reported to occur in every body organ system.

Cat scratch fever disease usually occurs as a self-limited infection with enlarged and inflamed lymph

Rickettsia, Ehrlichia, and Bartonella *(Continued)*

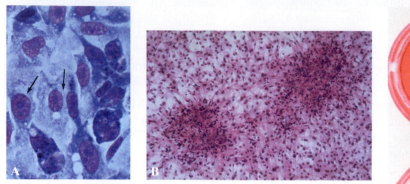

Figure 89

Ehrlichia chaffenis. (A) Infected cells with a large number of intracellular morulae (425×). (B) Overlapping infected tissue culture cells showing concentrated cytopathic effects (70×). (C) Neutral red staining of areas showing cellular destruction (plaques) in a tissue culture. The control tissue culture system shows no clear areas or evidence of infection. (2.2×). (From S.-M.Chen, V.L. Popov, H.-M. Feng, J. Wen, and D.H. Walker. *Inf. Immunol.* 63(1995):647–55.)

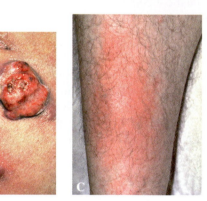

Figure 90

Clinical features of bacillary angiomatosis. (A) A magnetic resonance image showing a soft-tissue mass in front of femoral blood vessels and taken 8 months before diagnosis. (B) Multiple tender vascular, oozing skin lesions. (C) An erythematous, tender, raised lesion, called a plaque on the leg. (From J.E. Koeheler *et al., NEIM* 327(1992):1625–31.)

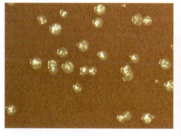

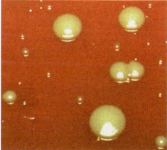

Figure 91

Microscopic features of *Bartonella henselae.* (A) Areas of tissue destruction in the liver of patients. (B) A higher magnification showing pleomorphic rods. (3,300×). (From D.A. Relman et al. *NEIM* 323(1990):1573–80.)

Figure 92

Primary culture of *Bartonella henselae* colonies on chocolate agar. (Courtesy of Drs. J.E. Koehler and J.W. Tappero. *NEJM* 337(1997):1876–1883.)

Figure 93

Primary culture of *Bartonella quintana* colonies on chocolate agar. (Courtesy of Drs. J.E. Koehler and J.W. Tappero. *NEJM* 337(1997):1876–1883.)

nodes in an area draining from a site of a cat scratch or bite. Some nodes may ulcerate.

Bartonella quintana

Bartonella quintana was first identified as an important pathogen during World War I and as the cause of trench fever. This disease is transmitted by the body louse (see Figure 87B). *Bartonella quintana* also has been identified as a cause of bacillary angiomatosis, endocarditis, and bacteremia.

Pathology and Clinical Features. The incubation period of *B. quintana* infections varies from 5 to 20 days. Clinical features vary but generally include nonspecific signs and symptoms including fever, malaise, headache, bone pain, and a temporary macular rash.

▲ The Beta Proteobacteria

The beta proteobacteria include a number of medically important species among their ranks. The subgroup overlaps to a certain degree with the alpha proteobacteria, and some species are noted for obtaining nutrients spreading from areas of anaerobic decomposition of organic substances.

▲ Microaerophilic Gram-Negative Helical Bacteria

Aquaspirillum

Aquaspirillum magnetotacticum

This highly motile spirillum usually is found in freshwater habitats. The bacterium is noted for its directed movement in a (geo)magnetic field, known as magnetotaxis. This movement is due to chains of 5 to 40 magnetic particles called magnetosomes present within cells of the organism (Figure 94A). The magnetosomes are generally uniform in shape and arrangement, and contain single crystals of the iron oxide magnetite ranging in size from 400 to 1,000 angstroms. They function as internal magnets that orient organisms along a specific magnetic field. *A. magnetotacticum* and related species are nonpathogens.

Morphology and Cultural Properties. *A. magnetotacticum* is a Gram-negative helical cell having a diameter ranging from 0.2 to 0.4 μm. It is the only member of the genus with magnetosomes. The turns of its helical shape are oriented in a clockwise direction, whereas most other *Aquaspirillim* species have a counterclockwise orientation (Figure 94B). *A. magnetotacticum* moves by means of flagella.

Aerobic, Gram-Negative Sheathed Bacteria

Leptothrix
Leptothrix cholodnii

Usually found in aquatic habitats such as lakes, springs, and streams. Some *Leptothrix* species may also be present in wastewater systems. *Leptothrix cholodnii* and related members of the genus are non-pathogens.

Morphology and Cultural Properties. *Leptothrix cholodnii* is a Gram-negative rod arranged in chains within a specialized covering known as a sheath (Figure 95A). This organism also occurs as free-swimming single cells, in pairs, or, in some *Leptothrix* species, as motile short chains of up to eight cells. Individual rods range in length from 0.6 to 1.4 μm, and are 1–1.2 μm in diameter. Free cells move by means of a single polar flagellum.

Sheaths are known for their pronounced tendency to become impregnated or covered with iron or manganese oxides (Figure 95B). Such encrusted sheaths are frequently empty because cells have left.

Leptothrix species are chemoorganotrophs, and are able to utilize a number of sugars as carbon and energy sources.

▲ Aerobic/Microaerophilic Gram-Negative Coccobacilli, Cocci, and Spirals

Bordetella

Bordetella pertussis

The causative agent of the respiratory disease **pertussis** (whooping cough) is *Bordetella pertussis*. The organism attaches to and grows in and on the ciliated cells of the mucous membranes of the human respiratory tract.

Transmission. *Bordetella pertussis* is spread by airborne droplets to individuals in close contact with infected persons in the early stages of illness. The human-to-human transmission is well established.

Morphology and Cultural Properties. *Bordetella pertussis* is a Gram-negative cocco-bacillus, 0.2–0.5 μm wide and 0.5–2.0 μm long. Cells are generally arranged singly or in pairs.

The organism is nonmotile and strictly aerobic. The optimum growth temperature ranges from 35° to 37°C. Colonies are glistening, smooth, convex, and pearl-like. They are grown on specially prepared media (Figure 96), such as charcoal blood agar medium, which may contain antibiotics such as cephalosporin to inhibit organisms from the normal microbial inhabitants of the respiratory tract. Colonies appear after 3 to 6 days.

Microaerophilic Gram-Negative Helical Bacteria

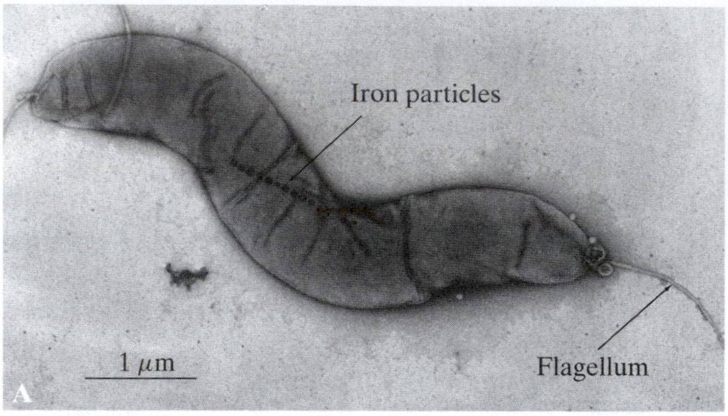

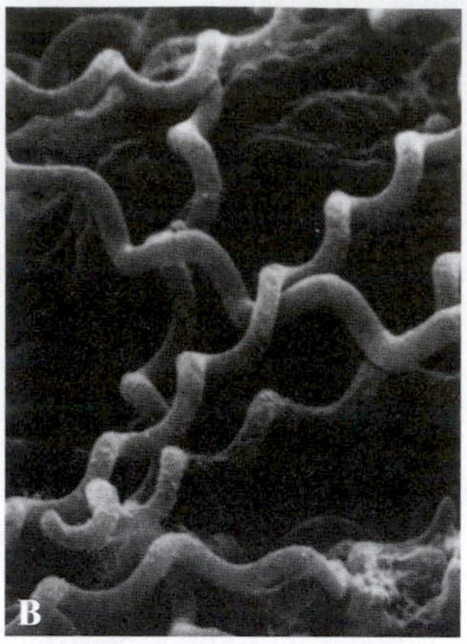

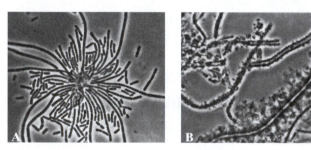

Figure 94

(A) Transmission electron micrograph of the magnetotactic cell of *Aquaspirillum magnetotacticum*. The intracellular chain of electron-dense, iron-rich magnetosomes is arranged in a reasonably straight line. (From D. Maratea, and R. P. Blakemore. *Internat. J. System Bacteriol.* 31(1981):452–55.) (B) Scanning electron micrograph of *Aquaspirillum peregrinum* showing typical left-handed spiral turns. (From H. Konishi et al., *J. Gen. Microbiol.* 132(1986):877–81.)

Aerobic Gram-Negative Sheathed Bacteria

Figure 95

Sheathed bacteria. (A) A phase-contrast view of the free-living cells and long chains of the sheathed bacterium Leptothrix lopholea. (B) *Leptothrix cholodnii* sheaths covered with deposits of ferric chloride. (From W.L. van Veen et al., *Microbiol. Revs.* 42(1978): 329–56.)

Pathology and Clinical Features. Pertussis follows a prolonged course consisting of three overlapping stages:

1. **Catarrhal stage.** Large amounts of watery discharge from the nose.
2. **Paroxysmal coughing stage.** Sudden and recurring violent coughing episodes up to 50 times a day for 2 to 4 weeks. The characteristic whoop, vomiting, and large amounts of mucous are typical of this stage.
3. **Convalescent stage.** A lessening of the signs and symptoms takes place over a 3–4-week period.

Diagnosis. Clinical diagnosis is confirmed by isolation of *Bordetella pertussis*. Specimens taken early in the course of the disease provide the best results. Fluorescent-antibody testing also is effective.

Neisseria

Most members of the genus *Neisseria* are Gram-negative diplococci, measuring 0.6–1.0 μm in diameter (Figure 97). The adjacent sides of cell pairs are flattened, giving the characteristic kidney bean or coffee bean appearance in microscopic preparations (Figures 97 and 98).

Neisseria species have an optimal growth temperature of 35° to 37°C, and they are aerobic and oxidase positive (Figure 99B).

Neisseria gonorrhoeae

The causative agent of the sexually transmitted disease gonorrhea (Figure 98) is *Neisseria gonorrhoea*. It is one of the most commonly reported bacterial infections in the United States. This organism is one of the most common causes of infertility in women worldwide.

Transmission. Infection is acquired by sexual contact, or an infected pregnant woman can transmit it to her newborn as it passes through the infected birth canal. Newborns not treated adequately may develop blindness of the newborn, or **ophthalmia neonatorum.**

Morphology and Cultural Properties. *Neisseria gonorrhoeae,* also commonly referred to as gonococcus, exhibits the typical Gram-negative diplococcus appearance (Figures 97 and 98). The organism grows best on media enriched with blood or on special substances such as chocolate agar (Figure 99A), and in the presence of 3% to 10% CO_2. Thayer-Martin chocolate agar containing specific antibiotics is widely used for isolation from clinical specimens, which include urethral, cervical, rectal, pharyngeal, and conjunctival discharges.

Pathology and Clinical Features. Gonorrhea has an approximate incubation period of 2 to 7 days. Infection without signs and symptoms occurs in about 10% of males and about 20 to 80% of females.

In males, the most common finding is an acute urethritis, which is accompanied by an abrupt and noticeable painful urination (**dysuria**) and a pussy urethral discharge (Figure 100A). Complications such as prostatitis, inflammation of the epididymis, and closure of the urethra (the canal for the elimination of urine extending from the bladder to the outside) are possible in untreated cases.

In females, the main site of urogenital disease is the lining of the cervix. Infection may be accompanied by a pussy discharge, painful urination, and a greater frequency of urination. Complications may lead to an inflammation of the uterine tubes and **pelvic inflammatory disease** (PID). Gonococci from cervical secretions may also contaminate the external region between the vaginal area and the anus (**perineum**), resulting in anorectal and neighboring gland infections.

In a small percentage of cases, *N. gonorrhoeae* may spread via the bloodstream to cause disseminated gonococcal infection (DGI). A sparse rash on the arms and legs and arthritis in one or more joints are typical of this condition.

Infants born to infected mothers may develop gonococcal ophthalmia (Figure 100C), also known as ophthalmia neonaturum, if preventive eyedrops or ointment is not administered at birth. This infection must be treated immediately or blindness can result. Non–sexually transmitted infections may occur in very young children, usually through accidental contamination with infectious discharges from infected individuals.

Diagnosis. The correct choice of specimens is critical to the successful isolation and identification of *N. gonorrhoeae*. The direct demonstration of the organism in white blood cells is diagnostic only when found in urethral discharges from males. Most other specimens require culture and biochemical testing.

Enzyme immunoassay, nucleic acid probes, and immunofluorescence tests (Figure 101) are available for the direct detection of gonococci.

Neisseria meningitidis

Bacterial meningitis (an inflammation of the coverings of the brain and spinal cord) is a relatively common and devastating disease. *Neisseria meningitidis* is one of the three most common causative agents of the disease.

Transmission. *Neisseria meningitidis* colonizes the nasopharynx and spreads from person to person by means of respiratory droplets released during breathing, coughing, or sneezing. Humans are the only reservoirs for this organism.

Morphology and Cultural Properties. *Neisseria meningitidis,* like other members of the genus, is a Gram-negative, nonmotile diplococcus with the appearance of a kidney or coffee bean. The organisms are capsulated and have pili.

Neisseria meningitidis is a strict aerobe and grows best on chocolate agar in the presence of 3% to 10% CO_2. The organism produces acid from glucose and maltose.

Pathology and Clinical Features. *Neisseria meningitidis,* also known as the meningococcus, can gain entrance into the bloodstream, from where it can progress to the cerebrospinal fluid to cause meningitis. Organisms spread from the nasopharynx through the bloodstream to produce meningococcemia, meningitis, or both. The disease may be mild, or it may progress rapidly, resulting in death within a few hours. Meningococcemia causes severe damage to blood vessels, the most visible effect being small hemorrhages (**petichiae**) in the skin (Figure 102).

Other signs and symptoms include fever, stiff neck, vomiting, severe headache, convulsion, and progression to a coma within a few hours.

Diagnosis. Laboratory diagnosis involves obtaining specimens for direct microscopic examination and culture. A serological test, such as latex agglutination, for the detection of capsular polysaccharides is essential.

Spirillum

Spirillum volutans

Frequently found in stagnant freshwater environments, *Spirillum volutans* is a nonpathogen.

Morphology and Cultural Properties. *Spirillum volutans* is a Gram-negative rigid helical cell, 1.4–1.7 μm wide and 14–60 μm long. The organism moves by means of large tufts of flagella located at both ends of a cell (Figure 103). *Spirillum volutans* is microaerophilic, but it can grow in special liquid media containing supplements. The organism is oxidase positive and catalase negative. It does not break down carbohydrates.

Aerobic/Microaerophilic Gram-Negative Coccobacilli, Cocci, and Spirals

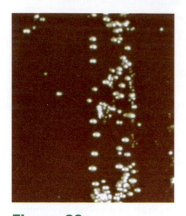

Figure 96
Colonies of *Bordetella pertussis* on charcoal blood agar.

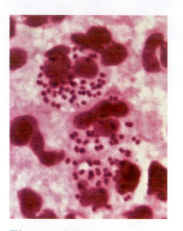

Figure 97
The "coffee bean" appearance of *Neisseria gonorrhoeae* from a urethral discharge. Note the presence of Gram-negative diplococci in leukocytes.

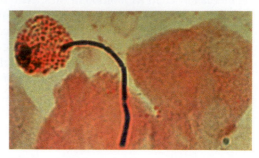

Figure 98
Neisseria gonorrhoeae in a cervical smear. Large cervical epithelial cells and a long Gram-positive rod are also evident.

Figure 99
(A) Growth of *Neisseria gonorrhoeae* on a chocolate agar medium. (B) A positive (dark colonies) oxidase test.

▲ The Gamma Proteobacteria

The gamma-proteobacteria is the largest of the subgroups. It contains most of the common chemoorganotroph, many of which are well-known pathogens.

▲ Anoxygenic Phototrophic Gram-Negative Vibroids and Rods

Purple phototrophic bacteria are noted for their **anoxygenic** form of photosynthesis, in which oxygen is not released. This is in contrast to the **oxygenic** photosynthesis of the cyanobacteria (the blue-green bacteria), in which free oxygen is a by-product. Purple bacteria have photosynthetic pigments which are contained in various types of intracytoplasmic membrane systems

(Figure 105). Suspension of these bacteria in various bodies of water exhibit a variety of colors, such as purple-red, rose red (Figure 104A), yellowish brown, and yellow.

Chromatium

Chromatium **and Related Species**
Chromatium and related species occur in sulfide-containing portions of freshwater, estuarine, marine, or highly concentrated salt environments (Figure 104A). All species are nonpathogenic.

Morphology and Cultural Properties. The cells of *Chromatium* species are Gram-negative and may be straight to slightly curved rods, 1.0–6.0 μm in diameter and 1.5–16.0 μm long. They occur singly or in multicellular filaments (Figure 104B). Purple bacteria form

Aerobic/Microaerophilic Gram-Negative Coccibacilli, Cocci, and Spirals *(Continued)*

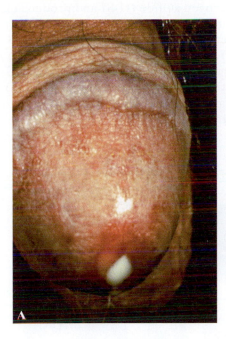

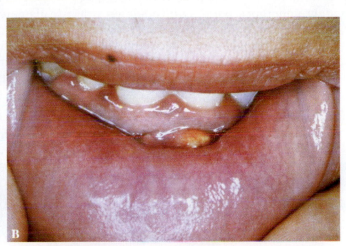

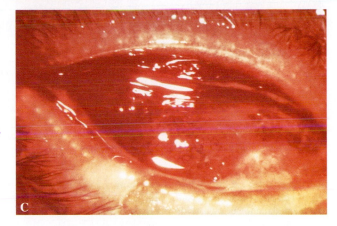

Figure 100

Clinical features of gonorrhea. (A) The typical discharge of this sexually transmitted disease in males. (B) An abscess (yellow growth) in the mouth of a patient. Such localized collections of pus in the mouth are not rare. (From D. Marini, S. Veraldi, and M. Innocenti. *Cutis.* 40(1987):363.) (C) A case of ophthalmia neonatorum caused by *Neisseria gonorrhoeae.*

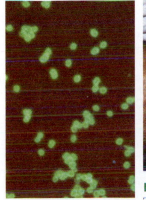

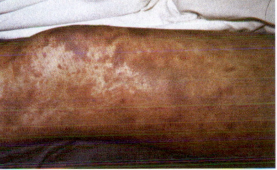

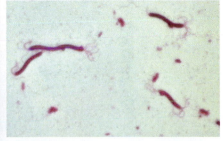

Figure 103

A photomicrograph of *Spirillum volutans.* In addition to the spiral morphology of this organism, the arrangement of two or more flagella at one or both ends of a cell is evident.

Figure 102

The petechial rash that commonly develops with the invasion (meningococcemia) of the bloodstream by *Neisseria meningitidis.* (From M. Barza. *NEJM* 328 (1996):34.)

Figure 101

The results of the fluorescent antibody technique showing *Neisseria gonorrhoeae.* The diplococcus arrangement can be seen.

intracytoplasmic membrane systems into which their photosynthetic pigments are inserted (Figure 105). Chromatia move by means of a single flagellum or a tuft of flagella located at one end of individual cells.

In the presence of sulfide and light, globules of sulfur appear within cells. Under anaerobic conditions, *Chromatium* undergoes photoautotrophic growth with sulfide or sulfur as the electron donor for CO_2 assimilation. Such organisms are called purple-sulfur bacteria. The optimum growth temperatures are 20–35° C., and the pH range is 6.8–7.5.

Purple-sulfur bacteria belonging to the genus of *Ectothiorhodospira* (Ek-tō-thī-ō-rō-dō-spī-ra) differ from other purple-sulfur organisms in that they oxidize hydrogen sulfide (H_2S) and produce elemental sulfur (S^0) outside of their cells. Cells of this species are vibroid in shape or rods, 0.5–1.5 μm in diameter, and have internal photosynthetic membrane arranged as lamellar stacks (Figure 105A). In addition, some species are able to tolerate environments containing extremely high salt concentrations.

Anoxygenic Phototrophic Gram-Negative Vibroids and Rods

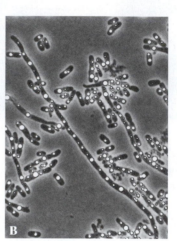

Figure 104

Phototrophic bacteria. (A) A small sulfide-containing freshwater inlet. A large accumulation of purple sulfur bacteria can be seen in the forefront of this view. (B) Phase-contrast micrograph showing the cells of the purple-sulfur bacterium *Chomatium gracile* containing sulfur deposits.

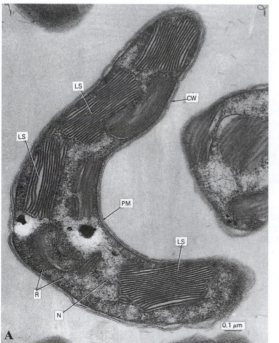
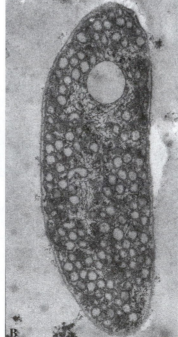

Figure 105

Transmission electron micrographs of the photosynthetic membrane machinery of phototrophic bacteria. (A) Thin section of *Ectothiorhdospira mobilis* showing its internal organization including a multilayered cell wall (**CW**), photosynthesizing membrane stacks (**LS**) or lamellae, nucleoid area (**N**), plasma membrane (**PM**) and ribosomes (**R**). From Remsen et al. *J. Bacteriol.* 95(1968):2374–92.) (B) Another example of a purple phototrophic bacterium *Chromatium* species with its photosynthesizing membranes organized into spherical-shaped vesicles (From N. Kaufman, et al. *Arch. Microbiol.* 131(1982):313–22.)

▲ Aerobic/Microaerophilic Gram-Negative Rods and Cocci

Francisella

Francisella tularensis is the causative agent of tularemia, or rabbit fever. The disease agents are harbored in the blood and other tissues of wild and domestic animals.

Transmission. *Francisella tularensis* can be acquired through the skin or the conjuctiva (lining of the eyelid) from handling of infected animals or animal products, from the bites of infected blood-sucking deer flies and wood ticks, by ingestion of improperly cooked contaminated meat or water, and by aerosol inhalation.

Morphology and Cultural Properties. *Francisella tularensis* is a Gram-negative rod, 0.3–0.5 μm wide and 0.2 μm long.

The organism is nonmotile and aerobic, and it requires special media for isolation. Small, smooth, grayish colonies appear within 48 hours. Growth may also occur later and may require 3 weeks. Optimum temperature is 37°C. Biochemical testing is of little value in identification.

Pathology and Clinical Features. Organisms establish infections locally and in regional lymph nodes. Such infections often result in ulcer formation and regional enlargement of lymph nodes (**lymphadenopathy**). Two forms of lymphadenopathy, ulceroglandular (Figure 106) and **oculoglandular**, are generally accompanied by fever, nausea, vomiting, or abdominal pain.

Diagnosis. Immunofluorescent or agglutination tests are commonly used for disease diagnosis.

Legionella

More than 30 species of *Legionella* have been identified. Of these, at least 12 species have been isolated from humans with respiratory disease. Attention here will be given to the most prevalent pathogen, *L. pneumophila*.

Two clinical forms of disease are caused by *L. pneumophila*: **Legionnaires' disease**, a pneumonia-like infection, and **Pontiac fever**, an influenza-like, self-limited illness. The organism was named after it was isolated from lung tissue (Figure 107) obtained from autopsies of individuals who had died during the epidemic at the 1976 American Legion Convention in Philadelphia.

Transmission. Airborne transmission by way of aerosols is the most common route of infection. Waterborne disease may occur, particularly in hospital and related environments. There is no known case of human-to-human transmission at this time.

Morphology and Cultural Properties. *Legionella pneumophila* is a Gram-negative rod, 0.3–0.9 μm wide and 2–20 μm long. The organism is aerobic and motile. It does not ferment carbohydrates and requires the amino acid cysteine for growth. (Figure 108). Among the preferred media used are Mueller-Hinton agar with hemoglobin and IsoVitaleX (Figure 108) and buffered charcoal-yeast extract agar (Figure 109). Optimum incubation temperature is 35°C. Colonies generally appear within 3 to 5 days and are convex, gray, and glistening (Figure 109).

Pathology and Clinical Features. Fever, chills, general discomfort, and muscle pain usually are the first symptoms that appear after an incubation period of 2 to 10 days following *L. pneumophila* infection. Legionnaires' disease involves many body systems and causes other notable clinical features, including coughing, breathing difficulties, chest and abdominal pain, vomiting, diarrhea, and mental confusion. Hospitalization is frequently required, and the illness may be fatal without antibiotic therapy.

Pontiac fever presents early symptoms similar to those found with Legionnaires' disease within 24 to 36 hours after exposure. This form of *L. pneumophila* infection is remarkable for the absence of pneumonia or fatalities.

Diagnosis. Diagnosis depends on isolation of the organism or its detection in respiratory secretions by direct immunofluorescence testing (Figure 110). Modern molecular methods involving nucleic acid probes also are of value.

Moraxella

Moraxella (Branhamella) catarrhalis

A common inhabitant of the human upper respiratory tract is the organism *Moraxella catarrhalis*. It is an important cause of otitis media (middle ear infection) in infants and children, lower respiratory tract infections in adults, and nosocomial respiratory tract infections. *Moraxella catarrhalis* has been isolated exclusively from humans.

Transmission. Aerosols are common sources of infection.

Morphology and Cultural Properties. *Moraxella catarrhalis* is a Gram-negative coccus, occurring as single cells or in pairs, with the adjacent sides flattened, giving a coffee-bean appearance (Figure 111). The cocci usually measure 0.6–1.0 μm in diameter.

Moraxella catarrhalis is aerobic, but it may grow poorly under anaerobic conditions. The organism will grow on media such as nutrient and blood agars. Blood agar colonies are small, circular, convex, and usually grayish white (Figure 112). Optimal temperature is 33° to 35°C.

Aerobic/Microaerophilic Gram-Negative Rods and Cocci

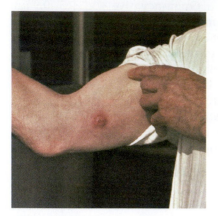

Figure 106
An enlarged lymph node in a patient with tularemia. (From Z. Cerny. *Internat. J. Dermatol.* 33(1994):468.)

Figure 107
Gram stain results with *Legionella pneumophila.*

Figure 108
Legionella pneumophila colonies after 6 days of incubation on Mueller-Hinton agar supplemented with 1% hemoglobin and other growth factors.

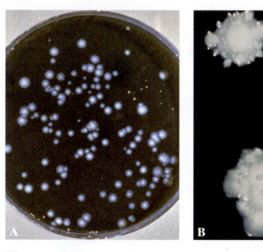

Figure 109
(A) *Legionella pneumophila* growing on buffered charcoal-yeast extract (BCYE) agar after 4 days of incubation. (B) A close-up of *Legionella pneumophila* isolated from a case of pneumonia after 6 days of incubation. (From M. Koide and A. Saito. *CID* 21(1995):199–201.)

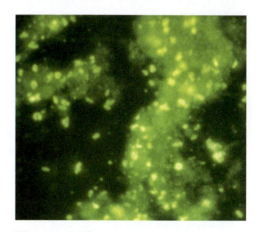

Figure 110
Brightly fluorescing *Legionella* in a lung tissue suspension stained by immuno-fluorescence.

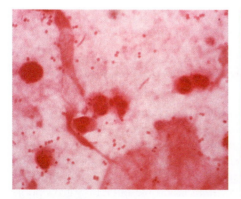

Figure 111
The results of a Gram-stained sputum specimen from a patient with chronic bronchitis caused by *Moraxella catarrhalis.* (From T.F. Murphy. *Microbiol. Revs.* 60(1996):267–79.)

Figure 112
Colonies of *Moraxella catarrhalis* on blood agar.

Figure 113
A positive color reaction for the enzyme butyrate esterase and a presumptive identification of *Moraxella catarrhalis.* (Courtesy of Remel Inc., Lenexa, KS.)

Moraxella catarrhalis is catalase and oxidase positive and does not produce acid from carbohydrates.

Diagnosis. Isolation and biochemical tests are necessary for diagnosis.

Pseudomonas

Pseudomonas aeruginosa

Among the most virulent opportunistic pathogens of humans is *Pseudomonas aeruginosa,* which colonizes and invades injured epithelial surfaces (including the lungs of intubated patients, corneal abrasions, and injured skin of burn patients). Seventy-five percent of all intensive care unit patients are colonized by this organism (Figure 114), and, not surprisingly, *P. aeruginosa* is the leading cause of mortality among cystic fibrosis patients. The treatment of *P. aeruginosa* infections is difficult, in part because this organism's antibiotic resistance is common and is becoming more widespread.

Pseudomonas aeruginosa is widely distributed in soil, water, sewage, intestinal tracts, and plants. The organism has been isolated from various materials, including disinfectants, cosmetics, and foods.

Transmission. *Pseudomonas aeruginosa* is spread in a number of ways, including by contaminated fingers and instruments such as urinary catheters, endoscopes, and respiratory therapy equipment. Bathing or soaking in contaminated water also can serve to transmit the organism.

Morphology and Cultural Properties. *Pseudomonas aeruginosa* is a Gram-negative rod, 0.5–1.0 μm wide and 1.5–5.0 μm long, usually occurring singly and in pairs (Figure 115).

The organism is motile and aerobic. It grows best at 37°C, but good growth can occur at 42°C. Large translucent, spreading colonies with irregular edges appear within 24 to 48 hours (Figure 116A). *Pseudomonas aeruginosa* produces a blue-green pigment that diffuses into media (Figure 116B). Special media such as *Pseudomonas* P and Pseudomonas F agars can be used to demonstrate pigment production (Figures 116B and 117).

Pseudomonas aeruginosa grows at 42°C, oxidizes glucose, and does not produce acid from dissacharides such as lactose. Colonial growth appears within 48 hours and exhibits a variety of forms, including smooth, rough, and mucoid and a small size. The variation exhibited by a single strain gives the false impression that several bacterial species are present.

Pathology and Clinical Features. *P. aeruginosa* has become a major cause of hospital-acquired infections. Infections with this organism rarely occur in persons with normal immune systems and defenses. Patients with extensive burns may become colonized with *P. aeruginosa* since burn injuries not only destroy the mechanical barrier of the skin but also seriously impair all other aspects of the immune system.

Skin lesions caused by *P. aeruginosa* appear as round, hardened, purple areas about 1 cm in diameter with an ulcerated center surrounded by a zone of redness. Inflammation of hair follicles, a skin condition and ear infections resulting from bathing in contaminated water and associated with hot tubs, and swimming pools and ear infections are among the few *Pseudomonas* infections occurring in healthy individuals.

Diagnosis. Most strains of *P. aeruginosa* are identified on the basis of its characteristic grapelike odor, colonial morphology, growth at 42°C, and the production of a water-soluble blue pigment, pyocyanin (Figures 116B and 116C).

P. fluorescens

Another species of this genus rarely associated with opportunistic infections is *P. fluorescens*. Most isolates are from respiratory tract specimens. The organism has also been isolated from various sources in hospital environments, such as water sources, sinks, floor, and contaminated blood. *P. fluorescens* is commonly associated with the spoilage of foods such as meats and fish.

Morphology and Cultural Properties. *P. fluorescens* is similar to *P. aeruginosa* in microscopic properties. It grows on most media, forms small colonies, and characteristically produces the fluorescent pigment, fluorescein (Figures 102C and 103).

Pathology. While *P. fluorescens* is mainly an environmental contaminant, it can be an opportunistic pathogen for humans. The organism has been associated with urinary tract infections, pelvic inflammatory disease, and postoperative infections.

Diagnosis. Most *P. fluorescens* strains are identified on the basis of colonial morphology and the production of the fluorescent pigment.

▲ Facultatively Anaerobic Gram-Negative Rods

Calymmatobacterium

Calymmatobacterium granulomatis

The causative agent of granuloma inguinale, also known as donovanosis and granuloma venereum, is *Calymmatobacterium granulomatis*. The disease is chronic and involves the skin and the subcutaneous tissue of the genital, inguinal, and anal regions (Figure 118).

Transmission. *Calymmatobacterium granulomatis* is thought to be sexually transmitted. However, several findings do not support this mode of transmission totally. Granuloma inguinale is mainly found in tropical areas of the world.

Aerobic/Microaerophilic Gram-Negative Rods and Cocci *(Continued)*

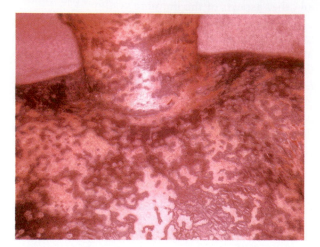

Figure 114
Generalized skin infection caused by *Pseudomonas aeruginosa*. Note the purple coloration of the skin.

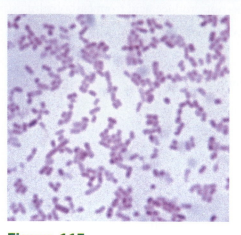

Figure 115
Gram-negative rods of *Pseudomonas aeruginosa*.

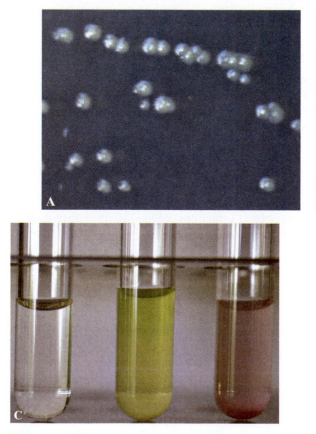

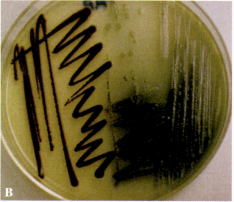

Figure 116
Pseudomonas aeruginosa cultural features. (A) A young culture of *P. aeruginosa* on nutrient agar. (B) *P. aeruginosa* growing on *Pseudomonas* P agar. This medium enhances the production of pyocyanin (a bluish pigment) and inhibits the production of the greenish yellow pigment fluorescein. The red-pigmented *Serratia marcescens* is shown for purposes of comparison. (C) A 72-hour broth culture of *P. aeruginosa* (center) and the appearance of its typical pigment.

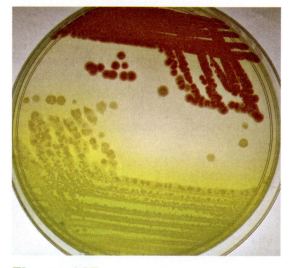

Figure 117
P. fluorescens growing on *Pseudomonas* F agar and producing its characteristic fluorescein (yellow fluorescent) pigment. Colonies of the red-pigmented *Serratia marcescens* are shown at the top of the culture plate.

Morphology and Cultural Features. *Calymmato-bacterium granulomatis* is a Gram-negative, pleomorphic rod, 0.5–1.5 μm wide by 1.0–2.0 μm in length (Figure 119). Material from specimens stained with Wright's, Giemsa, or other stains shows characteristic intracellular organisms within large mononuclear cells known as **Donovan bodies** (Figure 120).

Routine culture in embryonated eggs is neither practical nor highly successful. Specially prepared media containing growth factors have been used to grow *C. granulomatis*.

Pathology and Clinical Features. Granuloma inguinale begins as subcutaneous nodules that eventually erode and produce a clear, tumorlike, sharply defined lesion (Figure 118). Without treatment, the disease progresses by extension to neighboring skin areas and frequently spreads along the groin. Although such lesions generally are not painful, they can be mutilating.

Diagnosis. Diagnosis is usually based on finding the typical intracellular organisms (Donovan bodies) with large mononuclear cells in tissues known as histiocytes (Figure 120).

Citrobacter

Citrobacter freundii

Occurring in the feces of humans and lower animals, *Citrobacter freundii* is probably a normal intestinal inhabitant. The organism also is found in soil, water, sewage, and food. *Citrobacter freundii* often is isolated from clinical specimens as an opportunistic pathogen.

Morphology and Cultural Properties. *Citrobacter freundii* is a Gram-negative, straight rod, measuring about 1.0 μm in width and 2–6.0 μm in length and occurring singly or in pairs (Figure 121). The organism is motile, facultatively anaerobic, oxidase negative, and catalase positive.

It grows on various media, including nutrient, blood, and MacConkey agars (Figure 122). Optimal growth temperature is 39°C. On blood agar, colonies are large, white, raised, smooth, and entire (Figure 122A).

Citrobacter freundii are positive for methyl red, nitrate, and citrate tests and negative for indole and Voges-Proskauer. A variety of sugars, including glucose, mannitol, and maltose, are fermented with the production of acid.

Diagnosis. Isolation and biochemical tests are necessary for diagnosis. Lactose positive (fermenting) colonies are produced on media such as eosin-methylene blue or MacConkey agars. (Figure 122B).

Enterobacter

Enterobacter species are widely distributed in nature, occurring in soil, fresh water, sewage, and animal feces and on plants including vegetables. Several species are opportunistic pathogens. *Enterobacter aerogenes* and *E. cloacae* are the species most frequently isolated from clinical specimens and are important causes of hospital-acquired cases of bacteremia and contamination of wound infections in burn patients.

Transmission. *Enterobacter* species are found as contaminants of intravenous fluids and hospital equipment. Bacteremia can result from the use of these materials.

Morphology and Cultural Properties. Enterobacter species are small, straight, Gram-negative rods measuring 0.6–1.0 μm in diameter and 1.2–3.0 μm long (Figure 123). These organisms are motile.

Enterobacter species are facultatively anaerobic and grow on most agar media (Figure 124). Optimal growth temperatures range from 30° to 37°C. Colonies are generally cream to tan in color, entire, glistening, convex, and circular. Lactose-positive (fermenting) colonies are produced on media such as eosin-methylene blue or endo agars (Figure 124B).

Glucose and other carbohydrates are catabolized with the production of acid and gas. The I (indole), M (methyl red), VP (Voges-Proskauer), C (citrate) pattern for *Enterobacter* species usually is −, −, +, + (see Figure 70), which is the opposite of the reaction pattern produced by *Escherichia coli*. Hydrogen sulfide, DNAse, and lipase are not produced.

Escherichia

Escherichia species occur as normal members of the intestinal microbiota in the large intestine of most mammals. The most important of these is *E. coli*. This organism and several other related species commonly cause infections outside of the gastrointestinal tract. Urinary tract infections, primarily of the bladder, are the most common, followed by respiratory, wound, bloodstream (sepsis), and central nervous system infections. Many of these infections, especially sepsis (Figure 125) and meningitis, are life-threatening and are often hospital-acquired.

Escherichia coli

Transmission. *Escherichia coli* types, which include 0157:H7, associated with gastrointestinal and related infections are generally acquired by consuming contaminated water or food.

Morphology and Cultural Properties. *Escherichia coli* is a short, straight, Gram-negative rod, measuring 1.10–1.5 μm in diameter and 2.0–6.0 μm in length (Figure 126). Cells occur singly or in pairs. *Escherichia coli* is motile and does not form capsules.

Facultatively Anaerobic Gram-Negative Rods

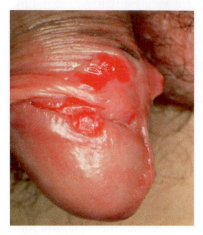

Figure 118
A case of granuloma inguinale, showing numerous ulcerated red nodules on the penis. (From P. Hacker and B.K. Fisher. *Internat. J. Dermatol.* 31(1992):696.)

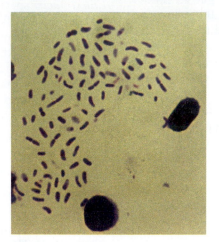

Figure 119
A rapid Giemsa (Rapidiff) stain of *Calymmatobacterium granulomatis* from a 48-hour monocyte culture (1,000×). (Courtesy of A. Kharsany, B. Housen, M. Housen, H.A. Housen, P. Kiepiela, T. Caicker, and A.W. Sturm, Faculty of Medicine, University of Natal, Durban.)

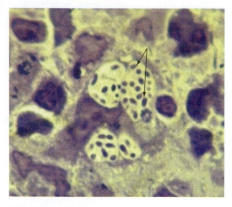

Figure 120
Microscopic view of a specimen from a case of granuloma inguinale showing Donovan bodies (arrows). (From P. Hacker and B.K. Fisher, *Internat. J. Dermatol.* 31(1996):696.)

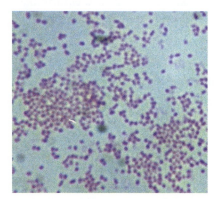

Figure 121
Small Gram-negative rods of *Citrobacter freundii*.

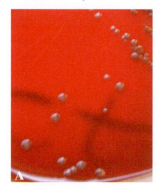
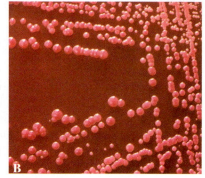

Figure 122
Growth of *Citrobacter freundii* on various media. (A) Blood agar. (B) The lactose-positive reaction on MacConkey agar.

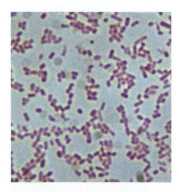

Figure 123
Enterobacter aerogenes Gram stain.

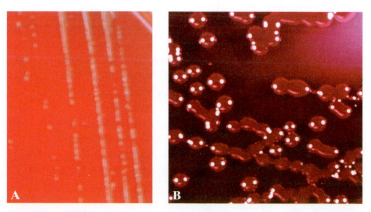

Figure 124
Enterobacter aerogenes on media. (A) Young culture on blood agar. (B) Colonies on Endo agar. Note the extension of the reaction (dark red) into the medium.

The organism is facultatively anaerobic and grows on a wide variety of agar media (Figure 127). Optimal temperature ranges between 30° to 37°C. Colonies on nutrient or blood agar are cream to tan in color, entire, convex, circular, and smooth. Dark purple colonies, typical of lactose fermentation results are formed on eosin–methylene blue agar (Figure 127C). The colonies of certain strains have a metallic sheen.

Escherichia coli is hydrogen sulfide, urease, and oxidase negative and catalase positive, and it catabolizes glucose and other carbohydrates with the formation of acid and gas. The I (indole), M (methyl red), VP (Voges-Proskauer), C (citrate) pattern of reactions is +, +, −, − (see Figure 70).

Pathology and Clinical Features. Several distinct types of *E. coli* are currently recognized to cause human diarrhea. These organisms can be grouped into several categories on the basis of pathogenic mechanisms: enteropathogenic *E. coli* (EPEC), enterohemorrhagic *E. coli* (EHEC), enteroinvasive *E. coli* (EIEC), enteroadherent *E. coli* (EAEC), and enterotoxigenic *E. coli* (ETEC).

ETEC strains are known to possess two pathogenic properties: an ability to adhere to intestinal tissue (Figures 128 and 129) and the production of two enterotoxins, one heat-labile and the other heat-stabile.

The signs and symptoms of mild infection include diarrhea, vomiting, chills, headache, and fever following an incubation period of 1 to 2 days. More serious infections produce severe abdominal cramps, toxemia, and watery stools consisting of blood and mucus. Severe dehydration, shock, and death also may occur.

Diagnosis. Isolation and biochemical testing are the main approaches used for identification. Specific selective and differential media and immunologic tests such as the particle agglutination procedure are used for the identification for *E. coli* 0157:H7 strains (Figure 127D).

Haemophilus

Haemophilus species are minute to medium-sized oval or rod-shaped Gram-negative cells (Figure 130). Organisms are nonmotile and facultatively anaerobic. They produce acid from glucose and other carbohydrates and reduce nitrates to and beyond nitrites.

Almost all species require preformed growth factors found in red blood cells, such as hematin (X factor) and NAD or nicotinamide adenine dinucleotide (V factor). The V factor is found in a variety of biological materials and is produced by bacteria such as *Staphylococcus aureus* and by yeast. *Haemophilus influenzae* will grow on blood agar in the vicinity of *S. aureus* colonies producing this growth factor. This phenomenon is called **satellitism** (Figure 131).

Haemophilus influenzae

Haemophilus influenzae is known to cause a variety of infections, including meningitis, especially in children, otis media, pneumonia, and epiglottitis. The organism is part of the normal upper respiratory tract microbiota of humans.

Transmission. *Haemophilus influenzae* is usually transmitted by contact with secretions produced by sneezing or coughing, or found on the hands.

Morphology and Cultural Properties. *Haemophilus influenzae* is a nonmotile, frequently capsulated, Gram-negative small rod, measuring less than 1.0 μm in width and variable in length. Growth is best on complex media such as chocolate agar (Figure 132).

Pathology and Clinical Features. *Haemophilus influenzae* is one of the common causes of an inflammation of the epiglottis **epiglottitis**. The infection often develops quickly, and if not diagnosed promptly, it may result in respiratory blockage and death. Typical signs and symptoms include sore throat, fever, difficulty in swallowing and breathing, and the appearance of a characteristic swollen epiglottis which looks like a bright-red cherry blocking the throat at the base of the tongue.

If *H. influenzae* colonizes the nasopharynx, infections from this site may extend to local tissues, the middle ear, invasion of the blood stream resulting in such conditions as meningitis, pneumonia, bronchitis, sinusitis, and osteomyelitis.

Diagnosis. Various specimens are used for direct microscopic examination or culture. These include blood, spinal fluid (Figure 130), bronchoalveolar lavage, and urine. Rapid immunologic tests for the detection of polysaccharide antigens also are used.

Haemophilus ducreyi

Haemophilus ducreyi is the causative agent of the sexually transmitted disease known as **chancroid** or **soft chancre** (Figure 133). Humans are the only known hosts for the disease. Chancroid was at one time limited in occurrence to tropical, undeveloped areas. However, since 1985, several large outbreaks have appeared elsewhere.

Morphology and Cultural Properties. *Haemophilus ducreyi* is a Gram-negative slender rod occurring in pairs or chains, and measuring 0.5 μm in width and 1.5–2.0 μm in length. They may appear in arrangements described as "schools of fish."

After 48 to 72 hours of incubation small, flat, smooth and yellow to grey colonies appear on selective media. Such colonies are translucent to opaque. Additional cultural properties of *H. ducreyi* include positive reactions for oxidase and nitrate reduction.

Facultatively Anaerobic Gram-Negative Rods *(Continued)*

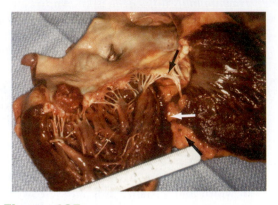

Figure 125
A large collection of fibrin and white and red blood cells known as a vegetation (arrow) on the mitral valve. This vegetation was found in a case of infective endocarditis caused by *Escherichia coli* (From C. Watanakuna Korn and J. Kim. *CID* 14(1992):501.)

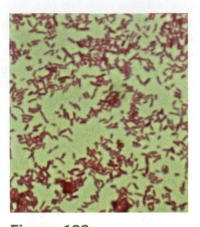

Figure 126
Gram-negative results with *Escherichia coli.*

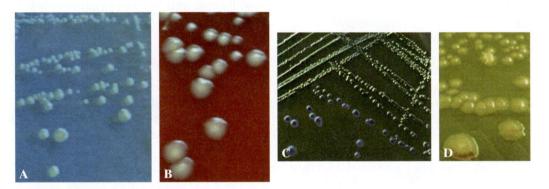

Figure 127
Escherichia coli on media after 48 hours of incubation. (A) Colonies on nutrient agar. (B) Colonies on blood agar. (C) The green metallic sheen produced by *E. coli* growing on eosin-methylene blue agar. Lactose fermentation also is shown. (D) Sorbitol-negative colonies (pale pink) typical of *Escherichia coli* 0157:H7. The medium here is Mac-Conkey sorbitol agar.

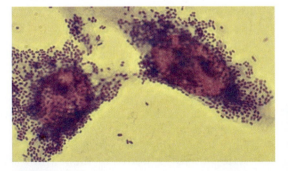

Figure 128
Light micrograph showing enteroadherent attached *Escherichia coli* (EAEC) to intestinal tissue cells (1,000×). (From A. Darfeuille-Michaud and Associates. *Inf. Immunol.* 58(1990):893–902.)

Figure 129
A scanning electron micrograph showing *E. coli* adhering to the surfaces of adult intestinal tissue. (From T. Yamamoto, et al. *Infection and Immunity.* 59(1991):3722–39.)

Pathology and Clinical Features. The incubation period of this disease is about 2 to 5 days. Typically a localized, red, elevated area or papule forms at the site of infection. The lesion develops into a painful ulcer with sharp margins (Figure 133).

Diagnosis. Specimens for diagnosis should be taken for microscopic examination and culture. *Haemophilus ducreyi* appears in characteristic parallel rows. Various media with growth factors are of value in isolating the organism (Figure 134).

Klebsiella

Klebsiella species are straight, Gram-negative rods measuring 0.3–1.0 μm in diameter and 0.6–6.0 μm in length. Cells occur singly, in pairs, or in short chains (Figure 135). Organisms are nonmotile and capsulated (Figure 136).

Klebsiella species are facultatively anaerobic, oxidase negative, and catalase positive. They reduce nitrates and ferment most commonly tested carbohydrates, producing acid and gas.

These organisms can be found in association with human feces, soil, water, fruits, vegetables, and grains. Certain species are opportunists and are known to cause urinary tract infections and a variety of noscocomial infections, bacteremia, and pneumonia.

Klebsiella pneumoniae

Klebsiella pneumoniae, the most common species in the genus, causes lobar pneumonia, a disease also associated with other encapsulated organisms.

Klebsiella pneumoniae forms large mucoid colonies, especially on carbohydrate media.

Diagnosis. Diagnosis involves isolation and biochemical testing.

Morganella

An opportunistic organism, *Morganella morganii* is found in the feces of several mammals and reptiles. The organism is known to cause infection in older patients with serious underlying disease. *Morganella morganii* has been isolated from respiratory and urinary tract infections, wounds, and cases of bacteremia.

Morphology and Cultural Properties. *Morganella morganii* is a Gram-negative rod 0.6–0.7 μm in diameter and 1.0–1.7 μm in length (Figure 137). The organism is nonmotile.

The organism is facultatively anaerobic and grows on a variety of media (Figure 138). Optimal temperature is 37°C. *Morganella morganii* is oxidase negative, catalase positive, and it hydrolyzes urea. Hydrogen sulfide is not produced. The organism's IMViC pattern is +, +, −, −. The only carbohydrates catabolized are glucose and mannose, with the production of acid and usually gas.

Pasteurella

Pasteurella multocida

One of several species found in the normal respiratory tract microbiota of some animals is *Pasteurella multocida.* This organism is by far the most common cause of an infected dog or cat bite (Figure 139).

Transmission. Humans are usually infected by the bite or scratch of a domestic cat or dog.

Morphology and Cultural Properties. *Pasteurella multocida* is a Gram-negative, ovoid or rod-shaped cell measuring 0.3–1.0 μm in diameter and 1.0–2.0 μm in length (Figure 140). Cells occur singly or in pairs.

Organisms are nonmotile and facultatively anaerobic. *Pasteurella multocida* are oxidase and catalase positive, produce acid but no gas from most carbohydrates, and reduce nitrates. Methyl red and Voges-Proskauer reactions are negative.

Pathology and Clinical Features. Infection develops at the site of a bite or scratch, often within 24 hours. The typical infection involves the subcutaneous tissue **(cellulitis)** with the development of a well-defined reddened border.

Diagnosis. Culture of pus taken from a lesion provides the best approach to diagnosis.

Photobacterium species

Several *Photobacterium* species have an established mutualistic relationship involving various ocean fish and other forms of aquatic life. Much of the light in ocean depths originates in bacteria nurtured in fish structures called light organs (Figure 141A). The flashlight fish swims above reefs in the Indian Ocean, shining its beam of light much like a car on a dark country road. The luminous organs of these marine forms are used to attract and capture prey, as well as to communicate with other members of their species. Members of the genus are found in marine environments, on the surfaces of, as well as in specific structures and the intestinal contents of marine animals.

Morphology and Cultural Properties. *Photobacterium* bioluminescent species (Figure 141B) are Gram-negative straight rods, 0.8–1.0 μm wide, and 1.8–2.4 μm long. They are generally motile by means of polar flagella (Figure 141C). Bioluminescence occurs only when oxygen is present. In addition, several other factors, including the enzyme luciferase, are necessary for this reaction to occur.

Facultatively Anaerobic Gram-Negative Rods (Continued)

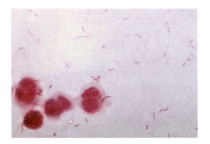

Figure 130
Gram-negative small rods or cocco-bacilli of *Haemophilus influenzae* in a spinal fluid specimen (1,000×).

Figure 131
Haemophilus influenzae growing on a blood agar plate containing *Staphylococcus aureus* (central white growth) and exhibiting satellitism.

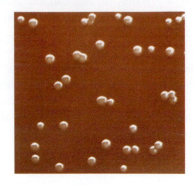

Figure 132
Colonies of *Haemophilus influenzae* on chocolate agar.

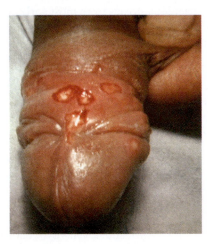

Figure 133
Clinical appearance of chancroid, a genital ulcer, that may be accompanied by painful, pus-filled enlarged lymph nodes.

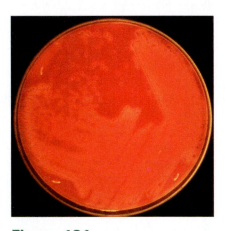

Figure 134
Entire plate and closer view of individual *Haemophilus ducreyi* colonies surrounded by zones of beta-hemolysis. (From P.A. Totten, D.V. Norn, and W.E. Stamm, *Inf. Immunol.* 63(1995):4409–16.)

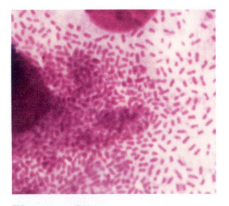

Figure 135
Rod-shaped *Klebsiella pneumoniae* (1,000×). (Courtesy of A. Faurse-Bonte, A. Darefeuille-Michaud, and C. Forester, Université d'Auvergne, France.)

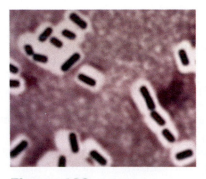

Figure 136
Klebsiella pneumoniae and its capsules.

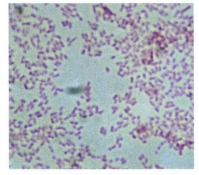

Figure 137
Gram stain of *Morganella morganii.* (1,000×).

Proteus

Proteus species, which include **P. vulgaris** and **P. mirabilis**, occur in the intestines of a variety of mammals and also can be found in manure, soil, and polluted waters.

Transmission. *Proteus* species may be acquired by the ingestion of contaminated food or water.

Morphology and Cultural Properties. *Proteus* species are Gram-negative, straight rods, $0.4–0.8\ \mu m$ in diameter and $1.0–3.0\ \mu m$ in length (Figures 142). These organisms occur singly and in pairs and are motile by means of peritrichous flagella. They generally produce concentric zones of growth (swarming) over most agar surfaces (Figure 143 and 144).

Proteus species are facultatively anaerobic and are positive for urea hydrolysis (see Figure 145B) and phenylalanine deaminase (see Figure 65). These organisms are oxidase negative and catalase and methyl red positive, and they give variable results with indole, Voges-Proskauer, and citrate tests. Glucose, but not lactose, and a few other carbohydrates are catabolized, with the production of acid and usually gas. Hydrogen sulfide is produced by *P. vulgaris* and *P. mirabilis* (Figure 145A).

Pathology and Clinical Features. *Proteus vulgaris* and *P. mirabilis* are associated with gastrointestinal infections, and they are often isolated from urinary tract and other extraintestinal infections such as septic lesions in burn patients. Urease production by *Proteus* species and *Morganella morganii* is thought to play a major role in the pathogenicity of these organisms, especially in the case of urinary tract infections.

Diagnosis. Culture and biochemical tests are used for identification.

Salmonella

Salmonella species other than *S. typhi* (the causative agent of typhoid fever) can be found in the gastrointestinal tracts of both warm- and cold-blooded animals and in the environment. These microorganisms either cause disease or are carried without any apparent harmful effects on the host.

Transmission. Infections with *Salmonella* species for the most part are acquired through the ingestion of contaminated food or drink. Contaminated meat and dairy products are the most likely sources of disease agents, although uncooked eggs also may be responsible for foodborne infection. Infected food handlers and other persons can transmit *Salmonella* infection as well.

Morphology and Cultural Properties. *Salmonella* species are straight, Gram-negative rods, $0.7–1.5\ \mu m$ in diameter and $2.0–5.0\ \mu m$ in length. The organisms are usually motile by peritrichous flagella (Figure 146).

Salmonella species are facultative anaerobes. Optimal growth temperature is 37°C. Organisms are oxidase negative and catalase positive, and they reduce nitrates. The IMViC pattern of reactions is $-, +, -, +$. Glucose and other carbohydrates, with the exception of lactose, usually are fermented. *Salmonella* forms distinctive colonies on various selective and differential media such as Hektoen enteric, *Salmonella-Shigella,* and XLT4 agars (Figure 147). Hydrogen sulfide is typically produced by various species (Figure 147).

Pathology and Clinical Features. From a clinical standpoint, most *Salmonella* infections can be divided into gastroenteritis (inflammation of the stomach and intestines), enteric fever, bacteremia, and the more serious typhoid fever.

Gastroenteritis has an incubation period of about 24 to 48 hours after ingestion of the causative agents. Episodes of nausea and vomiting are followed or accompanied by abdominal cramps and diarrhea. These signs and symptoms, especially diarrhea, may persist for 3 to 4 days. Fever does occur in about half of patients. *Salmonella typhimurium* and *S. enteritidis* are the major causes.

Enteric fever involves several organs, particularly the reticuloendothelial system, liver, spleen, and mesenteric lymph nodes. A prolonged fever is typical. *Salmonella typhi* and other *Salmonella* species are associated with enteric fever.

In typhoid fever, organisms penetrate the lining of the small intestine, enter the local lymph nodes, where they multiply, and eventually gain access to the bloodstream. By means of the bloodstream, the salmonellae are carried to most organs of the body, where they continue to multiply.

Typhoid fever has an incubation period ranging from 7 to 14 days. Common early signs and symptoms include headache, chills, fever, loss of energy, and generalized aches and pains. Constipation rather than diarrhea is typical. Organisms continue to be spread throughout the body, causing a bacteremia and a very high fever. Small purplish areas of bleeding (**petchiae**) called **rose spots** may develop in the skin around the trunk.

Systemic *Salmonella* infections are known to spread and may involve the heart and parts of the respiratory, skeletal, and central nervous systems.

Diagnosis. Various specimens are used for culture and diagnosis. These include stool for gastroenteritis, blood for enteric fever and bacteremia, stool and blood for typhoid fever, and sputum for pneumonia.

Facultatively Anaerobic Gram-Negative Rods (*Continued*)

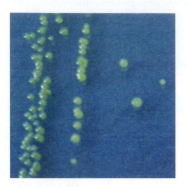

Figure 138
Morganella morganii colonies on nutrient agar.

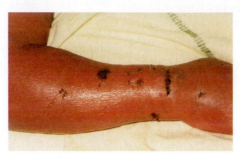

Figure 139
Wound infection and spreading inflammation involving subcutaneous tissue of the arm caused by *Pasteurella multocida*. (From O. Chosidow, *Internat. J. Dermatol.* 33(1994):471.)

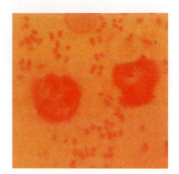

Figure 140
Bipolar Gram-negative coccoid to rod-shaped *Pasteurella multocida*. (From O. Chosidow, *Internat. J. Dermatol.* 33(1994):471.)

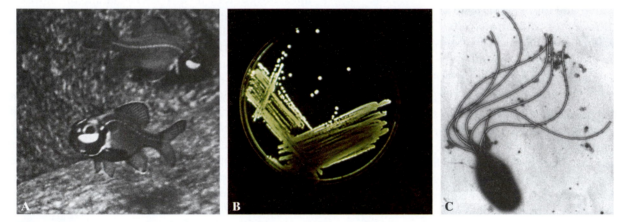

Figure 141
Photobacteria (luminous bacteria). (A) The location of microbiological light-generating organ of the flashlight fish (*Photobelpharon palpebratus*). (From J.G. Morin et al., Science. 190(1975):74–76.) (B) Appearance of bioluminescent bacteria on an agar medium. The colonies were photographed using only their own light. (Courtesy Kenneth Nealson.) (C) Transmission electron micrograph of *Photobacterium fischeri*, one of several species known to colonize the luminous organs of fish and other aquatic life. (From J.L. Reichelt, and P. Bauman. *Arch Microbiol.* 94(1973):283.)

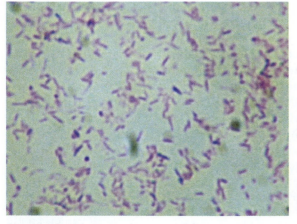

Figure 142
Gram stain of *Proteus vulgaris* (1,000×).

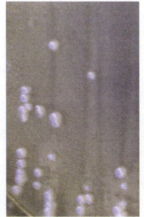

Figure 143
Proteus vulgaris colonies on nutrient agar.

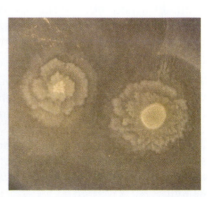

Figure 144
Proteus vulgaris exhibiting its characteristic swarming effect on trypticase soy agar.

Serratia

Serratia species are found in soil, in water, on plant surfaces, in the digestive tracts of rodents and insects, and in human clinical specimens. The red-pigmented *S. marcescens* (Figure 148) is a major opportunistic pathogen for hospitalized persons, causing urinary tract infections and septicemia.

Transmission. The means of transmission include aerosols.

Morphology and Cultural Properties. *Serratia* species are Gram-negative rods measuring 0.5–0.8 μm in diameter and 0.9–2.0 μm in length (Figure 149). These organisms move by means of peritrichous flagella.

Serratia species are facultative anaerobes. They grow well at temperatures ranging from 30° to 37°C, and they catabolize glucose and other carbohydrates, with the production of acid and often of gas. Pigment production is enhanced at lower temperatures. Most strains produce DNAse (see Figure 57), hydrolyze gelatin, and reduce nitrates. Voges-Proskauer and citrate tests are positive; indole is generally negative. The methyl red test is variable.

Shigella

The four species of *Shigella*—*S. dystenteriae* (Shiga bacillus), *S. flexneri*, *S. boydii*, and *S. sonnei*—cause bacillary dysentery. These organisms are found only in humans and certain other primates. The Shiga bacillus causes the severest form of the disease.

Transmission. Most *Shigella* infections are acquired by the fecal-oral route, by the ingestion of contaminated food or water.

Morphology and Cultural Properties. *Shigella* are straight, short, Gram-negative rods, 0.5–0.7 μm in diameter and 2.0–3.0 μm in length (Figure 150). Cells occur singly or in pairs. These organisms are nonmotile and do not form capsules.

Shigella species are facultative anaerobes. Optimal growth temperature is 37°C. Most species produce smooth, colorless, circular colonies. *Shigella* species are oxidase negative, catalase positive, and they reduce nitrates. Methyl red, Voges-Proskauer, and citrate reactions are negative. Indole production is variable. H_2S and urease are not produced. Glucose and certain other carbohydrates, with the exception of lactose, are catabolized, with the production of acid.

Pathology and Clinical Features. Once organisms gain access to and invade the cells of the small intestine, they multiply and bring about the destruction of the intestinal mucous lining, causing ulcer formation, bleeding, and diarrhea. The combination of diarrhea with blood and mucus constitutes dysentery. Infected persons may experience abdominal pain and severe cramping. Shigellosis is usually self-limiting in about 2 to 5 days.

Diagnosis. Stools or rectal swabs obtained during bowel examination are used for the inoculation of selective and differential media. Biochemical testing and serological tests are used for identification.

Vibrio

Morphology and Cultural Properties. *Vibrio* species are straight or curved Gram-negative rods, 0.5–0.8 μm in width and 1.4–2.6 μm in length (Figure 151). These organisms are motile by one or more polar flagella.

Vibrio species are facultative anaerobes. Optimal growth temperatures vary from 20° to 30°C. Glucose and most other carbohydrates are fermented, with the production of acid but not gas.

Vibrio species are commonly found in marine and estuarine environments and in the intestinal contents of marine animals. Some species are also found in freshwater environments.

Several species are pathogenic for marine vertebrates and invertebrates and for humans. The most notable of human pathogens are *V. cholerae*, the causative agent of cholera, *V. parahaemolyticus*, a major causative agent of food poisoning associated with contaminated fish or shellfish, and *V. vulnificus*, a cause of fatal septicemia.

Vibrio cholerae

Vibrio cholerae is the cause of worldwide pandemics resulting in extremely dehydrating diarrheal disease.

Transmission. Cholera is acquired through the ingestion of contaminated food or water. Bathing, playing, and related activities in water contaminated by sewage are high-risk activities.

Cultural Properties. *Vibrio cholerae* grow on a variety of media. Optimal growth temperature is 35°C. Most species form 1- or 2-mm-wide colonies within 24 hours (Figure 152). *Vibrio cholerae* is oxidase positive.

Pathology and Clinical Features. *Vibrio cholerae* multiplies rapidly on the mucosal surface of the small intestine and produces its toxin, **choleragen.** The action of the toxin causes the release of fluids and electrolytes into the intestinal lumen. Within several hours to 3 days, infected persons experience a sudden onset of explosive, watery diarrhea with vomiting and abdominal pain. Up to 7 liters of liquid stool can be released by one person in a 24-hour period. Another effect can be the shredding of the intestinal lining, which appears as small white flecks resembling rice grains in stools. The term "rice-water stools" is used to describe the effect.

Facultatively Anaerobic Gram-Negative Rods (Continued)

Figure 145

(A) Hydrogen sulfide (H$_2$S) production by *Proteus vulgaris* in a triple sugar iron agar slant. (B) H$_2$S production (top), and a positive urease reaction in the bottom of the Entero-Screen 4 tube (Courtesy Hardy Diagnostics.)

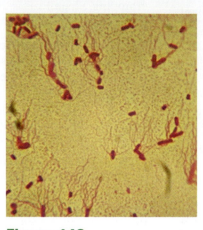

Figure 146

Salmonella typhi with peritrichous flagella.

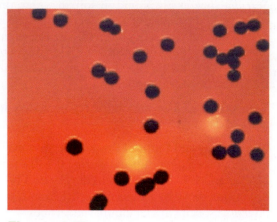

Figure 147

Salmonella species colonies (black) on XLT4 agar, showing hydrogen sulfide production. (Courtesy of Difco Laboratories Detroit, MI.)

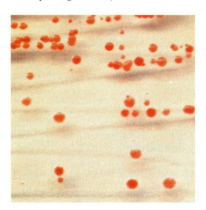

Figure 148

Red-pigmented colonies of *Serratia marcescens* on nutrient agar.

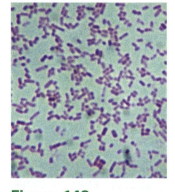

Figure 149

Gram-negative rods of *Serratia marcescens* (1,000×).

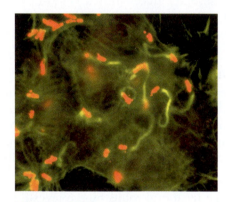

Figure 150

Tissue culture cells infected by *Shigella flexneri* shown by fluorescence labeling. (Courtesy of Dr. P.J. Sarsonetti, Unité de Pathogenie, Institut Pasteur.)

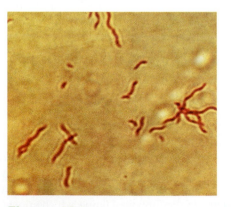

Figure 151

Gram-negative stain of *Vibrio cholerae* (1,000×).

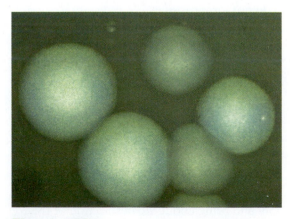

Figure 152

Vibrio cholerae colonies grown on meat extract agar and incubated overnight at 37°C. (From R.A. Finkelstein, M. Boesman-Finkelstein, Y. Chang, and C.C. Hase. *In. Immunol.* 60(1992):472–78.)

Diagnosis. Stool specimens are the only ones used for microscopic examination and culture. Serological tests are used for definitive identification.

Yersinia

Yersinia species are primarily animal pathogens. The genus contains several important human pathogens including *Y. enterocolitica,* the cause of **enterocolitis** and other diseases, and *Y. pestis,* the cause of plague. Another species, *Y. pseudotuberculosis,* causes pseudotuberculosis, a disease resulting in inflammation and damage to lymph nodes, the spleen, and the liver. The clinical features of fever and abdominal pain often mimic acute appendicitis. In most cases, wild animals are possible sources of infection.

Morphology and Cultural Features. Yersinia species are small, Gram-negative rods, sometimes appearing as cocco-bacilli. Cells measure 0.5–0.8 μm in width, and 1.0–3.0 μm in length. Retaining stain at the ends of cells produces a phenomenon called **bipolar staining**, a common finding (see Figure 140A). Most species are nonmotile when grown at temperatures above 30°C.

Yersinia species are facultative anaerobes and have an optimal growth temperature range from 28° to 30°C. Organisms are oxidase negative and catalase and urease positive. They reduce nitrates and produce negative Voges-Proskauer and citrate tests when grown at 37°C. Glucose and other carbohydrates are fermented, with the production of acid only. Hydrogen sulfide is not produced.

Yersinia enterocolitica

Yersinia enterocolitica (Figure 153) along with *Y. pseudotuberculosis* is associated with a condition known as yersiniosis, a clinical condition that mimics acute appendicitis. This organism causes a wider variety of infections than other members of the genus.

Transmission. Infection is usually acquired by the ingestion of contaminated food or water.

Pathology and Clinical Features. The most common infection caused by *Y. entercolitica* is an enterocolitis, usually in children. Typical signs and symptoms include fever, abdominal pain, and diarrhea. This infection generally is self-limiting.

Diagnosis. Stool specimens can be used for diagnosis, but attempts to isolate the organisms have limited success.

Yersinia pestis

Yersinia pestis (Figure 154) causes plague in both lower animals such as rodents (Figure 155A) and small mammals and humans.

Transmission. Plague has two major cycles, known as **urban** and **sylvatic**, and two major clinical forms, **bubonic** and **pneumonic**. The disease is transmitted to humans by the bite of infected fleas (Figure 155B).

Pathology and Clinical Features. The incubation period for bubonic plague ranges from 2 to 7 days after the flea bite. The appearance of fever and the painful bubo (Figure 154), usually in the groin and less often in the armpit, signal the onset of the disease. If the person is not treated, the organisms gain access to the bloodstream and cause a fatal septic shock. Pneumonic plague develops in about 5% of victims. Fever, general loss of energy, and a tight feeling in the chest are early symptoms. The production of sputum, difficulty in breathing, cyanosis, and death occur later, as early as the second or third day of illness.

Diagnosis. Gram stains of material taken from a bubo or other specimens such as blood, sputum, and spinal fluid usually are used for identification (Figure 156). Similar specimens are used for culture.

▲ The Delta Proteobacteria

The delta proteobacteria are a distinctive group of microorganisms. Included in their ranks are some predatory species of other bacteria and important contributors to natural biogeochemical cycles, such as the one involving sulfur. The sulfur cycle involves various oxidation and reduction stages of sulfur in the environment.

▲ Aerobic, Motile, Gram-Negative Predatory Curved Rods

Bdellovibrio

Bdellovibrio bacteriovorus

Several bacteria need living cells for their development and reproduction. Some species require eukaryotic animal and plant cells, while others use prokaryotes. One well-known intracellular predator of bacteria is *Bdellovibrio bacteriovorus* (del-lō, VIB-rē-ō-bak-tē-rē-Ō-vō-rus). This bacterium strikes and sticks to the surface of its prey before entering the new host. The generic name of the particular attacking bacterium reflects this behavior, *Bdello* comes from the Greek word meaning "leech." *Bdellovibrio* species are widely distributed in soil, sewage, and freshwater and marine environments.

Morphology and Cultural Properties. *Bdellovibrio bacteriovorus* cells are Gram-negative comma-shaped rods, 0.2–0.5 μm in diameter, and motile by means of a single sheathed flagellum. *B. bacteriovorus* is an obligate

Facultatively Anaerobic Gram-Negative Rods *(Continued)*

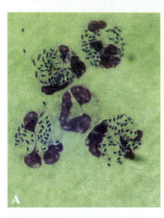
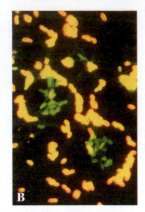

Figure 153

Two microscopic views of *Yersinia enterocolitica.*
(A) Giemsa stain showing *Y. enterocolitica* ingested by polymorpholeukocytes. (B) Similar view of phagocytosis shown by double immunofluorescence. Intracellular pathogens are green, and extracellular forms are yellow. (From J.H. Ewald, J. Hessemann, H. Riidiger, and I.B. Autenrieth. *J. Inf. Dis.* 170(1994):140–50.)

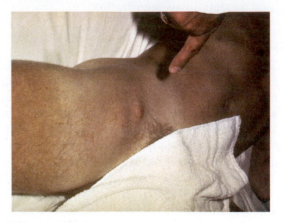

Figure 154

The appearance of a bubo, the first sign of the plague in humans.

Figure 155

Important players in the transmission of plague. (A) The rat, the source of *Yersinia pestis* among wild rodents. (B) The flea, the arthropod that transmits the disease agents among lower animals and to humans.

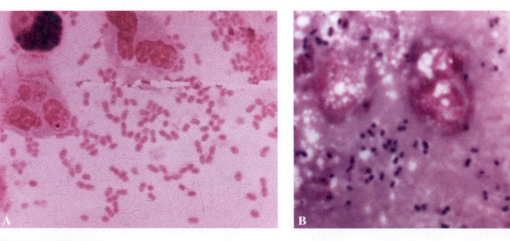

Figure 156

Two microscopic views of *Yersinia pestis*. (A) A Gram stain of spinal fluid from a patient with plague meningitis showing bipolar and pleomorphic Gram-negative cells. (B) A Wayson stain of bubo material taken from a patient with bubonic plague. (Courtesy of T. Butler, Division of Infectious Diseases, Texas Tech University Health Sciences Center, Houston, TX.).

aerobic chemoorganotroph. The highly motile *Bdellovibrio* locates its prey by means of chance collisions. After forcibly striking and attaching to a generally larger prey cell, the parasitic bacterium penetrates through the prey's cell wall and reproduces in the area between the wall and the plasma membrane known as the **periplasmic space** (Figure 157). The prey cell containing the invading bdellovibrio usually rounds up, swells, and eventually forms a spherical structure known as a bdelloplast. Soon after being attacked, the prey cell is killed and functionally becomes the substrate for *Bdellovibrio* development. The developing vibrio elongates into a spiral-shaped nonmotile cell that fragments into new small motile vibrios. The newly formed predacious offspring leave the prey cell to repeat the predatory cycle in susceptible bacteria. Interestingly, only Gram-negative cells are generally attacked.

Stigmatella

Stigmatella aurantiaca

Stigmatella is an example of the fruiting myxobacteria. Species also belong to the group of gliding bacteria, which are usually long rods or filamentous forms that lack flagella but are still able to move when in contact with surfaces. *Stigmatella* and related species have the interesting property of not only being able to form multicellular structures known as **fruiting bodies** (Figure 158), but also exhibiting the most complex life cycle found among bacteria.

Morphology and Cultural Properties. The vegetative cells of fruiting myxobacteria are Gram-negative, non-flagellated rods, 0.6–0.8 μm in diameter and 4–10 μm long. As these cells glide along surfaces they prey upon other bacteria as sources of nutrition, and leave a slime trail behind them. Under appropriate conditions, usually when nutrients are in short supply, large numbers of the vegetative cells come together and form a visible stalked fruiting bodies. These fruiting bodies, or *sporangia,* contain large numbers of resting structures known as *myxospores* (Figure 158). As the life cycle of myxobacteria continues, the myxospores germinate and give rise to new vegetative gliding cells.

Compared to vegetative cells, myxospores are more resistant to a number of physical factors, including the effects of drying, heat, ultraviolet radiation, and sonic vibration. Such resistance is much less than is found with the endospores of bacterial genera such as *Bacillus* and *Clostridium*.

Microaerophilic, Motile, Gram-Negative Predatory Curved Rods

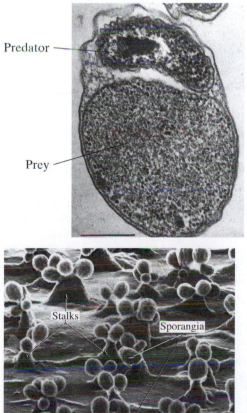

Predator

Prey

Figure 157

Transmission electron micrograph showing the presence *Bdellovibrio* after it has penetrated its prey. The prey cell will serve as the nutrient source for the predator. Eventually the prey cell will break open, releasing newly formed bdellovibrios. (From Abram et al. *J. Bacteriol.* 118(1974):663.)

Stalks

Sporangia

Figure 158

Scanning electron micrograph showing the multicellular fruiting bodies of the gliding mxyobacterium *Stigmatella aurantiaca*. Specific parts of a typical fruiting body include the characteristic myxospore-containing sac and the supporting stalk. (From D. White, J.A. Johnson, and K. Stephens. *J. Bacteriol.* 144(1980):400–405.)

▲ The Epsilon Proteobacteria

The epsilon proteobacteria include many species that are pathogenic to humans and other animals.

▲ Aerobic/Microaerophilic, Motile, Helical/Vibroid, Gram-Negative Bacteria

Helicobacter

Helicobacter pylori

A common pathogen of humans, *Helicobacter pylori* is the principal cause of chronic inflammation of the stomach (gastritis) and peptic ulcer disease (Figure 159). It is also a risk factor for gastric cancer, even though most infections are without symptoms. Once established, most infections last for years and rarely cure spontaneously. They usually can be cured by antimicrobial therapy. Infection also increases the risk of other diseases.

Transmission. *Helicobacter pylori* infections occur in human populations throughout the world and in all populations studied. Occurrence of infections increases with age. Person-to-person transmission appears to account for infections within families. Although the exact means by which infection is acquired is not well understood, fecal-oral or oral-oral transmission is likely.

Morphology and Cultural Properties. *Helicobacter pylori* is a Gram-negative, spiral, curved or straight rod, 0.5–1.0 μm wide and 2.5–5.0 μm long (Figures 160 and 161). The organism is motile and microaerophilic and grows best in an atmosphere of 5% O_2 with 5–10% CO_2 and on enriched media such as blood agar. The organism also can grow under aerobic conditions with high humidity (Figure 162). Optimal temperature is 37°C. Colonies are nonpigmented, translu- cent, and 1–2 mm in diameter. *Helicobacter pylori* rapidly hydrolyzes urea in laboratory tests by means of an unusual urease enzyme. There is speculation that *H. pylori*'s urease production serves as a significant virulence factor by neutralizing the acidity in the stomach to allow the pathogen to reproduce more easily in the tissues of the stomach. Isolated *H. pylori* are oxidase and catalase positive.

Pathology and Clinical Features. *Helicobacter pylori* is a highly successful microbe, infecting most of the world's population. Once established in the stomach, the organism persists for decades, seemingly unaffected by the acid and gastric movement (**peristalsis**) in its environment. In most cases, *H. pylori* silently coexists with its host, causing no symptoms or signs. But this relation is not always harmless. Gastric cancer develops in approximately 1% of infected persons, and 20% have duodenal ulcers.

Gastric cancer is a leading cause of death from cancer in the developing world, where infection with *H. pylori* in early childhood is the rule. In infected children, duodenal ulcer is rare. However, by early adulthood, the infection acquired in childhood has frequently progressed to a pathologic condition in which there is a patchy loss of gastric glands that secrete protein and acid (**multifocal atrophic gastritis**). The decrease in acid secretion opens the door to both stomach (gastric) ulcers and cancer. Childhood infection predisposes patients to gastric ulcers and stomach cancer, but it inhibits the development of duodenal ulcers.

In industrialized countries, deaths due to gastric cancer, gastric ulcers, and duodenal ulcers are decreasing. This decrease is believed to be due to improvements in hygiene and economic conditions that have interfered with the transmission cycle of the organism to the point where *H. pylori* is now hard to find in children.

Diagnosis. The urea-breath test, a clinical procedure, can be used to identify infected individuals. However, tissue specimens obtained by biopsy are necessary for tissue and microbiological examination and confirmation of *H. pylori* infection. Several serological tests, including the ELISA procedure, have been introduced to aid in diagnosis. Newer diagnostic tests are under development.

▲ Gram-Positive Bacteria

Gram-positive bacteria can be divided into two subgroups on the basis of their deoxyribonucleic acid guanine (G) plus the cytosine (C) base ratios, usually referred to as the GC ratio. Such determinations place bacterial species into one of two major phylogenetic subdivisions, **low GC** and **high GC**. These designations refer to the fact that specific species have GC ratios either significantly below or above 50 percent. The GC ratios also can be used as a means of introducing this large group of primarily chemoorganotrophic bacteria. Because of morphological and certain other similarities that exist among the Gram-positives, some overlapping between the two subgroups will occur in the following presentations.

Aerobic/Microaerophilic, Motile, Helical/Vibroid, Gram-Negative Bacteria

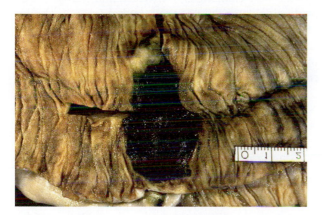

Figure 159
A large, well-defined ulcer. (From M. Koona and associates. *Acta Pathol. Jpn.* 42(1992): 267–71.)

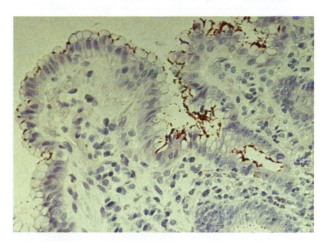

Figure 160
A cluster of *Helicobacter pylori* (red cells) along the lining of the stomach (1,000×). (Unpublished photo, courtesy of Dr. L.P. Andersen.)

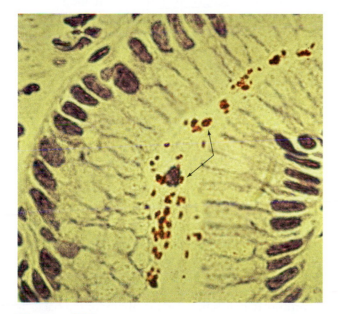

Figure 161
Helicobacter pylori (red cells) within the human intestine (1,000×.) (Courtesy of Dr. L.P. Andersen).

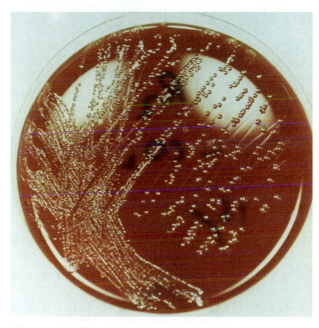

Figure 162
Colonies of *Helicobacter pylori* growing aerobically on an agar plate. (From H.X. Xia, C.T. Keane, and C.A. O'Morain. *European J. Clin. Microbiol. & Inf. Dis.* 13(1994):406.)

▲ Low GC, Endospore-Forming Gram-Positive Rods

Bacillus

Bacillus species are widely distributed in nature, particularly in soil, from where they are spread in dust, in water, and on animal or plant materials. The majority of *Bacillus* species are nonpathogenic and rarely are associated with diseases in humans or lower animals. Exceptions include *B. anthracis,* the causative agent of anthrax, *B. cereus,* the causative agent of human food poisoning, meningitis and burn infections, and species such as *B. larvae, B. popilliae,* and *B. thuringiensis,* known pathogens of specific insect groups.

Transmission. *Bacillus* species are spread through aerosols and by contact with infected animals or animal products such as hides and various types of brushes made with or contaminated by infectious animal materials.

Morphology and Cultural Properties. *Bacillus* species are Gram-positive, endospore-forming rods, 0.5–2.5 μm wide and 1.2–10.0 μm long. Cultures may become Gram-negative with age. Only one endospore is formed in each cell (Figures 163 and 164). Endospores are generally oval or sometimes round and are resistant to unfavorable environmental conditions and related factors such as extreme heat, cold, radiation, the effects of drying, and disinfectants. *Bacillus anthracis* forms capsules.

Bacillus species are aerobic or facultatively anaerobic and exhibit a wide range of physiological abilities. Several species are motile. Most members grow on simple media (Figures 164 and 165) and over a broad temperature range. Some, such as *B. circulans,* form unusual colonies (Figure 165C). The discussion of disease caused by members of this genus will be limited here to *Bacillus anthracis.*

The name *anthrax* comes from the Greek word for "coal." It refers to the typical black slough or lesion (*eschar*) that is seen on affected areas (Figure 166). The various forms of the disease are caused by the bacterium *Bacillus anthracis.*

Anthrax holds an important place in the development of modern medicine, and has long been intertwined with the history of humankind. The disease is believed to have been one of the Egyptian plagues at the time of Moses, and actual cases were clearly recorded by the ancient Romans. Today, human anthrax is not common. The disease mostly affects plant-eating animals, especially cattle, goats, and sheep. Moreover, human infection usually is of an occupational nature and is found mainly among veterinarians, farmers, commercial meat-handlers, and mill workers.

Unfortunately, the threat of bioterrorism has brought anthrax beyond the point of scientific inquiry and has created new challenges to medical and public health professionals. In light of the fact that large stockpiles of smallpox vaccine are on hand to treat exposed persons in the United States, anthrax is considered to be the major global biological threat. This is largely because of the following three reasons:

1. *Bacillus anthracis* spores (resistant form) are highly stable;
2. The spores of the pathogen can infect through the respiratory route;
3. The resulting inhalation disease (inhalation anthrax) has a high case-fatality rate.

Morphology and Cultural Properties. *Bacillus anthracis* is a Gram-positive, spore-forming, nonmotile rod with typical square ends, the appearance of which has been likened to that of bamboo rods. The organism occurs singly, in pairs, or in long chains (Figure 168).

In infected tissues spores are not present and cell chains are shorter and are surrounded by capsules, (a prominent feature). Capsules are typical of virulent (disease-causing) forms of *B. anthracis* (Figure 167B). Spores are readily formed in soil and in cultures (Figure 168).

Colonies of *B. anthracis,* which consist of parallel interlacing chains of cells, form colonies having a rough, uneven surface with numerous curled extensions at their edges. Such colonies resemble the mythical "Medusa head" (Figure 169).

B. anthracis is a facultative anaerobe that grows best aerobically on blood agar, at 37°C. and does not destroy red blood cells in a culture medium (nonhemolytic). The microscopic and cultural properties, combined with other important bacteriologic features, help to differentiate *B. anthracis* from other *Bacillus* species.

Anthrax depends both on the ability of the bacterium's capsule to interfere with host's immune response, phagocytosis, and on the bacterium's production of a powerful exotoxin.

Transmission. Human infections result from contact with contaminated animals or animal products, such as hides, wool, bone, or hair. Although moderately more resistant to anthrax than plant-eating animals, humans are still susceptible to the disease in industrial and non-industrial settings. While there have been no known cases of human-to-human transmission, a few cases of laboratory-acquired infections have been reported. The actual means of transmission is through the skin by contact with contaminated animal products or by inhaling *Bacillus anthracis* spores.

Endospore-Forming Gram-Positive Rods and Cocci

Figure 163

Spore-stained preparation of *Bacillus subtilis*. Carbol fuchsin and methylene blue were used as the primary and secondary stains, respectively. Spores appear as red swollen areas in rods (1,000×).

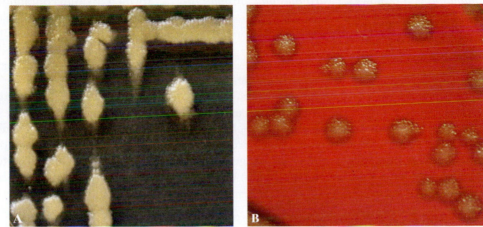

Figure 164

Bacillus subtilis on nutrient media. (A) On nutrient agar. (B) On blood agar.

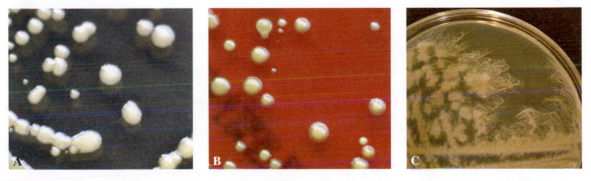

Figure 165

Bacillus species on nutrient media. (A) *B. megaterium* on nutrient agar. (B) *B. megaterium* on blood agar. (C) The unusual filamentous growth of *B. circulans*.

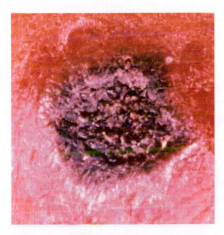

Figure 166

An eschar, a typical anthrax skin lesion. (Courtesy of the Centers for Disease Control and Prevention.)

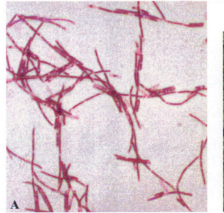

Figure 167

Microscopic views of *Bacillus anthracis*. (A) Gram-stained preparation. (Courtesy of the Centers for Disease Control and Prevention.) (B) The presence of a capsule surrounding a long chain of cells. (Courtesy of T. Abshire and J. Ezzell. USAMRIID, Ft. Detrick, MD.)

Currently five forms of anthrax are known: *cutaneous, inhalation* (pulmonary), *gastrointestinal, oropharyngeal,* and *meningitis.*

Clostridium

Clostridium species are widespread in the environment. They can be found in the soil as well as in the normal intestinal microbiota of humans and other mammals. The genus includes several saprophytes and pathogenic species. Pathogens are known to cause diseases such as botulism, tetanus, gas gangrene, and pseudomembranous colitis, a complication of antibiotic therapy (Figure 173).

Transmission. *Clostridium* species can be spread in a variety of ways, including aerosols, insects, and contaminated foods and objects.

Morphology and Cultural Properties. Clostridia are Gram-positive, commonly pleomorphic rods, $0.3-2.0$ μm wide and $1.5-2.00$ μm long (Figure 170). Cells occur in pairs or in short chains, with rounded or sometimes pointed ends. Endospores vary from oval to spherical forms and usually cause the cell to distend (Figures 171 and 172A).

Clostridia are strictly (obligately) anaerobic to aerotolerant and usually are motile. Species are metabolically diverse and may be saccharolytic, proteolytic, neither, or both. Growth temperatures range from 10° to 65°C.

Pathology. The discussion of clostridial agents will be limited to only those primarily causing disease in humans.

Diagnosis. Several methods are available for the diagnosis of different clostridial infections. These include tissue Gram stains (if appropriate), bacteriological culture, tissue culture cytotoxicity, enzyme immunoassay, and polymerase chain technology.

Clostridium botulinum

Clostridium botulinum (Figure 172B) is responsible for the paralyzing disease botulism. This species is divided into eight types (A, B, C alpha, C beta, D, E, F, G) on the basis of the specific toxins produced.

Four clinical forms of botulism are recognized. These are food-born (an intoxication), infant (infection and intoxication), wound (infection and intoxication), and adult intestinal colonization (infection and intoxication). Botulinum toxins A, B, E, and F are the agents usually associated with these conditions.

Foodborne botulism occurs after ingestion of preformed toxin in contaminated food. Infant botulism results from the ingestion of *C. botulinum* spores, followed by the germination of the spores, and toxin production in the infant's intestine. Wound botulism may occur after contamination of a deep wound by *C. botulinum.*

Adult intestinal colonization occurs following the ingestion of *C. botulinum* spore-contaminated foods. Exactly when the bacterium is introduced to an adult is largely unknown. In addition, adult intestinal colonization has been associated with gastrointestinal infections in adults after broad-spectrum antibiotic treatment and either intestinal surgery or inflammatory bowel disease. Once introduced into the gastrointestinal system, the condition follows a sequence of events similar to that occurring in case of infant botulism.

Other botulinum toxin-producing clostridia, primarily causing infant botulism, have been described in recent years. These include *C. butyricum,* producing botulinum toxin type E, and *C. baratii,* producing botulinum toxin type F.

Clostridium difficile

Clostridium difficile causes most cases of antibiotic-associated pseudomembranous colitis. In this condition, raised whitish yellow or greenish yellow plaques form on the colon (Figure 173A). These plaques often combine with one another to form a pseudomembrane and cause substantial tissue destruction (Figure 173B), dehydration, and electrolyte imbalance.

Clostridium perfringens

Various clostridial species can cause wound infections, including gas gangrene. *Clostridium perfringens* is the most commonly associated species. It causes the infection after invading normal healthy muscle surrounding the wound. Fermentation of muscle carbo–hydrate produces gas in subcutaneous tissues that can be felt on touch.

Clostridium tetani

Clostridium tetani is the cause of tetanus, or lockjaw. When spores (Figure 171) are introduced into wounds by contaminated soil or foreign objects such as nails or glass splinters, they germinate into vegetative cells that subsequently produce exotoxins. The tetanus toxins cause muscular dysfunction and allow uncontrolled muscle stimulation.

Endospore-Forming Gram-Positive Rods

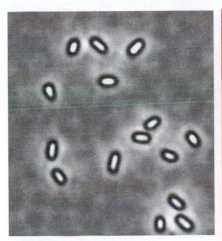

Figure 168
A wet mount preparation of
B. anthracis under phase contrast.
(Courtesy of T. Abshire and J. Ezzell,
USAMRIID, Ft. Detrick, MD.)

Figure 169
B. anthracis colonies on 5% sheep blood agar.
Note the dry ground glass surface and irregular
edges of the colonies. (Courtesy of T. Abshire
and J. Ezzell, USAMRIID, Ft. Detrick, MD.)

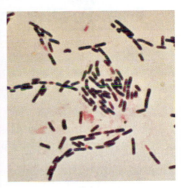

Figure 170
Gram-positive rods of
Clostridium species.

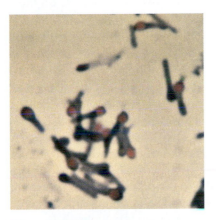

Figure 171
Microscopic view of *Clostridium
tetani* spores. The red spores appear at
the ends of cells.

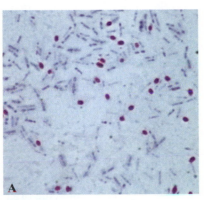

Figure 172
Clostridium botulinum. (A) Spore stain. (B) Colonies.

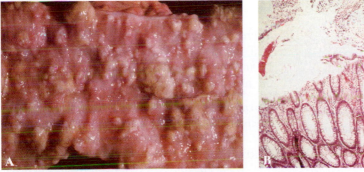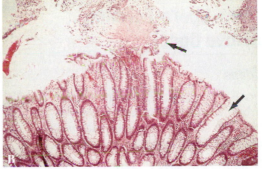

Figure 173
Pseudomembranous colitis. (A) A large intestine specimen showing yellow plaques
(enlarged areas). (B) A microscopic view of a specimen showing local destruction
(lower arrow) and pseudomembrane material and destroyed tissue (upper arrow).
(From C.P. Kelley, C. Pothoulakis, and J. T. LaMont. *NEJM* 330(1994):257.)

▲ Low GC, Regular, Non-Spore-Forming Gram-Positive Rods

Listeria

Listeria species are widely distributed in the environment. Certain species are pathogenic.

Listeria monocytogenes

Listeria monocytogenes causes listeriosis and related conditions primarily in immunocompromised hosts.

Transmission. Listeriosis is primarily transmitted by contaminated food.

Morphology and Cultural Properties. Listeria monocytogenes is a Gram-positive, non-spore-forming and nonencapsulated short rod, 0.4–0.5 μm in width and 0.5–2.0 μm in length (Figure 175). Cells have rounded ends and occur singly or in short chains.

The organism is a facultative anaerobe and is found intracellularly (Figure 175A). It is motile when grown at 20° to 25°C. However, its optimal temperature range is from 30° to 37°C. Listeria species are catalase positive and oxidase negative, and they ferment glucose.

Pathology and Clinical Features. Pregnant women, newborns, and immunocompromised individuals such as persons with AIDS are particularly susceptible to infections. The most common infections include bacteremia, spontaneous abortion, meningitis, and meningoencephalitis.

Listeria monocytogenes has created great interest because of its ability to reorganize host cell protein (actin) and to use this ability to spread to neighboring cells.

Diagnosis. Cultures of blood, cerebrospinal fluid, or wounds are used for diagnosis (Figure 174). Immunofluorescence and related methods also are of value (Figure 175C).

▲ Low GC, Non-Spore-Forming Gram-Positive Cocci

Aerococcus viridans

Aerococcus viridans is a rare cause of human infection. It is a common airborne organism in various environments including hospitals and has been associated with nosocomial infections in immunosuppressed patients.

Morphology and Cultural Properties. Aerococcus viridans is a Gram-positive coccus measuring 1.0–2.0 μm in diameter. Tetrads are commonly formed (Figure 176).

Aerococcus viridans is nonmotile and facultatively anaerobic. It grows best under reduced oxygen tension and at an optimal temperature of 30°C (Figure 177).

The organism is catalase negative and does not liquefy gelatin or reduce nitrate. Acid without gas is produced from various carbohydrates. Unusual cultural features include the ability to grow at pH 9.6 and in 10% NaCl and in 40% bile.

Enterococcus faecalis

Enterococcus faecalis is widely distributed in the environment, especially in the fecal matter of vertebrates. This organism is associated with urinary and biliary tract infections. In addition, it participates in mixed microbial infections involving the lungs, brain, abdomen, and pelvis.

Transmission. Transmission includes contaminated food or water, fomites, and aerosols.

Morphology and Cultural Properties. Enterococcus faecalis is a Gram-positive coccus measuring 0.6–2.5 μm (Figure 178). Cells occur in pairs or in short chains. The organism is nonmotile, does not form capsules, and is facultatively anaerobic. Acid without gas is produced from a wide range of carbohydrates.

Enterococcus faecalis grows best at 37°C on most media (Figure 179) but can also grow at both 10° and 45°C, at a pH of 9.6, and in media containing 6.5% NaCl (see Figure 191B) or 40% bile (see Figure 191A).

Pathology. Enterococci are frequently associated with urinary tract infections and bacteriuria following urologic, intraabdominal, or hepatobiliary surgery. Enterococcal bacteremia also is often found in elderly patients with serious underlying medical problems. The prevalence of enterococcal infections has increased dramatically to the point that enterococci are now the organisms second most commonly recovered from patients with nosocomial infections.

Diagnosis. Identification involves isolation and biochemical tests of E. faecalis.

Micrococcus luteus

Micrococcus luteus is primarily found on mammalian skin and in soil. It is also commonly isolated from the air and certain foods. M. luteus is actually a member of the high GC subgroup.

Morphology and Cultural Properties. Micrococcus luteus is a Gram-positive coccus measuring 0.5–2.0 μm in diameter (Figure 180). Cells occur in pairs, tetrads, or irregular clusters but not in chains.

Micrococcus luteus is strictly aerobic and grows best at temperatures ranging from 25° to 37°C on simple media (Figure 181). Colonies are usually yellow. The organism is catalase and oxidase positive (see Figures 68 and 67, respectively) and produces little or no acid from carbohydrates.

Regular, Non-Spore-Forming, Low GC, Gram-Positive Rods and Cocci

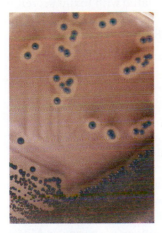

Figure 174

The appearance of *Listeria monocytogenes* colonies on BBL CHROMagar *Listeria* medium. This is a selective and differential medium specifically for Listeria isolation and identification. (Courtesy of Becton, Dickinson, and Company.)

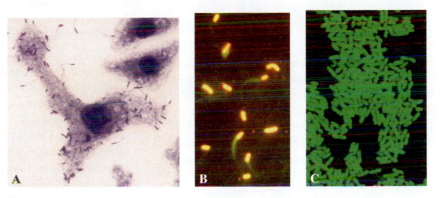

Figure 175

Microscopic views of *Listeria monocytogenes*. (A) Cells infected with *L. monocytogenes*. (Courtesy of Drs. S. Jones and D.A. Portroy.) (B) *L. monocytogenes* as shown by immunofluorescence. (From E. Gouin, P. Dehaoux, J. Mengaud, C. Kocks, and P. Cossart, *Inf. Immunol.* 3(1995):2729–37.) (C) Fluorescing cells of *Listeria* exposed to specific nucleic acid probes. (From B. Brehm-Stecher, J. J. Hyldig-Nielsen, and E. A. Johnson. *Appl. Environ. Microbiol.* 71(2005):5451–67.)

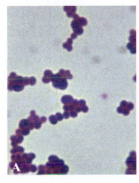

Figure 176

(A) Gram-stain preparation of *Aerococcus*. Note the tetrad arrangements (1,000×). (B) A methylene blue stain preparation showing tetrads (1,000×).

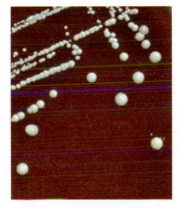

Figure 177

Aerococcus colonies on blood agar.

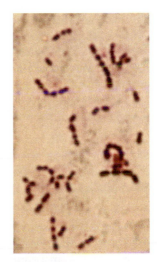

Figure 178

Gram stain of *Enterococcus faecalis*.

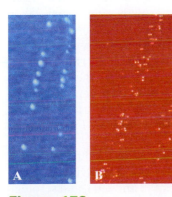

Figure 179

Enterococcus faecalis on culture media. (A) Colonies on nutrient agar. (B) Nonhemolytic colonies on blood agar.

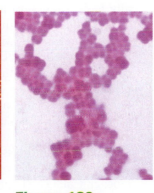

Figure 180

Gram stain preparation of *Micrococcus luteus* (1,000×).

Figure 181

Micrococcus luteus on culture media. (A) Colonies on nutrient agar. (B) Nonhemolytic colonies on blood agar.

Sarcina

Sarcina species are widely distributed in the environment and commonly are isolated from mammalian intestinal tracts.

Morphology and Cultural Properties. *Sarcina* species are Gram-positive cocci measuring 1.8–3.0 μm in diameter. Cells occur singly or in pairs, tetrads, or cubical packets of eight or more. Spore formation has been reported but is not usually seen.

Sarcina species are anaerobic and grow best at temperatures ranging from 30° to 37°C. Organisms are catalase negative and ferment carbohydrates.

Staphylococcus

Staphylococci are widespread in nature, though they are commonly found on the skin and in the mucous membranes of mammals and birds. Attention here will be given to only one species, *Staphylococcus aureus.*

Certain strains of *S. aureus* have developed a resistance to the antibiotic methicillin and have subsequently been designated as methicillin-resistant *Staphylococcus aureus* (MRSA). Hospital-associated strains associated with invasive diseases such as bloodstream, surgical site, and wound infections are designated as healthcare-associated MRSA or HAMRSA. Initially MRSA strains were confined to health care settings, but they have now spread to surrounding communities. Such strains are quite prevalent and are referred to as community-associated MRSA in order to distinguish them from other pathogenic strains.

Transmission. Staphylococci can be spread in a variety of ways, including aerosols, fomites, contact with infectious material from wounds, contaminated foods or water, and insects such as flies.

Morphology and Cultural Properties. Staphylococci are nonmotile, non-sporing, Gram-positive cocci, 0.50–1.5 μm in diameter (Figure 183). Cells occur singly, in pairs, and in grapelike clusters.

Staphylococcus aureus is facultatively anaerobic and grows best at a temperature from 30° to 37°C. Colonies are usually opaque, smooth, circular, cream, and sometimes yellow (Figure 184) or yellow-orange. Blood agar colonies normally show beta-hemolysis (Figure 184A). The organism is catalase positive (see Figure 68) and oxidase negative (see Figure 67). Nitrate is often reduced to nitrite (see Figure 66).

Pathology. *Staphylococcus aureus* causes a bewildering array of infections, including folliculitis (Figure 182A), superficial skin lesions (Figure 182B), and localized abscesses (Figure 182D); deep-seated infections involving bone (osteomyelitis), the central nervous system (meningitis), and the heart (endocarditis); toxic shock syndrome; pneumonia; and food poisoning. It, as well as *S. epidermidis,* is a major cause of nosocomial infections of surgical wounds and in-dwelling medical devices such as urinary catheters. Many of the staphylococcal infections are associated with toxins secreted by the organisms, one of which is staphylococcal scalded skin sydrome (Figure 182C).

Diagnosis. Several key characteristics of *S. aureus* can be used to identify the organism and to distinguish it from others. These include fermentation of mannitol on mannitol-salt agar (Figure 184B), DNAse production (see Figure 57), and a positive coagulase test (Figure 185). Coagulase is an enzyme that clots blood plasma. Commercially available molecular probes and identification systems, immunological tests, and bacterial viruses (phage typing) also are used for identification and related purposes.

Selective and differential can be used to detect the presence of MRSA in various types of specimens. Media of this type incorporate an antibiotic such as cefoxicin to discourage the growth of other bacteria. MRSA form pale purple-colored colonies (Figure 184E).

Streptococcus

The genus *Streptococcus* has great significance in medicine and industry. Various streptococci are important ecologically as members of the normal microbiota of humans and other animals. Some species are well known for the broad range of diseases they cause (Figures 186 and 187). Human diseases associated with the streptococci occur mainly in the respiratory tract or bloodstream or as skin infections. Industrially, several species are essential in various dairy processes and as indicators of pollution.

Transmission. The means of transmission include fomites, aerosols, and contaminated food and water.

Morphology and Cultural Properties. Streptococci are Gram-positive, nonsporing, nonmotile cocci, measuring 0.50–2.0 μm in diameter (Figure 188). Cells occur in pairs or chains (Figure 188A). Some specie are encapsulated (Figure 188B).

Streptococci are facultatively anaerobic and generally have a temperature growth range from 25° to 45°C. Optimal growth occurs at 37°C. These organisms are catalase negative and commonly attack media containing red bood cells (hemolysis). Streptococci can be divided into three groups by the type of hemolysis produced on blood agar: α hemolysis, producing green zones around colonies; β hemolysis, producing clear areas around colonies (Figure 189A); and γ hemolysis, producing no zones.

Streptococci are classified and placed into groups on the basis of colonial properties, hemolytic reactions, biochemical tests (Figures 190 and 191), and serological specificities based on antigenic differences in cell wall carbohydrates, proteins, and polysaccharide capsules. Examples of biochemical tests include bacitracin sensitivity, which differentiates group A streptococci

Gram-Positive Cocci

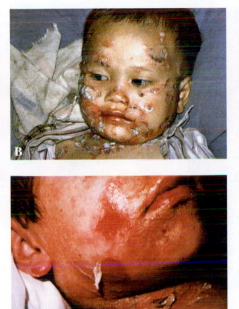

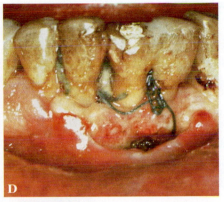

Figure 182

Clinical features of *Staphylococcus aureus* infections. (A) Folliculitis, inflammation of a hair follicle. (B) A child with impetigo. (C) A case of staphylococcal scalded skin syndrome. The outer skin layer has already separated from the basal layer. (From L.A. Schenfeld, *NEJM*. 342(2000): 1178.) (D) Gum tissue injury (necrosis) associated with penicillin-resistant staphylococci. (From H. Helovuo, K. Kakkarainen, and K. Pannio. *Oral Microbiol. Immuno.* 8(1993):75–79.)

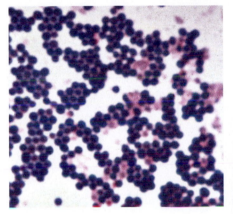

Figure 183

Gram stain of *Staphylococcus aureus*.

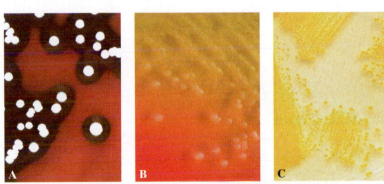

Figure 184

Staphylococcus aureus on culture media. (A) Beta-hemolytic colonies on blood agar. (B) Mannitol-salt agar reaction. (C) Reaction on *Staphylococcus* medium 110. (Courtesy of Difco Laboratories, Detroit, MI.) (D) *S. aureus* reaction on Vogel-Johnson. The medium is used in the early detection of coagulase positive, mannitol fermenting strains. (Courtesy of Remel, Lenexa. KS.) (E) A MRSA strain reaction on BBL CHROMagar MRSA medium This agar is a selective and differential medium for the direct detection of nasal methicillin resistant *S. aureus* isolates. (Courtesy of Becton, Dickinson and Company.)

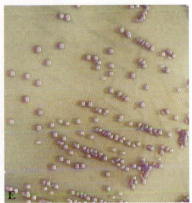

Gram-Positive Cocci *(Continued)*

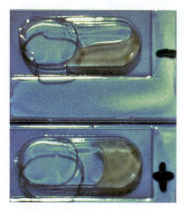

Figure 185
Coagulase reactions. A positive reaction is generally considered to be the best single indicator of potential pathogenicity. The formation of a plasma clot is a positive result; the absence of coagulation is a negative one.

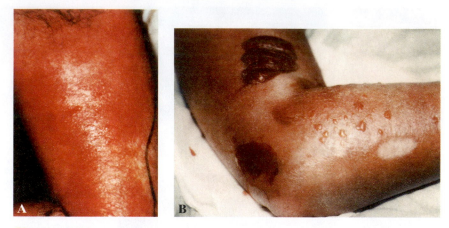

Figure 186
Examples of skin and tissue infections caused by group A streptococci (GAS). (A) Cellulitis of the arm, showing swelling and numerous thin, surface blisters oozing with bloody fluid. (B) Necrotizing fasciitis with blisters containing bloody fluid and a sloughing of skin from the elbow. (From B. Demers, A.E. Simor, H. Vellend, P.N. Schlievert, S. Byrne, F. Jamieson, S. Walmsley, and D.E. Low. *CID* 1(1993):792–800.)

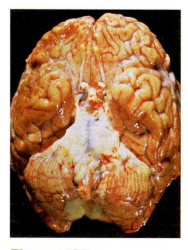

Figure 187
A view of the brain of a patient who died of *Streptococcus pneumoniae* meningitis. A thick, pussy accumulation can be seen covering a portion of the organ. (From J.A. Golden and D.N. Louis. *NEJM* 331(1994):34.)

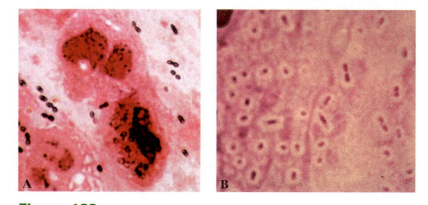

Figure 188
Microscopic views of streptococci. (A) A sputum specimen showing Gram-positive diplococci and an occasional chain. (B) A capsule stain of *Streptococcus pneumoniae*.

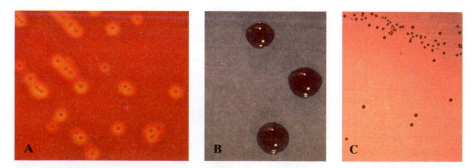

Figure 189
(A) Beta-hemolytic colonies of *Streptococcus pyogenes.* (B) Viridans streptococci *S. salivarius* colonies on *Mitis-Salivarius* agar. (C) *S. mitis,* another viridans streptococcus on the same medium.

Gram-Positive Cocci *(Continued)*

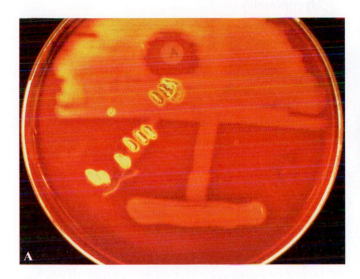

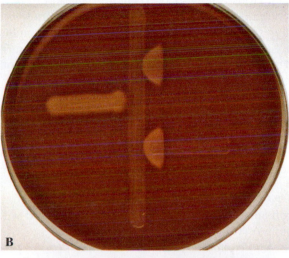

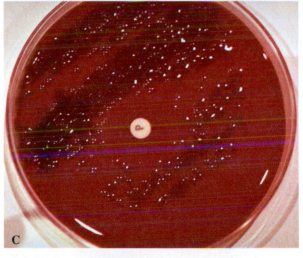

Figure 190

Streptococcus identification tests. (A) Typical characteristic growth inhibition by bacitracin. (The disk contains two units of the antibiotic.) Beta-hemolysis is evident from the agar stabs on the blood agar. The Christie, Atkins, and Munch-Peterson (CAMP) test is negative. (B) Positive CAMP reactions. The positive CAMP reaction is indicated by a triangle (arrowhead) perpendicular to the growth of a beta–lysin producing *Staphylococcus aureus* strain. (C) A positive growth inhibition optochin reaction. The *P* disk contains ethylhydrocupreine hydrochloride.

Figure 191

Tests for group B streptococci and certain enterococci. (A) The bile-esculin test. A positive reaction is a blackened slant. (B) Salt tolerance. A positive reaction is indicated when the brom-cresol purple indicator changes from purple to yellow. (C) Reactions in the Strep B carrot broth medium. The development of an orange color after incubation is a positive test. This medium is used as one-step screening for the presence of Group B streptococci. (Courtesy of Hardy Diagnostics.)

(GAS) from other hemolytic streptococci (Figures 190A and 190B), bile solubility, and optochin sensitivity, which differentiates group D and enterococci from other viridans streptococci (Figure 190C and 191A).

The viridans group includes *S. salivarius* (Figure 189B), *S. mitis* (Figure 189C) and *S. mutans,* which are among the most common causes of infective endocarditis. *Streptococcus mutans* plays an important role in tooth decay (caries).

Pathology. An enormous number of streptococcal species have been identified over the years from a wide variety of human and other animal sources. Attention will be given only to a few representatives of the genus *Streptococcus*.

Group A
Human disease is most commonly associated with group A streptococci (GAS). *Streptococcus pyogenes* and *S. pneumoniae* are members. *Streptococcus pyogenes* has long been recognized as an important cause of purulent diseases, such as pharyngitis (strep throat), impetigo, erysipelas, and, less frequently, severe generalized diseases such as sepsis and toxic shock–like syndrome. Certain strains cause scarlet fever. In addition, immunologic-related diseases including acute glomerulonephritis, acute rheumatic fever, and rheumatic heart disease have been linked to complications of group A pharyngitis.

Streptococcus pneumoniae (Figures 188B and 190C) is the most common cause of community-acquired pneumonia and lobar pneumonia in the United States and of fatal bacterial pneumonia in developing countries. It also causes otitis media, sinusitis, meningitis, and bacteremia.

Pneumococci are frequently isolated from the nasopharynx of healthy people. Virtually all humans are colonized by pneumococci at some stage, and pneumococcal carriage rates are higher in young children and where people are living in crowded conditions. An important feature of *S. pneumoniae* is its capacity to produce a polysaccharide capsule, which is structurally distinct for each of the 84 currently known serotypes of the organism.

Group B
Group B streptococci (GBS) are leading causes of neonatal sepsis, pneumonia, and meningitis in the United States. Infection of newborns exhibiting early-onset neonatal disease occurs via vertical transmission during passage through the birth canal or by exposure to infected amniotic fluid.

Diagnosis. Diagnosis of streptococcal infections requires the isolation and culture of the likely agent from clinical specimens (Figure 186). Gram staining and biochemical and serological tests are of major value (Figure 191C).

▲ Low GC Cell-Wall-less Bacteria

Mycoplasma
Mycoplasmas are found worldwide. A significantly large number of distinct species have been described. They can be found in humans (Figure 192A), in other animals including arthropods, and in plants. Some species, such as those found in the genera *Mycoplasma* and *Ureaplasma,* are pathogenic for humans, and those in the genus *Spiroplasma* commonly colonize and infect arthropods (Figure 192B). Spiroplasmas cause honeybee spiroplasmosis and suckling mouse cataract disease.

Morphology and Cultural Properties. Mycoplasmas are the smallest known free-living microorganisms, intermediate in size between bacteria and viruses. These organisms lack a cell wall (Figure 193), which is their single most distinguishing feature. Many of the biological properties of mycoplasmas are due to this feature. Even though the mycoplasmas exhibit a Gram-negative staining reaction, phylogenetically they are considered to be related to Gram-positive bacteria.

The various mycoplasmal species are very different organisms and have unique metabolic properties and cultivation requirements. No single medium will adequately support the growth of all organisms. Many, such as *M. pneumoniae,* require cholesterol, and ureaplasmas require urea for growth. The presence of sterols such as cholesterol in the cytoplasmic membranes of mycoplasma make them more resistant to the destructive effects of osmotic imbalances.

Several mycoplasmas such as *M. fermentans* and *M. pneumoniae* exhibit unusual colonies that look like fried eggs (Figure 192A).

Mycoplasma pneumoniae
Mycoplasma pneumoniae causes disease of the upper and lower respiratory tract. The infection is called **primary atypical pneumonia** to distinguish it from other types of pneumonia.

Transmission. *Mycoplasma pneumoniae* is spread by close personal contact and by aerosols.

Pathology and Clinical Features. *Mycoplasma pneumoniae* infection is most prevalent in colder months and mainly affects children from ages 5 to 9 years. The incubation period is generally long and usually presents signs and symptoms such as a persistent cough, fever, and headache.

Diagnosis. Culture of *M. pneumoniae* from specimens and serological tests are mainly used for diagnosis.

Ureaplasma urealyticum
Ureaplasma urealyticum is known to cause nongonococcal urethritis (NGU) in men and pelvic inflammatory disease (PID) and postabortal fever in women.

LowGC Cell-Wall-Less Bacteria—The Mycoplasma

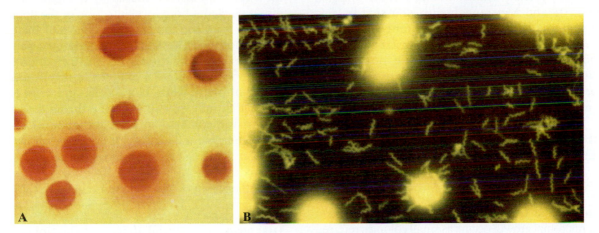

Figure 192
Mycoplasma. (A) Seven-day old stained *Mycoplasma fermentans* colonies on Dienes agar. Note the depressed centers surrounded by thin films of growth on the agar surfaces. Colonies range in size from 1 to 3 mm in diameter. (B) A fluorescent stained preparation of *Spiroplasma* species.

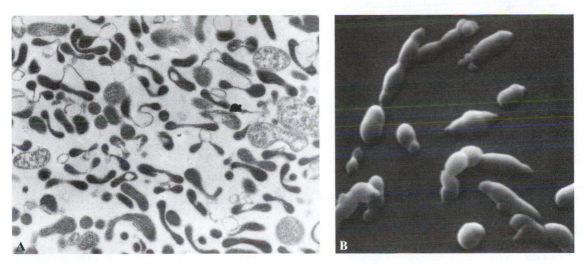

Figure 193
The mycoplasmas. (A) Transmission electron micrograph showing the wall-less nature of mycoplasmas. (B) Scanning micrograph of mycoplasma showing their variety of shapes.

Transmission. *Ureaplasma urealyticum* is primarily spread through sexual contact.

Pathology and Clinical Features. Persons infected with *U. urealyticum* usually experience painful or difficult urination (dysuria), frequent urination, and an unusual urethral discharge.

▲ High GC, Non-Spore-Forming Gram-Positive Rods

Actinomyces

Actinomyces species are normal inhabitants of certain portions of the gastrointestinal tract of warm-blooded vertebrates. They are commonly found in the mouth and on mucous membranes.

Actinomyces israelii

Most cases of human actinomycosis (Figure 194) are caused by *A. israelii*, although other species also have been isolated from lesions.

Transmission. *Actinomyces* species are highly adjusted to mucosal surfaces and do not cause disease unless they are introduced into deeper tissues. Conditions contributing to establishing infection include those causing tissue injury, such as tooth extractions, or some other form of trauma to the mouth or jaw.

Morphology and Cultural Properties. *Actinomyces israelii* is a Gram-positive, slender, straight, or slightly curved rod, 0.2–1.0 μm wide and 2.0–5.0 μm long. Organisms typically branch and may occur singly, in pairs, in V and Y arrangements, and side-by-side arrangements resembling a picket fence. Cells also may exhibit irregular staining and appear swollen.

Actinomyces species are nonmotile, nonsporing, and non-acid-fast. These organisms are facultative anaerobes and require CO_2 for good growth. Optimal growth temperature ranges between 35° and 37°C. Mature colonies appear after 7 to 14 days of incubation and are usually rough and crumbly in texture.

Actinomyces are catalase positive and indole negative. Acid but no gas is produced from carbohydrates.

Pathology and Clinical Features. Once an infection starts, microbial growth occurs as small colonies referred to as sulfur granules because of their color. Such colonies consist of branching cellular filaments (Figure 194).

Infections are most common in the mouth or jaw and usually give the face a swollen appearance.

Diagnosis. Tissue specimens (biopsies) are taken for culture and staining. Demonstrating the presence of the sulfur granule in tissue is of value. A longer than usual incubation period is necessary for *Actinomyces* species.

Corynebacterium

Corynebacteria are found in the environment on plants and as members of the microbiota of humans, lower animals, and plants. Several are commonly isolated in clinical laboratories.

Corynebacterium diphtheriae

Corynebacterium diphtheriae is the major pathogen of the genus and is the causative agent of diphtheria.

Transmission. *Corynebacterium diphtheriae* is spread by aerosols, direct contact with skin infections (Figure 195), and to some extent by fomites.

Morphology and Cultural Properties. Corynebacteria are Gram-positive, straight or slightly curved, slender rods 0.3–0.8 μm wide, and 1.5–8.0 μm long. Cells are usually found singly or in pairs and are arranged in a V or palisade formation. Some cells also can exhibit clubbed ends. Metachromatic granules or stored energy material (polymetaphosphate) can be demonstrated by special staining methods (Figure 196).

Corynebacteria are nonmotile, facultative anaerobes. Organisms are catalase positive, and they often reduce nitrate and tellurite. Most species produce acid but no gas from glucose and certain other carbohydrates. The optimal growth temperature is 35°C. Colonies of *C. diphtheriae* on tellurite-containing media appear gray or black (Figure 197).

Pathology and Clinical Features. Diphtheria is the most common disease caused by *C. diphtheriae*. The disease includes both local infection of the upper respiratory tract and the systemic effects of the organism's exotoxin. The signs and symptoms generally associated with the disease are the formation on the mucous membrane of a gray-white membrane consisting of fibrin, white blood cells, and dead cell remains that may extend from the throat area to the larynx and trachea; sore throat; general discomfort; headache; and nausea. Death can occur from respiratory blockage or destruction of heart tissue by the toxin.

Another form of the disease is cutaneous diphtheria (Figure 195).

Diagnosis. The initial diagnosis of diphtheria is based on clinical signs and symptoms. Definitive diagnosis is based on the isolation and culture of *C. diphtheriae* from the infection site and demonstration of its toxin production. Selective media containing potassium tellurite are used for isolation (Figure 197).

Mycobacterium

The mycobacteria are widely distributed in soil and water. Several species are pathogenic and can cause disease involving various vertebrates (Figure 198). The most commonly encountered pathogen is *Mycobacterium tuberculosis* (Figures 200 and 201), followed by *M. avium* complex (Figures 204 and 205), *M. kansasii* (Figures 203 and 204), *M. scrofulaceum* and *M. marinum* (Figures 198, 199). Unclassified, atypical, or nontuberculous organisms are frequently referred to as mycobacteria other than tubercle (MOTT) bacilli.

Transmission. *Mycobacterium tuberculosis, M. avium* complex (MAC), *M. kansasii,* and others causing respiratory disease are spread primarily through the inhalation of airborne droplet nuclei. Nosocomial transmission from patients or specimens is of major concern to health care workers.

Contaminated food and fomites also can spread mycobacteria.

Morphology and Cultural Properties. Mycobacteria are straight or slightly curved acid-fast rods, 0.20–0.7 μm wide and 1.0–10.0 μm long (Figures 198B and 205). Cells are poorly stained by the Gram method and are weakly Gram-positive.

Mycobacteria are nonmotile, non-spore-forming, and aerobic. Organisms grow slowly, with visible colonies appearing within 2 to 60 days depending on the species. Colonies of some species are often pigmented, especially when exposed to light (Figure 203). Mycobacteria are catalase positive.

High GC, Non-Spore-Forming Gram-Positive Rods

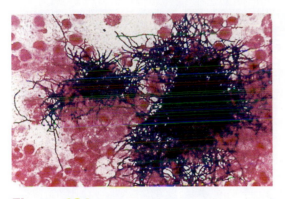

Figure 194
Gram-stained specimen from a patient with actinomycosis. The lesion, which included sulfur granules contained *Actinomyces israelii*. (Reprinted by permission of *Infections in Medicine* 9(1992):13. SCP Communications.)

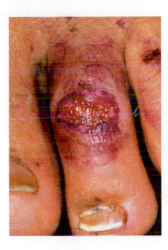

Figure 195
The most common form of cutaneous diphtheria, the ulcer or ecthyma diphthericum. (From W. Hofler. *Internat. J. Dermatol.* 30(1991):845.)

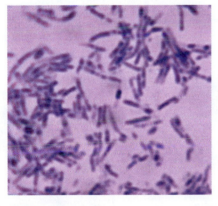

Figure 196
Metachromatic granules (enlarged and darkly stained areas) of *Corynebacterium diphtheriae*. (1,000×).

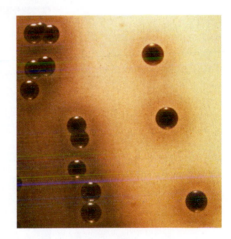

Figure 197
Colonies of *Corynebacterium diphtheriae* on a tellurite-containing medium.

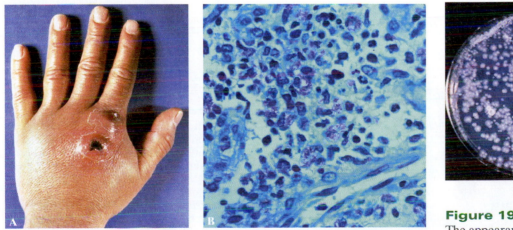

Figure 198
(A) Skin ulcers and disease caused by *Mycobacterium marinum*. (B) The results of acid-fast staining of the discharge from a skin ulcer caused by *Mycobacterium marinum*.

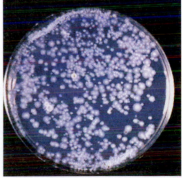

Figure 199
The appearance of colonies of *Mycobacterium marinum* when grown in the dark. This species is an example of a chromogenic bacterium, which when exposed to light will turn a bright yellow.

High GC, Non-Spore-Forming Gram-Positive Rods *(Continued)*

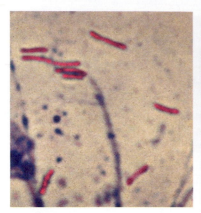

Figure 200
The acid-fast rods of *Mycobacterium tuberculosis.*

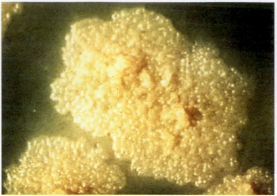

Figure 201
Colonies of *Mycobacterium tuberculosis.*

Figure 202
The BBL™ *Mycobacterium* Growth Indicator Tube (MGIT)™. Growth of mycobacteria is indicated by an orange fluorescent glow upon exposure to a long-wave ultraviolet light source. (Courtesy of Becton Dickinson Microbiology Systems.)

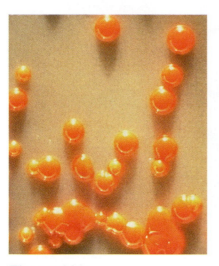

Figure 203
Mycobacterium kansasii colonies.

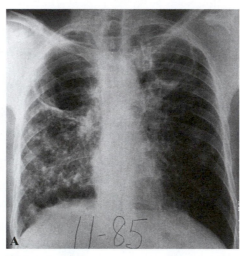

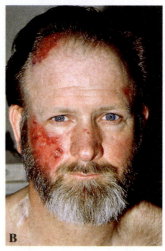

Figure 204
Mycobaterium avium complex (MAC) clinical states. (A) An x-ray showing extreme involvement of the lungs (large dense area). (From E. Wolinsky. *CID* 15(1992):1–12.) (B) A human immunodeficiency virus-infected person with a nontuberculous, atypical mycobacterium skin infection caused by MAC. (From S.T. Nedorost, B. Elewski, J.W. Tomford, and C. Camisa. *Internat. J. Dermatol.* 30(1991):491–97.)

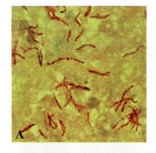

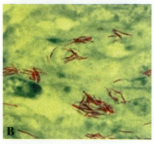

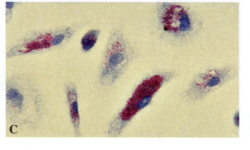

Figure 205
(A) Fite-stained lymph node specimen containing *Mycobacterium kansasii.* The rods are long and beaded (1,000×). (From F.S. Jannotta and M.K. Sidaway, *Arch. Path. Lab. Med.* 113(1989):1120.) Microscopic views of *Mycobacterium avium-intracellulare* (MAC). (B) A Fite-stained lymph node specimen. The rods are short and beaded (1,000×). (From F.S. Jannotta and M.K. Sidaway. *Arch. Path. Lab. Med.* 113(1988):1120.) (C) A skin specimen showing intracellular acid-fast rods. (From S.E. Hoffner, G. Kallenius, and S.B. Svenson. *Res. Microbiol.* 143(1992):391–98.)

Mycobacterium tuberculosis complex
On the basis of cultural and biochemical properties, *M. tuberculosis, M. bovis, M. africanum,* and *M. microtii* are members of the *M. tuberculosis* complex. Other species are grouped together as **nontuberculosis mycobacteria.** These include *M. kansasii* and *M. avium intracellulare* (Figures 203–205) which cause respiratory disease similar to tuberculosis, and *M. ulcerans* and *M. marinum,* which cause skin and soft tissue infections (Figure 198).

Pathlogy and Clinical Features. Tuberculosis primarily affects the lower respiratory system. A chronic productive cough, fever, night sweats, and weight loss are typical of the disease. Tuberculous mycobacteria are able to avoid destruction by the host's defenses and form the primary lesion known as the **tubercle.** Organisms spread to regional lymph nodes, enter the blood, and reseed the lungs. Tissue injury results from cell-mediated hypersensitivity.

Diagnosis. Active disease is diagnosed on clinical grounds, an abnormal chest radiograph, acid-fast organisms in sputum or other specimens (Figure 193), and isolation of the organism. Various approaches are used to isolate or to detect mycobacteria, including the BBL *Mycobacterium* Growth Indicator Tube (Figure 196).

Mycobacterium avium-intracellulare
(M. avium complex, MAC)
The majority of disease states caused by *M. avium-intracellulare* (Figure 204) and other MOTT bacilli appear as chronic pulmonary disease (Figure 204), local lymph node inflammation, and diseases of skin, soft tissue, bones, and joints. Overwhelming MOTT-disseminated disease also occurs in patients with AIDS. MAC causes the third most common opportunistic bacterial diseases affecting adults with HIV infections.

Mycobacterium leprae
Mycobacterium leprae is the causative agent of leprosy (Hansen's disease), a chronic tumor-associated (granulomatous) disease of the peripheral nerves and mucous membranes (mucosa) of the nose. For centuries, and in certain locations perhaps still today, leprosy-infected individuals (lepers) were kept isolated from others in society (Figure 206).

Transmission. Transmission of *M. leprae* is believed to occur most commonly by contamination of the nasal mucosa or skin lesions with infectious secretions. The reservoir of the disease is the infected human.

Cultural Properties. Growth of *M. leprae* in artificial media or in tissue cultures is generally limited to only a few generations. The organism can be successfully grown in the foot pads of normal mice, irradiated thymectomized mice, and the armadillo, which may also be infected by natural means.

Pathology and Clinical Features. Two major forms of leprosy are recognized; **tuberculoid** and **lepromatous.** Intermediate forms also occur. The incubation period for the disease, which in general is very long, is measured in years.

Individuals with tuberculoid leprosy (Figure 207) can exhibit large flattened patches with raised or elevated red edges and dry, pale, hairless centers on any body surface. Loss of sensation on the skin also may develop as a result of invasion of peripheral sensory nerves.

Individuals with lepramatous leprosy exhibit extensive skin involvement, with thickening of looser skin parts of the lips, forehead, and ears. The classic lion face is typical of such infections. Extensive penetration of *M. leprae* into body tissue can cause severe body damage resulting in the loss of facial bones, fingers, and toes.

Diagnosis. Tuberculoid leprosy is diagnosed on clinical features. The lepromatous form involves demonstrating *M. leprae* in stained infected tissues (Figure 208).

Mycobacterium smegmatis
M. smegmatis (Figure 209) is an acid-fast rod that can be isolated as a common contaminant from urine specimens taken from males. Finding this bacterium in urine does not necessarily indicate an infection.

Bacillus of Calmette and Guérin (BCG)

The bacillus of Calmette and Guérin (Figure 210) is the basis for the only available live vaccine used as a preventative measure against tuberculosis. The vaccine developed by Albert Calmette and Camile Guérin was originally derived from a *Mycobacterium bovis* strain that was weakened by repeated subculturing. *M. bovis* is the cause of bovine tuberculosis. Once a serious health threat, this disease is now rarely encountered.

▲ Nocardia

Nocardia

The nocardia are widely distributed in soil. Some species are pathogenic opportunists for humans and lower animals. *Nocardia asteroides* and *N. brasiliensis* are of medical importance. These organisms cause pulmonary infection (*N. asteroides*), and skin and subcutaneous infections (*N. brasiliensis*).

Transmission. Pulmonary nocardiosis is acquired by inhalation of the disease agent, which is present in dust and soil and on contaminated mucosal surfaces. Many individuals with the disease have poorly functioning immune systems.

Skin and related infections generally result from direct inoculation of *Nocardia* by a thorn or a wood sliver.

Morphology and Cultural Properties. Nocardia species are Gram-positive rods or are coccoid in shape and measure 0.50–1.20 μm in diameter. Cells are partially acid-fast (Figure 211) and give a beaded appearance.

Nocardia are aerobic and grow well on most enriched media, forming colonies within 2 to 7 days and exhibiting a variety of colors, including tan, gray, orange, purple, and white (Figure 212).

Diagnosis. Laboratory diagnosis includes direct smears from infected sites, culture, and biochemical tests.

Bifidobacterium

Bifidobacterium bifidum
Generally found in sewage and as natural members of the microbial populations in the mouths and gastrointestinal tracts of warm-blooded vertebrates, in insects the bifidobacteria are usually considered to be nonpathogenic.

Morphology and Cultural Properties. Bifidobacteria are generally Gram-positive, but often stain irregularly. They exhibit a range of rod-shaped forms, occurring singly or in pairs, arranged in chains, rosettes, or Vs. The cells are somewhat curved, or clubbed (Figure 213).

The bifidobacteria are anaerobic, chemoorganotrophic, nonmotile, non-acid fast, and non-spore-forming. The optimum growth temperature is between 37° and 41°C. Organisms grow within the pH range of 4.5–8.5, and actively ferment carbohydrates.

Gardnerella

Gardnerella vaginalis
Found in the human genital and urinary tracts, *Gardnerella vaginalis* is associated with bacterial vaginosis. The condition also involves various obligate anaerobic bacteria.

Transmission. *Gardnerella vaginalis* is generally acquired through sexual activity.

Morphology and Cultural Properties. *Gardnerella vaginalis* is a Gram-negative to Gram-variable pleomorphic rod, about 0.5 μm in width and 1.5–2.5 μm in length. The organism is nonmotile and does not form capsules.

Gardnerella vaginalis is facultatively anaerobic. Special media and a CO_2 incubator are necessary for culture. Small, opaque, convex colonies typically appear after 48 hours of incubation. Optimal growth temperature is 35° to 37°C.

Gardnerella vaginalis is catalase and oxidase negative, produces acid but no gas from glucose and certain other carbohydrates, and does not reduce nitrates.

Pathology and Clinical Features. *Gardnerella vaginalis* infection produces a foul-smelling vaginal discharge

with a pH ranging from 4.5 to 5.5. The organism is frequently isolated from women with infections of the mucous membrane lining the inner surface of the uterus.

Diagnosis. The direct microscopic examination of vaginal secretions is more relevant for the diagnosis of *G. vaginalis* infection than is the isolation of the organism from such specimens. A minimum diagnostic requirement from bacterial vaginosis usually includes the finding of (1) excessive vaginal discharge, (2) vaginal pH of greater than 4.5, (3) vaginal epithelial cells covered by small Gram-negative rods known as **Clue cells** (Figure 214B), and (4) a fishy chemical odor in the potassium hydroxide test.

The Chlamydia

The chlamydiae are a small group of coccoid cells that are obligate intracellular parasites of eukaryotic cells. Members of the genus cause several significant human diseases, including trachoma, the leading cause of infectious blindness worldwide (Figure 215A), several common sexually transmitted diseases (Figure 215B), and acute respiratory infections. Arthopods neither transmit nor serve as hosts.

Morphology and Cultural Properties. Chlamydiae are Gram-negative and are grown in cell cultures (Figure 217). These organisms have a unique biphasic life cycle that alternates between an infectious, rigid, metabolically inactive form, the **elementary body,** and noninfectious, fragile, metabolically active form, the **reticulate body** (Figure 216).

Diagnosis. The clinical properties of chlamydial infections are often diagnostic. Laboratory confirmation may include immunofluorescence tests (Figures 217C and 220), nucleic acid probes (Figure 218), and certain serological tests.

Chlamydia trachomatis
Chlamydia trachomatis causes trachoma (Figure 215A), inclusion conjunctivitis, and genital infections including inflammation of the urethra (nongonococcal urethritis) in men and acute inflammation of the uterine tubes (salpingitis) and inflammation of the cervix (cervicitis) in women. Specific strains known as serovars cause lymphogranuloma venereum (LGV), a sexually transmitted disease (Figures 215B and 217B).

Transmission. *Chlamydia trachomatis* eye infections are spread by flies, contaminated fingers, and inanimate objects (fomites).

Pathology and Clinical Features. The clinical properties of trachoma include the development of follicles and an inflamed inner eyelid lining (conjunctiva) as shown in Figure 215A. The cornea may also become

High GC, Non-Spore-Forming, Gram-Positive Rods (Continued)

Figure 206
A Scandinavian church which had a special entrance and section reserved specifically for individuals infected with leprosy in earlier centuries.

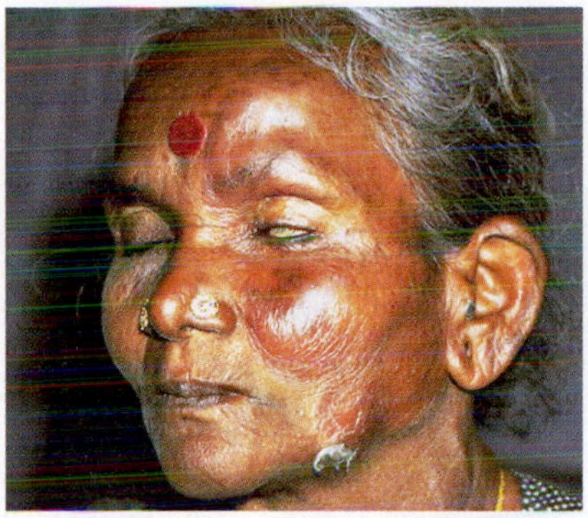

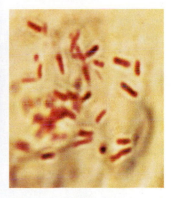

Figure 207
A case of early leprosy. (From M.M. Hogeweg. *Trop. Doctor.* Suppl. 1(1992):15–21.)

Figure 208
The acid-fast reaction of *Mycobacterium leprae.*

Figure 209
Colonies of *Mycobacterium smegmatis.*

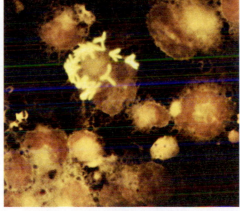

Figure 210
Bacillus of Calmette and Guérin (BCG) as shown by immunofluorescence. (From F.C. Bange, A.M. Brown, and W.R. Jacobs Jr. *Inf. Immunol.* 64(1996):1794–99.)

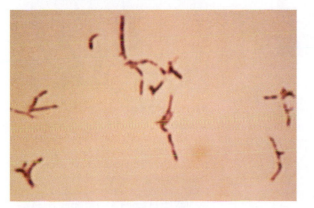

Figure 211
Acid-fast rods of *Nocardia asteroides* (1,000×).

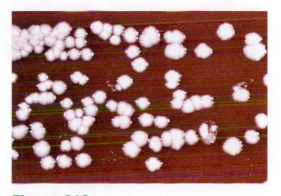

Figure 212
Nocardia asteroides colonies on blood agar.

Facultatively Anaerobic, High GC, Non-Spore-Forming, Irregular Gram-Staining Rods

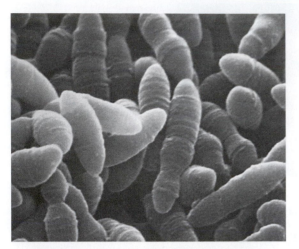

Figure 213

The bifidobacteria. These anaerobic organisms are important members of the natural microflora of the human gastrointestinal system. (H. Bauer, and H. Sigarlakie. *Can. J. Microbiol.* 21(1975): 1305–16.)

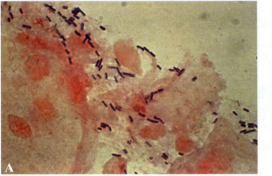

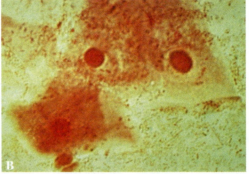

Figure 214

Gardnerella vaginalis. (A) Normal cervical smear showing epithelial cells and Gram-positive rods. (B) Clue cells covered with the Gram-negative rods of *G. vaginalis*.

The Chlamydia

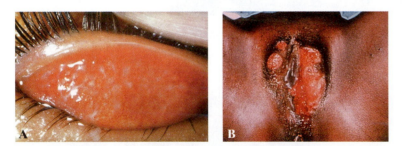

Figure 215

Clinical states caused by *Chlamydia trachomatis*. (A) Trachoma. The pronounced inflamed inner eyelid with many round, pale swelling (follicles) of the eye disease are shown. (Courtesy of Director-General and Programme Manager, Prevention of Blindness, World Health Organization.) (B) Clinical view of a lesion found on the genitalia of a female patient with lymphogranuloma venereum. This is one type of sexually transmitted disease caused by *C. trachomatis*.

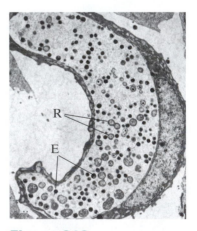

Figure 216

Transmission electron micrograph showing a membrane-bound vacuole in a cell infected with *Chlamydia trachomatis*. The vacuole contains both the larger reticulate bodies (R) and the smaller, dark elementary bodies (E) (4,664×). (From P.B. Wyrick and S.J. Richmond. *JAVMA* 195(1989):1509.)

cloudy. Inclusion conjunctivitis is a much milder infection with a pussy discharge.

Lymphogranuloma venereum is a chronic, invasive, sexually transmitted disease that causes swelling and ulceration of regional lymph nodes (buboes).

Chlamydia pneumoniae

Chlamydia pneumoniae is an important cause of human respiratory tract infections including sore throat, bronchitis, and pneumonia. Recently, it has been implicated in coronary artery disease.

Transmission. *Chlamydia pneumoniae* is spread from human to human by aerosols.

Pathology and Clinical Properties. *Chlamydia pneumoniae* typically causes hoarseness, prolonged coughing, and bronchial asthma, all of which damage the epithelial cells lining the respiratory system.

Chlamydia psittaci

Chlamydia psittaci is widely distributed in nature, usually causing respiratory infections in many mammals and birds (Figure 219). The organism also has been associated with various forms of heart disease (Figure 221).

Transmission. Psittacine (tropical) birds are considered to be the major reservoirs, but human cases have been associated with infected pigeons, sparrows, ducks, cockatiels, and many other birds. Close or prolonged contact with infected birds is not necessary. Infections can be acquired from contact with a contaminated environment or materials such as a cage or bird droppings. Aerosols are the principal means of transmission.

Pathology and Clinical Features. An influenza-like illness known as psittacosis is the usual clinical condition caused by *C. psittaci.*

▲ The Spirochetes

The spirochetes are helical (spring-shaped) motile organisms. Their morphological features and a unique type of movement distinguish them from other bacteria. The spirochetes are unicellular and Gram-negative. The size range of spirochetes is $0.1-3.0$ μm $\times$ $5-250$ μm. Because some spirochetes are very thin, they can be seen only by dark-field microscopy (Figure 232), electron microscopy (Figure 222) or by special staining techniques that increase their dimensions to bring them within the resolving power of the bright-field microscope. Dark-field microscopy depends on the reflection of light from the surfaces of particles or cells.

Spirochetes may be anaerobic, facultively anaerobic, or aerobic. Some spirochetes are free-living, whereas

others are members of the normal microbiota of humans and lower animals. Three genera — *Borrelia, Leptospira,* and *Treponema* — include causative agents of human and zoonotic diseases. A **zoonosis** is an infectious disease transmissible to humans from a lower animal host by natural means.

Borrelia

Borrelia burgdorferi

Lyme disease (Figure 221) has emerged as the leading arthropod-borne disease in the United States. *Borrelia burgdorferi* (Figure 222) is the causative agent of the disease and related disorders.

Transmission. Various *Ixodes* ticks including *I. dammini* (Figure 223), *I. pacificus,* and *I. scapularies* transmit the disease agent. Small mammals, particularly mice, are important hosts for these ticks and are critical for maintaining *B. burgdorferi* in nature. Deer are important hosts for adult tick stages.

Morphology and Cultural Properties. *Borrelia burgdorferi* is a large helical cell, $0.2-0.5$ μm wide and $30-20$ μm long, with $3-10$ loose spirals. It is Gram-negative but is most easily shown by other staining procedures such as Giemsa and immunofluorescence. *Borrelia burgdorferi* is microaerophilic and can be grown on various types of blood agar media (Figure 224). The organism exhibits beta hemolysis, which can be made more pronounced with hot-cold incubation (incubating first at 34°C and then overnight at 4°C).

Pathology and Clinical Features. Lyme disease has been divided into three stages:

1. **Early Localized Stage.** Develops within a few days to weeks after infection. **Erythema (er'-i-THĒ-ma) migrans,** the characteristic lesion, appears as a large round or oval reddened area with definite borders at the site of the tick bite (Figure 221). Fever, headache, enlarged local lymph nodes, muscle and joint pain, and fatigue are typical symptoms.

2. **Disseminated or Spreading Stage.** Occurs within days to months. The appearance of various-sized reddened areas is typical. The symptoms and signs found with early Lyme disease also can be present. Facial paralysis may occur.

3. **Late Stage.** Appears within months to years. The effects of the disease include long-lasting arthritis and central nervous system involvement.

Diagnosis. Finding spirochetes in blood specimens taken from individuals in the early stage of Lyme disease is the principal basis for diagnosis. Cultural

The Chlamydias

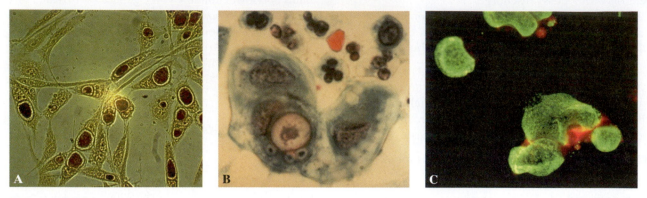

Figure 217

Microscopic features of *Chlamydia trachomatis*. (A) The organism in McCoy cells. (From R.L. Hodinka and P.B. Wyrick. *Inf. Immunol.* 56(1998):1456–63.) (B) Intracytoplasmic inclusions from a specimen taken from a patient with lymphogranuloma venereum (1,200×). (C) *C. trachomatis* inclusions shown by fluorescent-antibody. (Courtesy of S.T. Knight and P.B. Wyrick.)

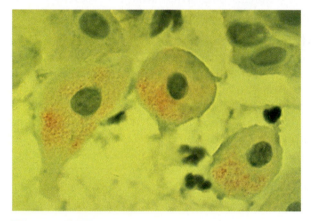

Figure 218

Chlamydial intracytoplasmic deposits revealed by a nucleic acid probe technique. (Courtesy of Drs. D. Raisi, C. Ghirardini, and M. Portolani. University of Modena and Mirandala, Hospital Cytology Lab, Italy.)

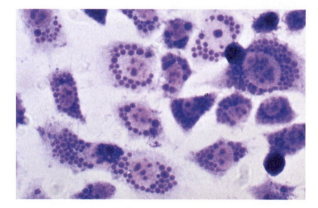

Figure 219

Chlamydia psittaci inclusions in McCoy cells. (From R.L. Hodinka and P.B. Wyrick. *Inf. Immunol.* 56(1998):1456–63.)

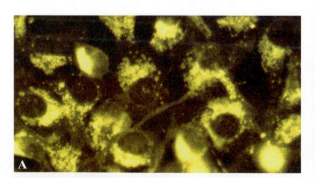

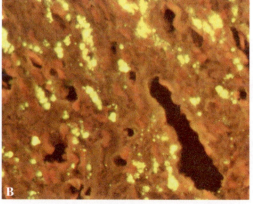

Figure 220

The presence of *Chlamydia psittaci* as shown by fluorescent antibody. (A) Tissue culture inoculated with a pharyngeal (throat) specimen. Chlamydial inclusions are shown. (B) Chlamydial antigen and inclusions (yellow glowing areas) in a mitral valve growth. (From D.S. Shapiro, S.C. Kenney, M. Johnson, C.H. Davis, S.T. Knight, and P.B. Wyrick. *NEJM* 326(1992):1192–95.)

isolation from typical lesions and immunological tests such as the enzyme-linked immunoabsorbent assay (ELISA) also are of value (see Section 11, "Immunology").

Borrelia recurrentis
The causative agent of **endemic relapsing fever** is *Borrelia recurrentis* (Figure 225).

Transmission. Endemic relapsing fever is transmitted to humans by soft-bodied ticks of the genus *Ornithodoros,* and **epidemic relapsing fever** transmitted by body lice of the genus *Pediculus* (Figure 226).

Morphology and Cultural Properties. *Borrelia recurrentis* is a large spirochete, 0.3 μm wide and 10–30 μm long. It is Gram-negative, but, as is the case with the other *Borrelia* the organisms can be seen more easily with other staining procedures. *Borrelia recurrentis* is microaerophilic and can be cultured on various types of artificial media containing long-chain fatty acids.

Pathology and Clinical Features. After an incubation period of about 7 days, a massive number of spirochetes appear in the blood, producing a high fever accompanied by a severe headache, muscle pain, chills, and general weakness. The fever period lasts approximately one week and ends suddenly with the development of an adequate immune (protective) response. Usually, less severe relapses (reoccurrences) occur 2–4 days later.

Diagnosis. Finding *B. recurrentis* among blood cells (Figure 225) in stained smears prepared from blood taken during the fever period is characteristic. Special immunological tests also may be helpful.

Leptospira

Leptospira interrogans
Leptospirosis (lep′-tō-spi-RŌ-sis), a worldwide disease of a wide range of animals, is caused by *Leptospira interrogans* (Figure 227).

Transmission. Humans acquire the disease mainly by ingesting water contaminated with infected animal urine.

Morphology and Cultural Properties. *Leptospira interrogans* is a thin spirochete, 0.15 μm wide and 5–15 μm long. It consists of fine, closely wound spirals with hook-like ends. Immunofluorescence staining and dark-field microscopy produce the best microscopic views. *Leptospira interrogans* is aerobic and can be cultured in certain enriched semisolid media.

Pathology and Clinical Features. Leptospirosis usually occurs in two phases:

1. **Phase 1.** Develops after an incubation period of seven days with the appearance of an influenza-like illness. Typical symptoms include fever, chills, headache, and muscle pain.

2. **Phase 2.** Usually lasts three or more weeks. Typical symptoms include muscle aches and headaches, accompanied by signs such as rash, yellowing of body tissues (jaundice), and laboratory evidence of liver and kidney damage. In the most severe form of the leptospirosis, known as Weil's disease, extensive damage to the liver, kidneys, and blood vessels occurs.

Diagnosis. *Leptospira* can be isolated from the blood or from other body fluids such as urine during the first weeks of the disease. Specimens must be used immediately for culturing.

Treponema

Treponema pallidum
The causative agent of syphilis, a sexually transmitted disease (STD), is *Treponema pallidum.* It was first recognized in the late fifteenth and early sixteenth centuries, when it spread through Europe, often with fatal effects. *Treponema pallidum* subspecies cause certain tropical diseases. These include *T. pallidum* subspecies *T. p. carateum,* the cause of pinta, and *T. p. pertenue,* the cause of the nonvenereal yaws (Figure 233). Victims of yaws in extreme cases exhibit a marked saddle nose deformity and an oval swelling on either side of the nose bridge.

Transmission. Syphilis is typically a sexually transmitted disease, but it can also be spread nonsexually through blood transfusions or nonsexual bodily contact, and a fetus can acquire it *in utero* or during the passage through the birth canal of an infected mother (**congenital syphilis**).

Morphology and Cultural Properties. *Treponema pallidum* is a slim spirochete, 0.15 μm wide and 5–15 μm long, with a characteristic corkscrew appearance (Figures 231 and 232). Dark-field microscopy and immunofluorescence techniques are routinely used to show the organisms in specimens. The usual bacteriological stains are not effective. However, the depositing of silver on spirochete surfaces can be used to demonstrate organisms in infected tissues (Figure 231). Special tissue culture methods can be used to grow *T. pallidum.*

Pathology and Clinical Features. Syphilis can be divided into four stages (Figures 228–230):

1. **Primary syphilis.** Appears within 21 days after infection as a hardened, swollen sore called a **chancre** (SHANG-ker) at the site of infection; if untreated, this lesion heals within 3 to 8 weeks (Figure 228A).

2. **Secondary (disseminated) syphilis.** Occurs 2 to 10 weeks after primary chancre has healed. Signs of infection include blisterlike eruptions on the

The Spirochetes

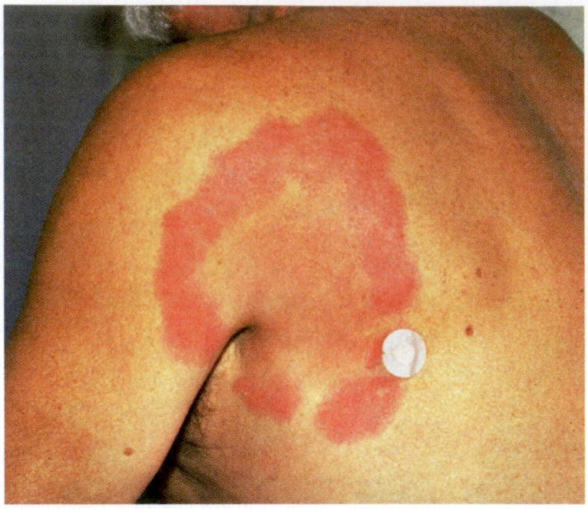

Figure 221

Erythema migrans, the typical lesion of early localized Lyme disease. The lesion develops at the site of the tick bite and occurs in 60–80% of cases. (From J. Cote. *Internat. J. Dermatol.* 30(1991):500–501.)

Figure 222

Scanning electron micrograph of *Borrelia burgdorferi*, the causative agent of Lyme disease and related disorders. (Courtesy of Dr. F.-R. Matsucha, Free University of Berlin.)

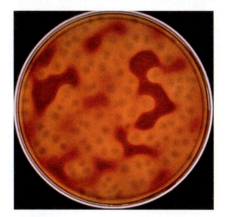

Figure 224

Hemolytic (blood agar plate) activity of *Borrelia burgdorferi* after incubation at 34°C and transfer to 4°C for overnight incubation. (From L.R. Williams and F.E. Austin. *Inf. Immunol.* 60(1992): 3324–30.)

Figure 223

A female deer tick, *Ixodes dammini*, which transmits *Borrelia burgdorferi*, the cause of Lyme disease. (Courtesy, Dr. F.-R. Matsucha, Free University of Berlin.)

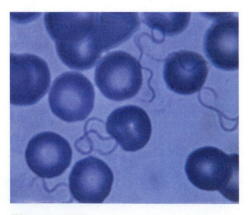

Figure 225

Microscopic view of *Borrelia recurrentis*, the causative agent of epidemic relapsing fever.

Figure 226

The human louse *Pediculus humanus*, known to transmit relapsing fever and other diseases.

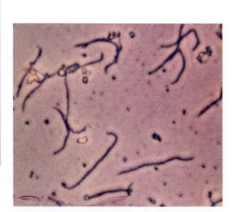

Figure 227

Microscopic view of *Leptospira interrogans*, the cause of leptospirosis.

body (Figure 228B), involving the face, palms, and soles of the feet, and painless warty growths on the genitalia and surrounding areas.

3. **Latent syphilis.** Occurs after the secondary stage. No evidence of infection is apparent other than positive blood tests.

4. **Tertiary stage.** Appears in the untreated as early as 5 years after infection, but generally after 15 to 20 years. Extensive tissue destruction occurs in the cardiovascular and nervous systems (Figure 229), but may involve any part of the body. The **gumma** (Figure 230) a firm, white lesion measuring up to 40 μm may be found in the body tissues and dramatically emphasizes the effects of *T. pallidum* infection.

Diagnosis. Dark-field and immunofluorescent microscopy (Figure 232) can detect *T. pallidum* in primary and secondary lesions. Most cases of syphilis, however, are diagnosed by tests that demonstrate the presence of antibodies in a patient's blood. These include the fluorescent treponemal antibody absorption test (FTA-ABS), and the micro-hemagglutination test for *T. pallidum* (MHAP-TP).

▲ Anaerobic/Facultative Anaerobic, Non-Motile, Gram-Negative Rods

Porphyromonas

This genus *Porphyromonas* was newly created in 1988. It includes three former *Bacteroides* species: *Porphyromonas asaccharolytica, P. gingivalis,* and *P. endodontalis.*

Morphology and Cultural Properties. Members of this genus are short, Gram-negative rods, 0.5–0.8 μm in diameter and 1.0–3.0 μm in length. Organisms are nonmotile and strict anaerobes. *Porphyromonas* species produce pigmented colonies ranging from buff to tan to black after 2 to 21 days incubation (Figure 234).

These organisms do not break down carbohydrates (asaccharolytic), a property that separates them from *Prevotella* species.

Certain *Porphyromonas* species are associated with gum disease and have been isolated from infected dental root canals.

Prevotella

The genus *Prevotella* was created in 1990. Fifteen former *Bacteriodes* species were transferred to *Prevotella,* including several human pathogens, such as *P. denticola* (Figure 235), *P. intermedia* (Figure 236), and *P. melaninogenica* (Figure 237).

Morphology and Cultural Properties. *Prevotella* species are Gram-negative, rods measuring 0.5–0.8 μm in diameter and 1.0–3.0 μm in length. These organisms are nonmotile and non-spore-forming. They tend to be pleomorphic.

Prevotella denticola is catalase and indole negative. It hydrolyzes esculin and ferments glucose, lactose, and maltose. *Prevotella intermedia* differs from *P. denticola* by being positive for indole and fermenting glucose and lactose only. *Prevotella melaninogenica* produces reactions identical to those of *P. denticola* with the exception of esculin hydrolysis. None of these species produces cellobiose.

Pathology. *Prevotella denticola* is commonly found in the human oral cavity, but its clinical significance is not known. *Prevotella intermedia* also has been isolated from the human oral cavity but also has been found in specimens from head and neck infections and from pleural infectious conditions.

Streptobacillus

Streptobacillus moniliformis

The only member of this genus is *Streptobacillus moniliformis.* It is found in the throat and nasopharynx of mice and of wild and some laboratory rats. *Streptobacillus moniliformis* causes one form of rat bite fever in humans.

Transmission. Most human cases are acquired from bites of infected rats, mice, or cats. Infections also can result from the ingestion of milk contaminated with rat feces and by environmental exposure of persons working or living in rat-infested buildings.

Morphology and Cultural Properties. *Streptobacillus moniliformis* is a Gram-negative to Gram-variable rod, 0.1–0.7 μm in diameter and 1.0–5 μm in length. Cells occur singly or in long, wavy chains referred to as a **string of beads** 10–150 μm long (Figure 238). This organism is nonmotile.

Streptobacillus moniliformis is a facultative anaerobe and requires enriched media for best growth. The optimal temperature is 35° to 37°C. On blood agar, the organism forms small gray-white colonies. It is one of the few bacterial species that spontaneously converts to L forms (cells without cell walls). The organism is oxidase and catalase negative, and does not reduce nitrates. Glucose and other carbohydrates are fermented, with acid production only. All media for biochemical testing require blood, serum, or other enriching substances.

Pathology and Clinical Features. Illness occurs quickly after an incubation of less than 10 days. Usually, signs and symptoms include chills, fever, vomiting, and severe headache. A rash also may appear on the palms

The Spirochetes *(Continued)*

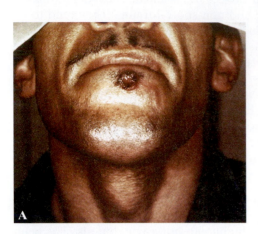

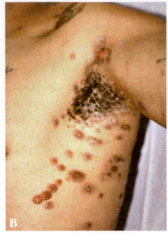

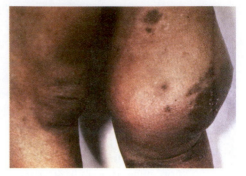

Figure 229
Charcot knee, a complication of tertiary syphilis involving the cardiovascular system.

Figure 228
Early stages of syphilis. (A) The primary sore, or chancre. This lesion contains hundreds of *Treponema*. (Courtesy of U.S. Public Health Service.) (B) An example of the rash that may develop in secondary syphilis. (From P. Lawrence and N. Saxe. *Clin. Exper. Dermatol.* 17(1992):44–48.)

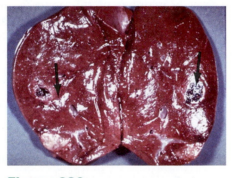

Figure 230
Gummas in the liver. Two gummas are evident (arrows). One shows extensive tissue injury.

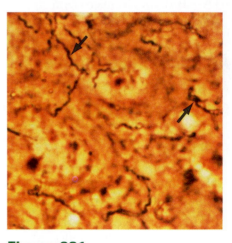

Figure 231
Treponema pallidum (arrows) in tissue.

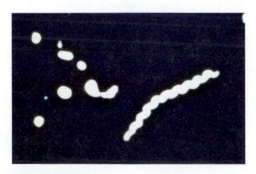

Figure 232
The spirochete *Treponema pallidum* revealed by dark-field microscopy. Organisms appear bright in a dark background.

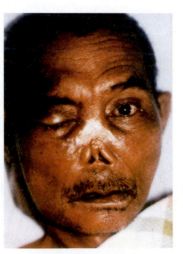

Figure 233
Nasal destruction in a case of yaws. (Courtesy of Drs. S. Elango and S.P. Palaniappan.)

Anaerobic/Facultative Anaerobic, Non-Motile, Non-Spore-Forming Gram-Negative Rods

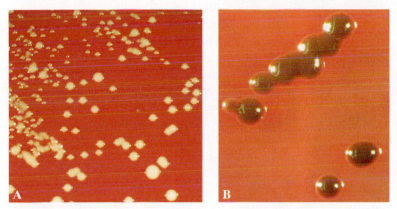

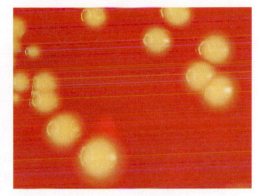

Figure 234
Porphyromonas endodontalis. (A) Colonies on *Brucella* agar. (B) Highly pigmented colonies on *Brucella* agar. (Courtesy of Dr. H.F. Somer, Anaerobe Laboratory, National Public Health Institute, Helsinki, Finland.)

Figure 235
Prevotella denticola colonies on a laked blood medium. (Courtesy of Dr. H.F. Somer, Anaerobe Laboratory, National Public Health Institute, Helsinki, Finland.)

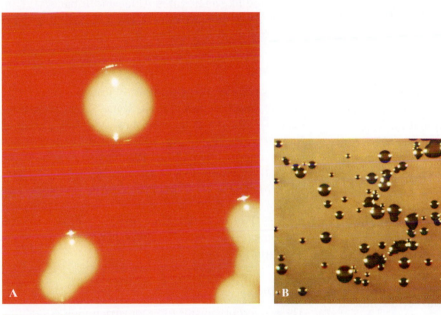

Figure 236
Prevotella. intermedia on two different agar media. (A) Colonies on *Brucella* agar. (B) Colonies on a laked blood medium. (Courtesy of Dr. H.F. Somer, Anaerobe Laboratory, National Public Health Institute, Helsinki, Finland.)

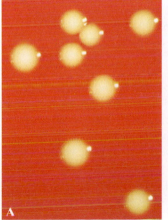

Figure 237
Prevotella melaninogenica on two different media. (A) Colonies on *Brucella* agar. (B) Colonies on a laked blood medium. (Courtesy of Dr. H.F. Somer, Anaerobe Laboratory, National Public Health Institute, Helsinki, Finland.)

and soles. Arthritic involvement of the elbows, wrists, knees, and ankles is quite common.

Diagnosis. Blood, other body fluids, pus, and discharges from skin lesions are used for microscopic examination and culture. Biochemical tests and serological tests such as a tube agglutination are of value.

▲ The Cyanobacteria

The cyanobacteria (blue-green bacteria) represent the largest, most diverse, and most widely distributed group of photosynthetic bacteria. They bear a resemblance to gliding bacteria in general morphology, to some photosynthetic bacteria in their ability to use both carbon dioxide and gaseous nitrogen with light energy, and to eukaryotic algae and higher plants in having chloroplast *a* and in being able to split water enzymatically (*photolysis*). All cyanobacteria have characteristic pigments, *phycobilins*, which function as accessory pigments in photosynthesis. The photosynthetic pigments are contained in **phycobilisomes.**

The cyanobacteria are found in a wide variety of environments. Some grow freely in snow and on high mountain tops. Others thrive in thermal springs, such as those found in Yellowstone National Park, where the water temperature may be as high as 85°C. Still others are found in marine and fresh waters, in soil, and even in wet flower pots. Certain blue-green bacteria can grow on volcanic rock, where most plant life fails to develop. The explanation lies in the ability of these microorganisms to use gaseous nitrogen (nitrogen fixation), carbon dioxide, and water vapor.

The cellular properties of the cyanobacteria are clearly unlike those of any eukaryotic algal group. These include cell wall composition, ribosome structure, features of protein synthesis, and certain nucleic acid properties.

The cyanobacteria may be spherical, rod-shaped, or spiral (Figure 239). Their cells are 1.5 to 2 μm in diameter, Gram-negative, with a multilayered petidoglycan cell wall, plasmids, and pili. The large number of cyanobacteria can be divided into five subgroups based on morphology and their means of asexual reproduction.

Group I consists of unicellular cells, but new cells produced by binary fission may remain connected to form aggregates.

Group II contains small spherical cells that reproduce by means of a type of endospore known as **baeocytes.** These reproductive cells are formed from multiple splitting of vegetative cells.

Group III is made up of filamentous cells that divide by binary fission in a single plane (Figure 239B).

Group IV includes filamentous forms that produce round **heterocysts.** Heterocysts arise from the differentiation of vegetative cells and are the sole sites of nitrogen fixation. In filaments, cells are bound together in a common sheath known as a **trichome** (Figure 239C). Reproduction occurs by breakage of trichomes into shorter chains of cells called **hormogonia.**

Group V consists of branching forms of cyanobacteria.

Myxosarcina

Myxosarcina (Figure 239A) species are found in fresh and marine water environments. The vegetative cells are spherical, measure 5–6.3 μm in diameter, and form cube-like aggregates. At least four baeocytes are released per vegetative cell. The temperature range for best growth is 30–44°C.

Spirulina

These cyanobacteria commonly are found in fresh, marine, and brackish waters. Some *Spirulina* species occur in inland salt lakes and in some hot springs at temperatures as high as 50°C. They are generally not found in terrestrial habitats. Pigments are variable and range from blue-green to extreme redness.

The *Spirulina* are filamentous forms that appear as tightly closed right-handed or left-handed springs or helixes (Figure 239B). They divide by binary fission. Movement is described as a type of spiraling or screwlike motion.

Calothrix

These microorganisms form long trichomes containing many cells in a common sheath. Heterocysts also are present (Figure 239C). The wide ends of individual trichomes measure approximately 2.5–18 μm in diameter. The sheaths of these organisms vary in color and may appear golden-brown to dark brown.

Anabaena

The trichomes formed by *Anabaena* species contain spherical or ovoid (barrel-shaped) vegetative cells and generally range in diameter from 2 μm to 10 μm. The trichomes may be straight, curved, or spring-like (Figure 239D). Heterocysts may appear within a long chain of cells or at the very end position. Reproduction is by fragmentation of parent, large trichomes into shorter forms.

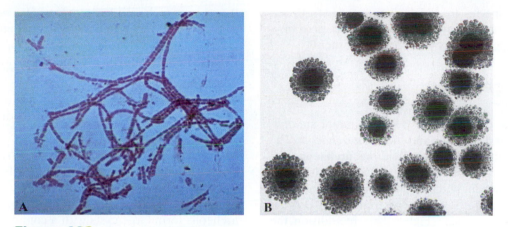

Figure 238

(A) Long, wavy chains of Gram-negative *Streptobacillus moniliformis* (1,000×).
(B) Colonies on Dienes agar (magnification 200×). (From M.R. Pins, J.M. Holden, J.M. Yang, S. Madoff, and M.J. Ferraro. *CID* 22(1996):471–76.)

The Cyanobacteria

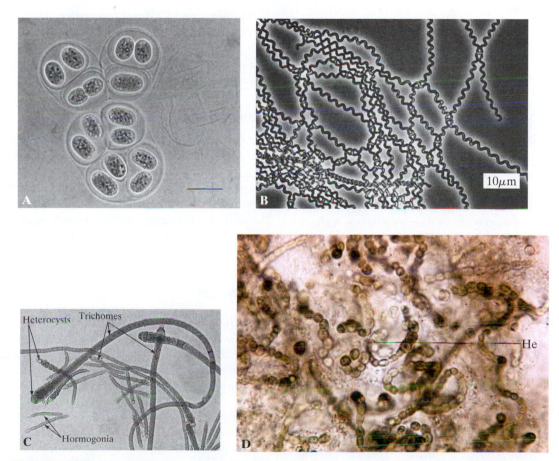

Figure 239

Examples of cyanobacteria. (A) Young aggregates of the coccal cyanobacterium *Myxosarcina*. Note the packets of cells. (B) The spiral, loosely coiled cells of *Spirulina*. (C) The long trichomes (**T**) of *Calothrix*. The trichomes contain many cells in a common sheath. Shorter cell chains, hormogonia (**Ho**), and heterocysts (**He**) also are evident. (Courtesy of R. Rippka.) (D) Light micrograph of a massive number *Anabaena* trichomes. Heterocysts (**He**) also can be seen.

Anabaena species are widely distributed in fresh and marine waters. They are important members of phytoplankton populations in these environments.

▲ THE ARCHEAE

The Archeae domain consists of two major subdivisions, the *Euryarchaeota* and the *Crenarchaeota*. The prokaryotes in these two subdividions, **archaeons**, are referred to as extremophiles because of their ability to grow in extremely harsh environments, including the highest temperatures and extremes of pH of all known microorganisms (Figure 240). In addition, the species in this domain have several properties that separate them from other prokaryotes. These include a different sequence of nucleotides in ribosomal ribonucleic acid (rRNA), a universal absence of muramic acid (a typical petidoglycan compound) as a component of cell walls, and differences in the translation process of protein synthesis.

Euryarchaeota

The Euryarchaeota include a physiologically diverse number of species. Among the ranks of this subdivision are the methanogens, obligate anaerobes prokaryotes whose metabolism is connected to methane production (Figure 240); the halophiles (salt-lovers), obligate aerobes whose metabolism and reproduction requires very large amounts of salt; and the high-temperature-requiring hyperthermophiles. The halophilic prokaryotes are a distinct group of organisms clearly able to adapt to the high salt concentrations and light intensities of their natural habitats. The halophiles can use light to produce adenosine triphosphate (ATP), but in a way that is different from the process used by other microorganisms. A characteristic property of extreme halophiles is the presence of red pigment in their cell membranes. These bacteria are known for their production of a typical brick-red color in seawater ponds, from which salt is recovered by evaporation, and on salted dry fish or hides treated with salt containing halophiles.

The thermoacidophiles, include cell-wall-less archaeons that grow best at relatively high temperatures and extremely low pH.

Methanothermus

Methanothermus fervidus
Methanothermus fervidus (Figure 241B) is well known for it ability to grow in environments in which temperatures reach 83°C.

Morphology and Cultural Properties. *M. fervidus* is a Gram-positive straight to slightly curved rod,

0.3–0.4 μm wide and 1–3 μm long. It occurs singly or in short chains. This archeon is a strict anaerobe and appears to be motile. The optimal temperature and pH are 83–88°C. and 6.5, respectively.

Methanosarcina mazei
Methanosarcina mazei is found in marine sediments, lake sediments, or anaerobic digestors devoid of any free oxygen.

Morphology and Cultural Properties. *Methanosarcina* species are irregular spherical bodies (Figure 241C). They generally exhibit a Gram-variable reaction and occur alone or in aggregates of cells. Occasionally the archeons form large cysts with a common outer wall surrounding individual cells. *Methanosarcina* species are nonmotile, and strictly anaerobic,

Pyrococcus

Pyrococcus furiosus
P. furiosus can be found in geothermally heated marine environments.

Morphology and Cultural Properties. *Pyrococcus furiosus* is a slightly irregular coccus, 0.8–2.5 μm in width. It occurs singly or in pairs, and is highly motile by means of a tuft of polar flagella (Figure 241D). Pyrococcus furiosus is strictly anaerobic and heterotrophic.

Crenarchaeota

The Crenarchaeota mainly contain hyperthermophilic species, including among their ranks prokaryotes able to grow at the highest temperatures of all known microorganisms (Figure 241 D). Moreover, many hyperthermophilic species whose energy is obtained from the oxidation of organic compounds (chemolithotropic) are the only primary producers in their respective extreme environments.

Pyrodictium

Pyrodictium species
Pyrodictium species occur in environments with extreme high temperatures, such as the submarine, hydrothermic sea floors of volcanoes, and the hot deep sea vents near Guaymus, Mexico.

Morphology and Cultural Properties. *Pyrodictium* species occur as disk- to dish-shaped ells, 0.3–2.5 μm in diameter, and approximately 0.2 μm in thickness (Figure 241A). The organisms grow in flocks and form networks of hollow fibers. *Pyrodictium* species are Gram-negative, nonmotile, chemolithotropic, strict anaerobes. The optimum temperature and pH for growth are 105° C. and 5.5, respectively.

Figure 240
A hot spring in Yellowstone National Park from which methogens were isolated. The gas bubbles on the surface contain methane and carbon dioxide.

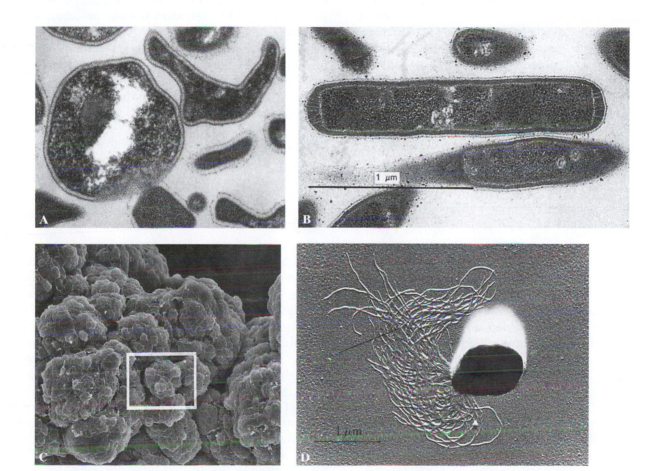

Figure 241
Represenatatives of the Archeae. (A) Electron micrograph of *Pyrodictium.*(Courtesy of K.O. Stetter, University of Regensburg.) (B) *Methanothermus fervidus,* an extremely thermophilic organism capable of growing at temperatures up to 97° C. (Courtesy of K.S. Stetter, University of Regensburg.) (C) Scanning electron micrograph showing the packaging of cells (sarcinae formation) in aggregates of *Methanosarcina mazei.* Courtesy of R.W. Robinson et al, Department of Microbiology and Cell Science, University of Florida.) (D) Shadow-cast preparation of the marine, heterotrophic Archaea species, *Pyrococcus furiosus* (From G. Fiala and K.O. Setter. *Arch. Microbiology* 145(1986):56–61.)

5

Mycology

There are more things in heaven and earth, Horatio,
Than are dreamt of in your philosophy.

— William Shakespeare, Hamlet

Mycology (the study of fungi) had modest beginnings in the eighteenth century, often as a Sunday hobby for physicians. Today, fungal forms of life are still of interest to physicians, but a number of other specialists have been added to those concerned with fungi. These include the geneticist, genetic engineer, molecular biologist, plant pathologist, ecologist, and commercial microbiologist. This list reflects the widespread distribution of fungi, their usefulness as research organisms, and their involvement in many aspects of everyday life. Here is a brief exploratory look at this group of microorganisms.

▲ Properties of Fungi

Like bacteria, the fungi are extremely diverse. Unlike bacteria, however, they have a eukaryotic form of cellular organization. They are also much larger and contain the organelles typical of eukaryotic cells. In contrast to algae (the next group to be considered), the fungi lack chlorophyll, and, consequently, despite the fact that several are green, do not carry out photosynthesis. Fungi also have cell walls, which generally contain **chitin**, a polysaccharide that is found in the skeletons of insects, the shells of crabs, and related forms of life.

Although the fungi are a large and diverse group of eukaryotic microorganisms, three fungal groups have major practical importance: the *microscopic* **molds** (Figure 242A) and **yeasts** (Figure 242B) and the *macroscopic* **mushrooms** (Figures 242C and D).

Fungi can be found in a wide variety of habitats. Most are found in soil or on dead plant matter. Here, these types play crucial roles in the mineralization of organic carbon in nature. Equipped with some of the most powerful digestive enzymes known, decomposing fungi are capable of reducing wood, fiber, and foods to their basic chemicals with staggering efficiency. Like animals, fungi are heterotrophic and must obtain preformed organic substances from their environment. Some fungi are aquatic, living primarily in fresh water, and a few marine forms are known.

Fungi also are recognized for their disease-producing capabilities. Plants and animals, including humans, are susceptible (Figure 243). These pathogenic fungi have the enzymes necessary to obtain nutrients directly from the living host.

Molds and Yeasts

Fungi are identified in the laboratory on the basis of their vegetative and reproductive structures (Figures 244–247). The most commonly seen fungi in the laboratory exist in one or both of two forms: molds and yeasts.

Molds, or filamentous fungi, form tubelike filaments called **hyphae** (singular, *hypa*). Some hyphae have cross-walls, or **septa**, to separate cells (Figure 244A), others lack septa and are called **coenocytic**, or **nonseptate**. Hyphae begin as single filaments and then branch repeatedly, forming a network known as a **mycelium** (plural, *mycelia*). Many are dry, cottony, raised masses of branching hyphae (Figures 258 through 262). Three basic types of mycelia are recognized: *vegetative*, *aerial*, and *reproductive*. The rhizoids of a vegetative mycelium penetrate the surface of nutrient materials and absorbs nutrients (Figure 244B); an aerial mycelium grows above the surface of a nutrient, such as an agar medium; and a reproductive mycelium gives rise to and bears the reproductive structures called **spores** (Figures 244A and 245).

Asexual reproduction may involve the formation of **conidia** (singular, *conidium*), or spores (Figure 244). Asexual spores, known as **sporangiospores** (Figure 244C), form within a saclike structure known as a **sporangium**. They are the result of cytoplasmic division in the sporangium. The sporangia are typically formed on

MYCOLOGY
Properties of Fungi

Figure 242

Examples of fungi. (A) The cottony mycelium (colony) of the mold *Phialiophora repens* on potato dextrose agar. (From M. Hironga and associates, *J. Clin. Microbiol.* 27(1989):394–99.) (B) Colonies of Brewer's yeast, *Saccharomyces cerevisiae.* (C) A group of common mushrooms, *Lepiota cristata,* growing on a lawn. (D) The shelf mushroom, *Polyporus versicolor,* growing on dead wood.

Figure 243

Fungus diseases of plants. (A) Peach brown rot. The fungus pathogen eventually completely penetrates the fruit, and its enzymes cause it to rot. The peach becomes a dry, distorted form called a "mummy." (B) Another example of fungus destruction. The strawberry on one side has been destroyed by fungus—caused fruit rot.

Properties of Fungi (Continued)

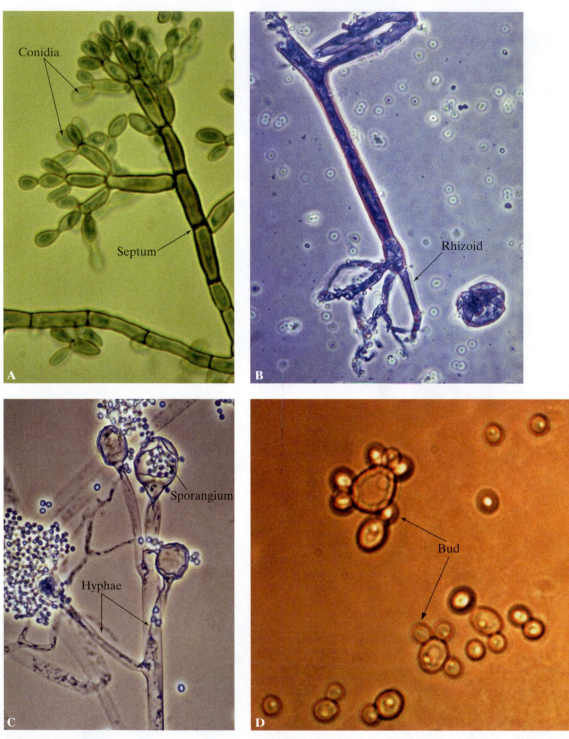

Figure 244

Microscopic views of fungi showing typical structures and certain differences between molds and yeast. (A) A microscopic view of a fungus showing septate hyphae and a round to oval type of spore called a *conidium*. (Courtesy of Dr. T. Matsumoto.) (B) The rootlike rhizoids. (C) The saclike sporangia and sporangiospores. Nonseptate hyphae also can be seen in these micrographs. (From G. St. Germain, A. Robert, M. Ishak, C. Tremblay, and S. Claveau. *Clin. Inf. Dis.* 16(1993):640–45.) (D) The yeast *Saccharomyces cerevisiae,* showing parent cells and buds.

Properties of Fungi *(Continued)*

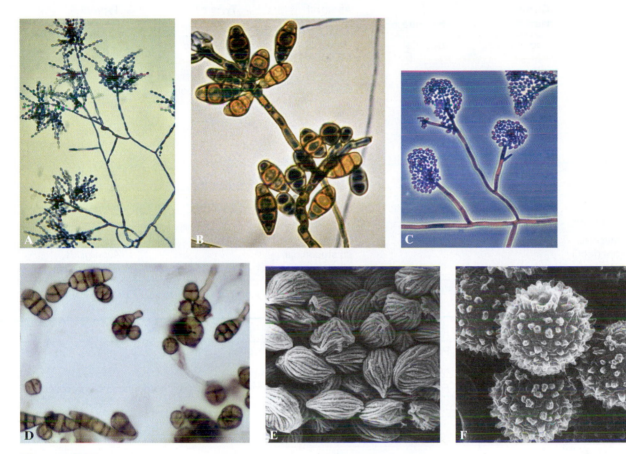

Figure 245

Microscopic views of conidia-producing fungi. (A) *Cladosporium carrionii.* The conidiophores produce long, branching chains of smooth-walled, oval, somewhat pointed conidia. (B) *Curvularia geniculata.* Conidia are large, usually contain four cells, and may appear curved owing to the swelling of a central cell. (C) *Exophiala spinifera.* Conidia are oval and gather in clusters at the end and sides of conidiophores and at points along filamentous hyphae. (Courtesy of Dr. T. Matsumoto.) (D) *Alternaria* species. Conidia are large and brown and have both transverse and longitudinal crosswalls. They are single or in chains. A comparison of fungal spore surfaces with scanning electron microscopy. (E) *Rhizopus stolonifer* (8,600×). (F) *Eurotium restrictus* (5,500×). (Courtesy of Professors T. Ohtsuki and M. Osumi.)

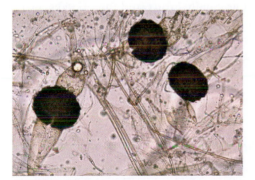

Figure 246

Representative thick black zygospores of *Rhizopus* species. Note the two hyphal branches connecting to the zygospores.

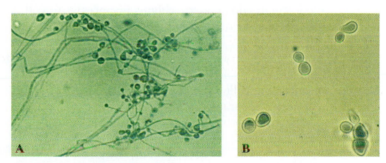

Figure 247

Dimorphism. (A) A temporary wet mount with lactophenol cotton blue stain preparation. *Blastomyces dermatitidis* in its mycelial phase growing at 25°C. Single conidia are directly attached to hyphae or are on small conidiophores (400×). (B) Yeast phase of the fungus growing at 37°C. Single-budding cells can be seen (400×). (From S.E. Hoy, and T.-Y. Chuang. *Cutis.* 48(1991):193–96.)

Table 13   Classification and General Properties of Fungi

Group	Common Name	Hyphae	Asexual Structures	Type of Sexual Spore	Typical Representatives
Ascomycetes (Ascomycota)	Sac fungi	Septate	Conidia, blastoconidia	Ascospore	*Neurospora, Saccharomyces, Morchella* (morels)
Basidiomycetes (Basidomycota)	Club fungi, mushrooms	Septate	Basidiospores	Basidiospore	*Amanita* (poisonous mushroom), *Agaricus* (edible mushroom)
Zygomycetes (Zygomycota)	Bread molds	Coenocytic	Spores, conidia	Zygospore	*Mucor, Rhizopus* (common bread mold)
Oomycetes (Chytridiomycota)	Water molds	Coenocytic	Spores, conidia	Oospore	*Allomyces*
Deuteromycetes (Fungi Imperfecti)	Fungi imperfecti	Septate	Conidia, blastoconidia, phialoconidia	None	*Penicillium, Aspergillus, Candida*

Table 14   Examples of Conidia

Conidium Type	Brief Description
Arthroconidium (arthrospores)	Produced by fragmentation of fertile hyphae
Blastoconidium (blastospores)	Formed by outgrowth of new cells
Chlamydoconidium (chlamydospores)	Formed by enlarging and developing thick walls; produced during unfavorable conditions
Macroconidium	Large multiseptate spores
Microconidium	Small spores
Phialoconidium	Formed at the tip of a flask-shaped specialized conidiophore, the phialide

specialized hyphae known as **sporangiophores**. A distinctive domelike internal structure, the **columella**, is formed during sporangium development. Identification of fungi belonging to the phylum Zygomycetes (Table 13) is based on the presence of these reproductive structures and the rootlike rhizoids (Figure 244B). Several fungi are able to form thick-walled, resistant spores called chlamydoconidia (**chlamydospores**) (Figure 278B). Table 14 briefly describes the features of conidia.

Conidia are nonmotile reproductive structures and typically are produced on aerial hyphae (Figure 245). Conidia can be single-celled or can appear as multicompartment forms separated by internal crosswalls. The crosswalled larger type are known as **macroconidia** (Figures 245D and 268C). A variety of conidia are formed by the fungi. Their development, arrangement, and microscopic appearance are of major importance to rapid and precise identification (Figure 245E). Spores and conidia often are pigmented and resistant to drying, and serve to spread fungi to new habitats. Their presence also gives a rather dry, dusty appearance to mycelial surfaces.

Some molds also produce sexual spores, which result from sexual reproduction (Figure 246). Fungal species belonging to different taxonomic groups produce different sexual spore types (see Table 13).

Yeasts generally form, smooth, moist colonies (Figure 242B). Microscopically, they are usually spherical or oval (Figure 244D). Yeasts reproduce by **budding**, a process by which a new cell (**blastoconidium**, or bud) forms as a small outgrowth of the parent cell. The bud gradually grows and then breaks away. Yeasts do not form filaments or mycelia. At times, yeasts may not separate from one another and instead form a chain called a **pseudomycelium**. Some yeasts also can undergo sexual reproduction, resulting in the formation of a fertilized cell called a **zygote**.

Dimorphism

Fungi that are able to grow as two different forms are referred to as being **dimorphic**. This ability is usually temperature-dependent, but it may be influenced by other factors as well. For example, *Blastomyces dermatitidis* (Figure 247), the fungus that causes blastomycosis, an infection of the lungs and other body tissues, grows as a yeast at 37°C and as a mold at 23° to 25°C (room temperature).

Classification

A number of fungal properties are used for purposes of identification and classification. These include the type of hyphae and spores produced, habitats, biochemical reactions and cultural characteristics. Table 13 compares the major phyla of fungi.

▲ Culture and Microscopic Techniques

Although, with few exceptions, the direct microscopic examination of specimens rapidly yields important information, the identification of a fungus depends on culture. Several types of media are available for a variety of purposes. As is the case for bacteria, media may be enriched, selective, or selective and differential (Figure 248). Sabouraud's dextrose agar is a commonly used selective agar for fungal cultivation. Other media containing different ingredients, such as bird seed (Figure 249A) and dyes (Figure 249B), are used for the isolation and identification of fungi.

Direct Microscopic Examination

Because certain fungi tend to grow more slowly than other microorganisms or are more difficult to isolate, significant attention is placed on direct microscopic examination and related techniques. Several methods are available for direct microscopic examination.

1. **Temporary wet mount**. Direct examination of a specimen taken from a mycelium can be achieved by placing it in a good-sized drop of lactophenol cotton blue, a mounting fluid containing both disinfectant (phenol) and stain (cotton blue). Lactic acid acts as a clearing

Culture and Microscopic Techniques

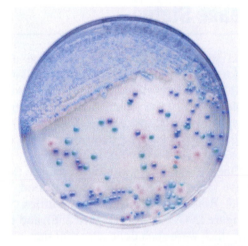

Figure 248
The reactions on CHROMagar *Candida*. A selective and differential medium used for the direct isolation and differentiation of different species of the yeast *Candida. C. albicans* forms green colonies, *C. krusei*, pinkish colonies, and *C. tropicalis*, dark blue to metallic blue colonies, some of which are surrounded by a halo. (Courtesy of Becton, Dickinson and Company.)

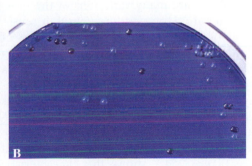

Figure 249
Culture techniques. (A) On the right, orange to brown colonies of *Filobasidiella* (*Cryptococcus*) *neoformans* on *Guizotia abyssinica*–creatinine agar are shown. On the left, nonpigmented colonies of the same organism growing on Sabouraud's glucose agar. (Courtesy of Dr. F. Staib, Former Chief, Mycology Unit, Robert Koch Institute.) (B) Colonies of *F. neoformans* (dark blue) and *Candida albicans* (light blue) on Sabouraud's glucose agar containing the dye trypan blue.

agent and aids in preserving fungal structures (Figures 247A and 250A).

2. **Potassium hydroxide (KOH) preparation**. This technique involves placing a specimen into a drop of KOH and then mixing and gently heating it. After cooling, the preparation is examined. The KOH digests protein and other types of tissue debris, thereby making it easier to see fungal structures (Figure 250B).

3. **India ink preparation**. India ink is used to highlight certain structures, such as capsules (Figure 251). The India ink does not stain the structure but provides a dark background against which the capsule can be seen.

A number of techniques also are available to demonstrate and study fungi in tissues. These include the acid-fast stain, modified Gram stain (Figure 253), and specific histopathologic stains, such as:

1. **Gomori-methenamine-silver nitrate stain**. This staining procedure sharply outlines fungi in black due to the depositing of silver on their cell walls (Figure 252A). The internal parts of hyphae stain deep rose to black, and the background appears green.

2. **Periodic acid-Schiff stain**. The hydroxyl groups in the carbohydrates of fungal cell walls are oxidized by the periodic acid to form aldehydes. The aldehydes, in turn, combine with basic fuchsin dye to produce a pink-purple color (Figure 272B).

Immunofluorescence microscopy and related techniques also are used to detect fungi directly in tissues and body fluids.

Representative Fungi, Their Distinctive Properties, and Selected Disease States

▲ Macroscopic Views of Fungi

Mushrooms

Mushrooms (Figures 254 and 255) represent a large group of filamentous fungi that typically form large structures called **fruiting bodies**, or **basidiocarps**. The most familiar of these visible forms include the gilled mushrooms and the fleshy fungi (Figure 254E).

The mushroom represents the reproductive phase of the fungus that formed it. It develops from a mycelium that is hidden from view in soil or another substrate. A mushroom is a specialized body that ensures the maximum production of spores (**basidiospores**). There are many variations of the basic mushroom structure. However, in general, a mushroom consists of a **cap**, or **pileus**, radiating layers of **lamellae** (gills) found on the underside of the cap and representing the spore-producing region, and a **stipe**, or **stalk**, on which the cap sits. Many mushrooms may have one or two layers of tissue, known as **veils**, on young fruiting bodies. Patches of veil material may be found on the cap and on portions of the stipe. A persistent ring of tissue on the stipe is called an **annulus**. Figure 254A shows the characteristic parts of a mushroom.

The identification of a particular mushroom involves noting the presence or absence of these parts and other properties, including the size, shape, cap and stipe, color, surface appearance and texture, odor, color changes of structures following injury, habitat, and spore color. Determination of spore color is one of the most essential steps in mushroom identification. A spore print (Figure 257) is the technique usually used to determine spore characteristics as well as certain features of gills.

One of the predominant interests that humans have in mushrooms is their use as food. The Greeks and Romans were fond of truffles, mushrooms (Figure 254A), morels (Figure 256), and puffballs, all of which were such delicacies that they were the only foods the wealthy insisted upon preparing themselves. As a group, mushrooms occur in a wide variety of habitats from the arctic to the tropics. In addition to mushrooms grown commercially (Figure 254A), many wild species are edible and are collected by mushroom lovers. Certain other species, such as those belonging to the genus *Amanita*, produce some of the most potent toxins known. The red or yellow *Amanita muscaria* (Figure 255A) is responsible for a number of mushroom poisonings every year. This species is popularly known as the fly-agaric because it was formerly used to kill flies. The cap of the mushroom was brought indoors and sprinkled with sugar to attract flies, which nibbled on it. Other mushrooms and related species of the group (the basidomycetes) cause agricultural losses affecting various types of plant life, including grains, trees used for lumber, and grasses (Figure 255C).

Culture and Microscopic Techniques *(Continued)*

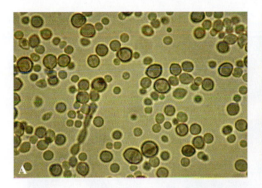
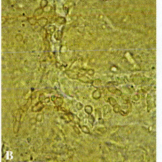

Figure 250
Microscopic techniques. (A) Temporary wet mount of a culture showing yeastlike cells and septate hyphae. (Courtesy of Drs. T. Matsumoto and T. Matsuda.) (B) A potassium hydroxide preparation of a nail specimen showing filamentous hyphae and spherical cells. (Courtesy of Drs. T. Matsumoto and T. Matsuda.)

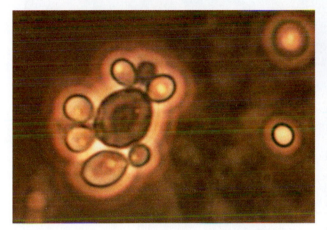

Figure 251
An India ink preparation of *Filobasidiella (Cryptococcus) neoformans.*

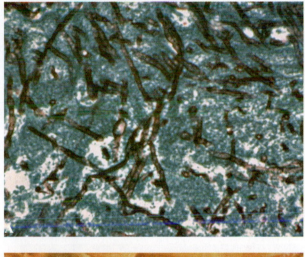

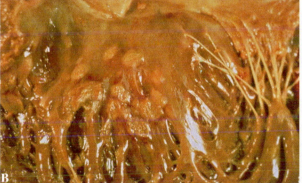

Figure 252
(A) A Gomori-methenamine-silver nitrate stained preparation of heart tissue showing the presence of hyphal elements. (B) A fungus-caused infective endocarditis. Specimens were taken from the irregular yellow-tan growths on the left ventricle. (Courtesy Dr. F. Welty, Harvard Medical School).

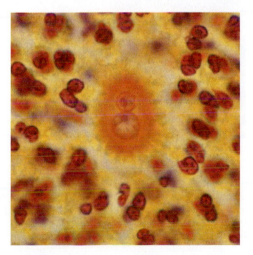

Figure 253
A Brown-Hopps Gram-stained preparation showing a budding yeast cell of *Sporotrichum (Sporothrix) schenckii* in a skin specimen (750×). (From R.C. Neafie and A.M. Marty. *Clin. Microbiol. Revs.* 6(1993):34–56.)

Macroscopic Views of Fungi

Figure 254

Examples of macroscopic common fungi. (A) The structures of a mature mushroom (*Agaricus* species). The specific parts shown include the cap (pileus) attached to the stalk (stipe), a ring (annulus) around the stalk, and the lower surface of a cap showing the radiating strips of tissue called gills (lamellae). (B) The oyster mushroom (*Pleurotus ostreatus*) and its small buttons. (C) The bush or hen of the woods mushroom, *Polyporus frondosus*. (D) Cup-shaped mushroom, *Peziza* species. (E) An example of a fleshy mushroom.

Macroscopic Views of Fungi *(Continued)*

Figure 255

Poisonous mushrooms. (A) *Amanita* species, a highly poisonous species. (B) Common earth-ball, *Scleroderma citrinum*. (C) Fairy ring disease is destructive to grasslands, including lawns, pastures, and parks. The destructive action is caused by the enzymes of growing mushrooms species that damage the roots of the host grass plants, thereby weakening them and permitting a massive fungus invasion. It is believed by some that anyone stepping into the mushroom ring will eventually become insane.

Figure 256

Morchella esculenta, the morel, a highly edible variety.

Figure 257

A spore print.

Puffballs

Puffballs (earth-balls) are fleshy (later dry) fruiting bodies found in rich soil and decaying wood. The inside of the developing puffballs contains numerous small chambers containing club-shaped structures that produce large numbers of basidiospores.

Morels

True morels (Figure 256) represent a group of tasty edible mushrooms, many of which are highly prized by connoisseurs of fine food. They are often found in forests around the world.

▲ Common Molds and Yeasts

A number of fungi are commonly studied in microbiology and related courses. Several of them are also considered as contaminants, but under certain circumstances they may be labeled as pathogens. Fungi such as *Rhizopus nigricans* (Figure 251), *Penicillium* species (Figure 260), *Fusarium* species (Figure 262A), and *Scopulariopsis* species (Figure 262C), though generally harmless to healthy persons, cause disease in individuals whose immune systems have been weakened by severe diseases such as AIDS or by immunosuppressive therapy. Some authorities refer to such fungi as **opportunists**.

The following descriptions of mycelia (both macroscopic and microscopic) are based on fungal growth on Sabouraud's dextrose agar.

1. *Alternaria* **species**. Mycelial surface is grayish white at first and later becomes greenish black or brown (Figure 261). Hyphae are septate and dark. Microscopic structures include conidiophores and brown macroconidia (Figure 245D).

2. *Aspergillus* **species**. Mycelial color, which varies widely, from white to shades of green, black, yellow, brown, and gray, usually is determined by the color of conidia. Microscopic structures include conidia (phialoconidia), conidiophores, vesicles, and small vessel-shaped structures (phialides) on which conidia are formed. (See Figure 259).

3. *Fusarium* **species**. Mycelia are cottony, often producing a diffusible pink, purple, or yellow pigment (Figure 262A).

4. *Penicillium* **species**. Mycelia are greenish blue, grow rapidly, and appear with a powdery or velvet texture (Figure 260). Some species secrete yellow surface droplets. Microscopically, the fungus appear as a broom or brush form with conidia (phialoconidia), conidiophores, and phialides also (Figure 260B).

5. *Rhizopus* **species**. Mycelia mature within four days and quickly cover media surfaces. Mycelial appearance is cottony and is white at first and then becomes dotted with black dots (sporangia). Microscopic structures include rhizoids, sporangiophores, sporangia, columellae, and sporangiospores (Figure 258B).

6. *Rhodoturula* **species**. Mycelia develop rapidly and are soft and pink to coral in color (Figure 262B). Microscopically, *Rhodoturula* appear as budding cells that are round or oval.

7. *Saccharomyces cerevisiae*. Colonies appear as smooth, moist, and white to cream-colored growths. *Saccharomyces* appear as budding yeast cells (Figures 242D and 244D).

8. *Scopulariopsis* **species**. Mycelia range in color from light tan to dull gray (Figure 262C). Microscopic structures include conidiophores and chains of pear-shaped, thick-walled conidia.

▲ Funguslike Protists

Slime molds are examples of funguslike protists. They are eukaryotic and have properties similar to both fungi and protozoa. Slime molds can be divided into two groups, the **cellular slime molds**, which are composed of single ameba-like cells, and the **acellular slime molds**, which consist of **plasmodia**, naked protoplasm masses of indefinite size and shape (Figure 263).

▲ Spectrum of Mycoses

Most fungi that infect humans and other animals and cause disease are categorized by the tissue or organ levels that are the primary sites of involvement. These categories include **superficial**, **cutaneous**, **subcutaneous**, and **systemic mycoses**.

Agents of Superficial and Cutaneous Mycoses (Dermatophytoses)

Superficial fungal infections (mycoses) involve only the outermost layers of the skin or parts of hair shafts. *Phaeoannellomyces werneckii* causes such skin infections, and *Trichosporon beigeli* and *Piedraia hortae* cause the hair shaft and follicle infections known as **white piedra** and **black piedra**, respectively (Figure 265). Potassium hydroxide preparations are helpful to diagnosis.

The cutaneous mycoses, or dermatophytoses, are infections of the hair, nail, or skin on a living host (Figures 264 and 265) and are caused by species of the

Common Molds and Yeasts

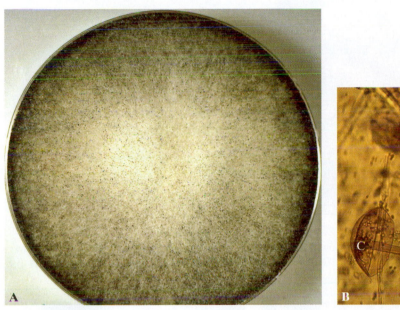

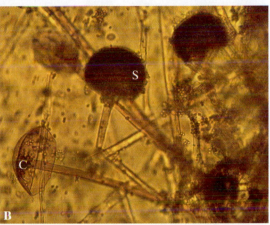

Figure 258
The bread mold *Rhizopus nigricans*. (A) The mycelium grown on Sabouraud's dextrose agar. Note the different pattern of growth. The black dotlike structures are sporangia (sacs of sporangiospores). (B) A microscopic view. The black sporangia (S) filled with spores, columella (C) and free spores can be seen.

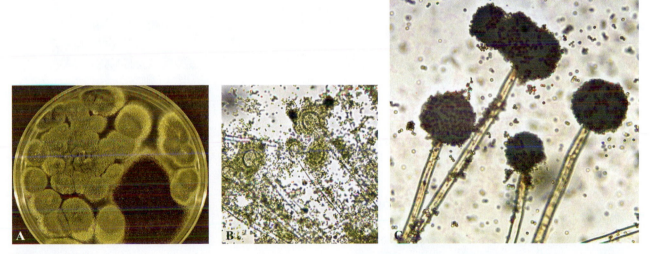

Figure 259
Aspergillus species. (A) Two different mycelia, *A. niger* (black) and *A. flavus* (greenish) are shown. Microscopic views of *A. niger* (B) and *A. flavus* (C) showing a conidiophore (lower supporting structure), the swollen vesicle connecting to the phialide area, and the oval conidia.

Common Molds and Yeasts *(Continued)*

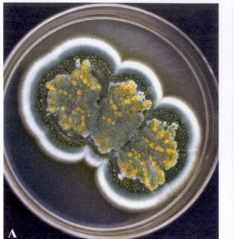
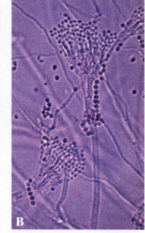

Figure 260
Penicillium species. (A) Mycelium. (B) Microscopic view showing lower supporting conidiophore, connecting to a phialide, and the oval conidia.

Figure 261
Alternaria species mycelium. See Figure 245D for the spores of the mold.

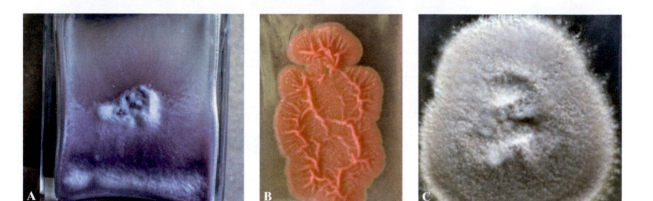

Figure 262
Common contaminants, all of which can cause infections in individuals with weakened immune systems. (A) *Fusarium* species. This mold forms a fluffy, cottony mycelium, which may be accompanied by a diffusible, pink, purple, or yellow pigment. (B) *Rhodoturula* species. This genus includes nonfermenting yeasts usually forming mucoid, yellow to red colonies. (C) *Scopulariopsis* species. This mold forms white to light beige or dull grayish mycelia.

Funguslike Protists

Figure 263
The scrambled egg–appearing slime mold *Fuligo septica*. Both well-formed yellow plasmodia and filamentous newly forming (whitish) plasmodia are shown.

Superficial and Cutaneous Mycoses (Dermatophytoses)

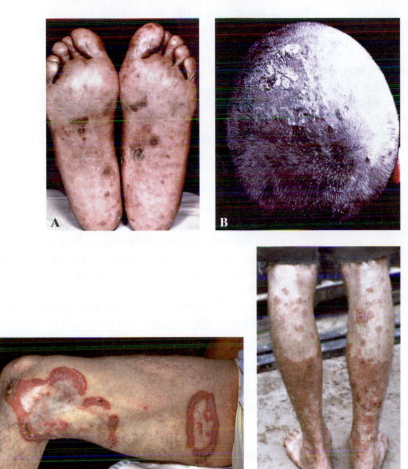

Figure 264
Examples of ringworm skin infections. (A) Athlete's foot, tinea pedis. (B) The scalp
lesion in tinea capitis (arrow). (C) Tinea corporis. (D) Disabling fungal infection
complicated by a secondary streptococcal infection.

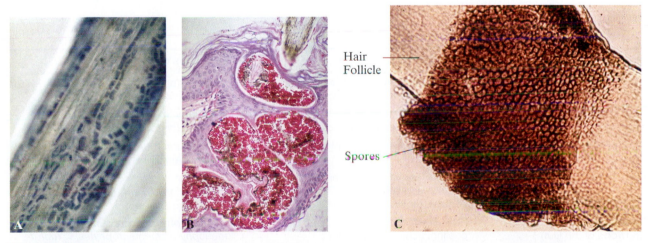

Figure 265
Infection of the hair. (A) An endothrix infection. Arthroconidia can be seen in the infected hair. (B) Spore-ridden hair shafts
forming darkly staining plugs in the surface layer of the scalp (stratum corneum). (C) Spores almost completely replacing
the contents of a hair. (From J. Lee, Y.-Y., and M.-L. Hsu. *J. Cutan. Pathol.* 19(1992):54–58.) (C) An ectothrix (external) hair
infection. Caused by *Piedra hontae* Note the large numbers of conidia on the hair follicle surface.

three genera *Epidermophyton, Microsporum,* and *Trichophyton* (Figures 266 through 271). These fungi attack and are limited to the keratinized tissues. They are referred to as **dermatophytes**. Occasionally, the terms *tinea* and *ringworm* are used to indicate the diseases (dermatophytoses) caused. Table 15 lists several examples. In general, *Microsporum* species attack hair and skin but not nails. *Trichophyton* species attack hair, skin, and nails. *Epidermophyton floccosum* infects skin and occasionally nails. The dermatophytoses are generally acquired through direct contact with infected persons or other animals and fomites.

Endothrix and ectothrix are the two known forms of hair invasion. In the **endothrix** infection, hyphae grow down the hair follicle and penetrate the hair shaft (Figure 265). The **ectothrix** infection is characterized by spores, arthroconidia (formed by frequent fragmenting in the area surrounding the hair), forming a sheathlike covering, and hyphae growing around the hair shaft and eventually destroying it.

Diagnosis of a dermatophytosis includes KOH preparations of appropriate specimens and culture. Mycelial characteristics and microscopic examination of cultures are also used for fungal identification (Figures 266 through 271).

Agents of Subcutaneous Mycoses

Subcutaneous mycoses affect the deeper layers of the skin (Figure 272). The causative fungi gain access to subcutaneous tissues through some form of injury. Examples of such diseases include chromoblastomycosis and sporotrichosis (Figure 273). The organisms causing these two diseases differ from other fungi that cause subcutaneous mycoses by being dimorphic.

Chromoblastomycosis

This mycosis begins after the causative agent is implanted into the skin. Lesions develop slowly and gradually increase in size to form a subcutaneous nodule or tumor (Figure 272A).

Fungi that cause chromoblastomycosis include *Fonsecaea pedrosoi* (Figure 272 C–E), *Cladosporium carrionii* (see Figure 245A), and *Exophiala spinifera* (see Figure 245C). These organisms appear in tissue as thick-walled spherical cells that are called **muriform cells** or **Medlar bodies** (Figure 272B).

Sporotrichosis

This disease also is caused through tissue injury. The classic form of the disease includes the formation of numerous nodules, abscesses, and ulcers that develop along the lymphatics that drain the site of entry. The disease generally does not extend beyond this area.

The causative agent *Sporotrichum* (Sporothrix) *schenckii,* grows as a mold in cultures at 25°C and as yeastlike cells in tissues (Figure 273). These properties are of value in diagnosis.

Agents of Systemic Mycoses

Several fungi are known to cause systemic mycoses. Only three are discussed here: *Blastomyces dermatitidis* (see Figure 247), *Coccidioides immitis* (Figure 274), and *Histoplasma capsulatum* (Figure 275). Infection due to these fungi is usually acquired by inhalation of conidia from an environmental source.

Blastomyces dermatitidis

Blastomycosis is a chronic, tumor-forming infection of the lungs that may spread to other tissues. The

Table 15   Representative Dermatophytoses

Disease	Description	Examples of Fungal Causes
Tinea barbae (ringworm of the beard)	Circular patches on bearded area; some loss of hair	*Trichophyton mentagrophytes T. rubrum, Microsporum canis*
Tinea capitis (ringworm of the scalp)	Involvement of the scalp and hair; crusty, scaly lesions on scalp	*Mycrosporum canis, Trichophyton tonsorans*
Tinea corporis (ringworm of the body)	Ringlike lesions with a central scaly area	*Trichophyton rubrum, T. mentagrophytes, Microsporum canis*
Tinea cruris (ringworm of the groin)	Ringlike, sometimes leathery, patches and itching in skin folds of pubic region	*Epidermophyton floccosum, Trichophyton rubrum, T. mentagrophytes*
Tinea pedis (ringworm of the feet)	Fluid-filled lesions, skin cracks, peeling, itching between toes	*Epidermophyton floccosum, Trichophyton mentagrophytes, T. rubrum*
Tinea unguium (ringworm of the nails)	Hardening and discoloration of the nails	*Epidermophyton floccosum, Trichophyton mentagrophytes, T. rubrum*

Superficial and Cutaneous Mycoses (Dermatophytoses) *(Continued)*

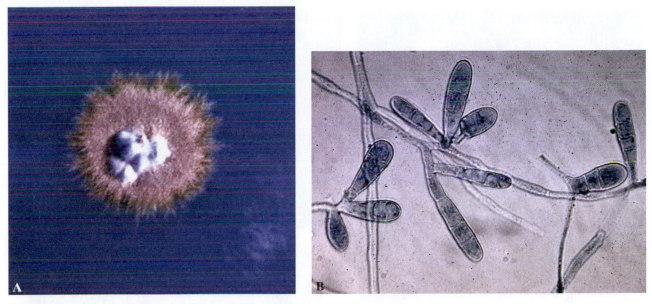

Figure 266
Epidermophyton floccosum. (A) Mycelium on Sabouraud's dextrose agar. (B) Microscopic view showing macroconidia.

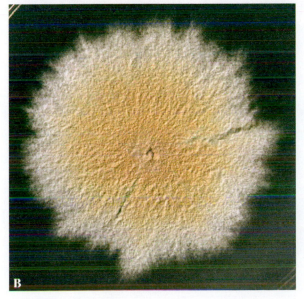

Figure 267
Microsporum species. (A) *M. canis* mycelial surfaces.
(B) *M. gypseum* mycelium. (C) Scanning electron micrograph
of *M. gypseum* macroconidia (800×). (Courtesy of N. Contet-
Audonneau and M. Miegeville.)

Superficial and Cutaneous Mycoses (Dermatophytoses) *(Continued)*

Figure 268
Trichophyton rubrum. (A) Mycelium on Sabouraud's dextrose agar. (B) The red to purplish pigment produced by the mold can be seen on the undersurface. (C) *T. rubrum* macroconidia.

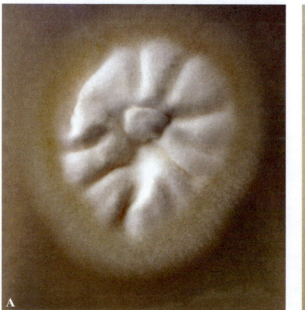

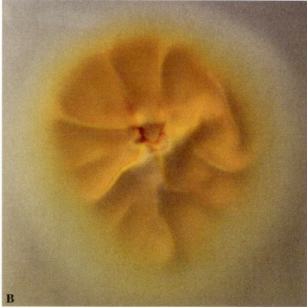

Figure 269
Trichophyton tonsorans. (A) Top view of mycelium. (B) Undersurface of mycelium.

Superficial and Cutaneous Mycoses (Dermatophytoses) *(Continued)*

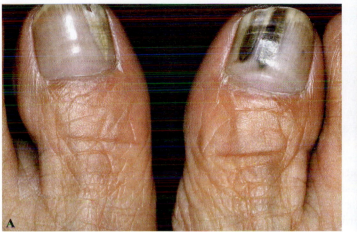

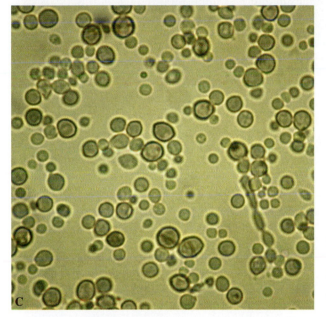

Figure 270
Nail infection, a case of fungal melanonychia (black pigmentation of the nail). (A) Black discoloration of toenails. (B) *Wangiella dermatitidis* mycelium, the causative agent. (C) Hyphae and conidia (600×). (From T. Matsumoto, T. Matsuda, et al. *Clin. Exp. Dermatol.* 17(1992):83–86.)

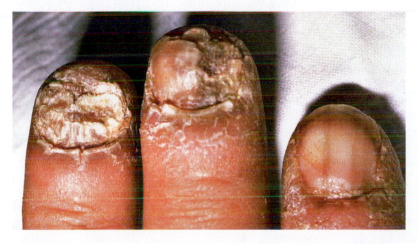

Figure 271
Trichophyton rubrum nail infections. A severe case of onychomycosis in a 37–year-old man with AIDS. (From N.S. Prose, K.G. Abson, and R.K. Scher. *Internat. J. Dermatol.* 31(1992):453.)

causative agent, *Blastomyces dermatitidis* is found worldwide and can be found in soil and wood.

Transmission. *Blastomyces dermatitidis* is inhaled. However, it is not spread from person to person or from lower animals to humans.

Morphology and Cultural Properties. *Blastomyces dermatitidis* is a dimorphic fungus. It appears as a mold (Figure 247A) at room temperature (25°C) and as a yeast in body tissues and on cultures grown at 37°C (Figure 247B)

Pathology and Clinical Features. Most infections are without signs or symptoms. Symptomatic disease usually appears as a mild respiratory infection, which usually is accompanied by fever, weight loss, general weakness, and a productive cough. If the fungus spreads, tissues such as skin, bones, joints, prostate gland, and testes are most commonly involved.

Diagnosis. The most rapid means of diagnosis is the direct demonstration of yeast cells in KOH preparations. Culture on routine mycologic media is also used but may take as long as 4 weeks.

Coccidioides immitis

Coccidioidomycosis, or valley fever, is the oldest of the major systemic mycoses on the basis of reported cases. The disease is also the most geographically restricted of this group of diseases because the causative agent, *Coccidioides immitis,* grows only in soil in semiarid climates (dry areas with high summer temperatures and mild winters).

Transmission. Infection usually results from the inhalation of arthroconidia (Figure 274G). However, introduction of this fungus into other sites also can result in a self-limited infection (Figure 274A and B). Domesticated, zoo, and wild animals are susceptible to infection.

Morphology and Cultural Properties. *Coccidioides immitis* is a dimorphic fungus. On routine culture at room temperature and at 37°C it grows as a mold (Figure 274E). In tissue, instead of a yeast phase, the fungus produces large, distinctive, round-walled structures, **spherules**, containing up to hundreds of endospores (Figure 274F). This structure is unique among pathogenic fungi.

Pathology and Clinical Features. In the self-limited form of the disease, general symptoms include cough, chest pain, loss of appetite, general discomfort, and fatigue. More than half of infections show no symptoms, or the disease is so mild it is not detected. Disseminated coccidioidomycosis may involve cutaneous and subcutaneous tissues (Figure 274D), the meninges, and visceral organs.

Diagnosis. The most rapid means of laboratory diagnosis is the direct KOH demonstration of the thick-walled spherules in sputum and other specimens. Culture from sputum and skin and other tissues and serological and skin tests also are useful.

Histoplasma capsulatum

Histoplasmosis is a disease of worldwide distribution. The causative agent, *Histoplasma capsulatum,* can be found in soil, chicken or pigeon coops, and bat caves.

Transmission. *Histoplasma capsulatum* is acquired through inhalation of conidia from an aerosol. Infected bats also are possible sources of infection. Person-to-person transmission does not occur.

Morphology and Cultural Properties. Histoplasma capsulatum is a dimorphic fungus that appears as a yeast in tissue and in cultures grown at 37°C. Cultures incubated at 22° to 25°C exhibit mold characteristics with the formation of a diagnostic structure, the **tuberculate macroconidium** (Figure 275C). *Histoplasma capsulatum* grows on standard media used for fungi, but it may take several weeks to produce a recognizable mycelium.

Pathology and Clinical Features. Histoplasmosis is an infection of the reticuloendothelial system, where *H. capsulatum* invades, grows, and reproduces in macrophages and giant cells. Most cases of the disease are asymptomatic or show only fever and cough for a few days or weeks. More severe cases may exhibit chills, chest pain, and general discomfort. Progressive disease develops along a path similar to pulmonary tuberculosis. Signs and symptoms include lung cavities, sputum production, night sweats, and weight loss.

Disseminated histoplasmosis may involve several organs, including the skin, central nervous system, lungs, gastrointestinal tract, and adrenal glands (Figure 275A and B).

Diagnosis. Direct microscopic examination of sputum in pulmonary histoplasmosis is rarely helpful. Culture of blood and biopsy specimens from a reticuloendothelial organs such as the spleen or lymph nodes are more likely to be successful. Staining of tissue specimens with methenamine silver (Figure 252) is also helpful to diagnosis.

Agents of Opportunistic Mycoses

Certain microorganisms regularly cause infection and disease when they enter a nonimmune host. Such organisms are known as primary pathogens. In contrast to these infectious disease agents, there are also opportunists. These organisms rarely cause disease in healthy humans. However, if an individual's defense

Subcutaneous Mycoses

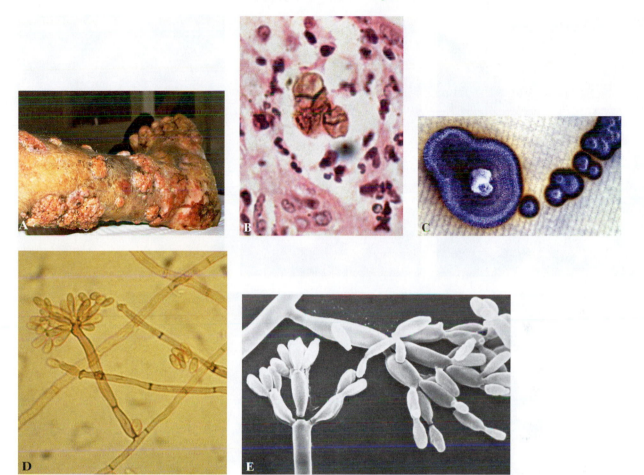

Figure 272

Chromoblastomycosis. (A) Clinical appearance. (B) A skin biopsy showing muriform cells (an intermediate form between a yeast and a mold) stained by the periodic acid-schiff technique. (C) *Fonsecaea pedrosoi* on Sabouraud's dextrose agar. (D) Microscopic view of *F. pedrosoi* showing septate hyphae and the typical fonsecaea-type conidia formation. (E) A scanning electron micrograph of *F. pedrosoi* conidia (oval cells). (From F. Queiroz-Telles, et al. *Internat. J. Dermat.* 31(1992):805–12.)

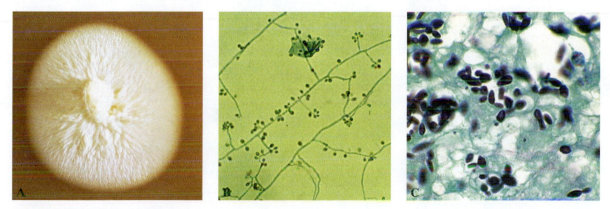

Figure 273

Sporotrichum (Sporothrix) schenckii. (A) Mycelium on Sabouraud's dextrose agar. (B) Microscopic view of cells grown at 25°C and stained with lactophenol cotton blue. Note the flowerlike arrangement of the oval conidia (400×). (From S.E. How and T.-Y. Chuang. *Cutis* 48(1991):193–96.) (C) Yeast phase in a Gomori-methenamine-silver nitrate stained tissue specimen (750×). (From R.C. Neafie and A.M. Marty. *Clin. Microbiol. Revs.* 6(1993):34–56.)

Systemic Mycoses

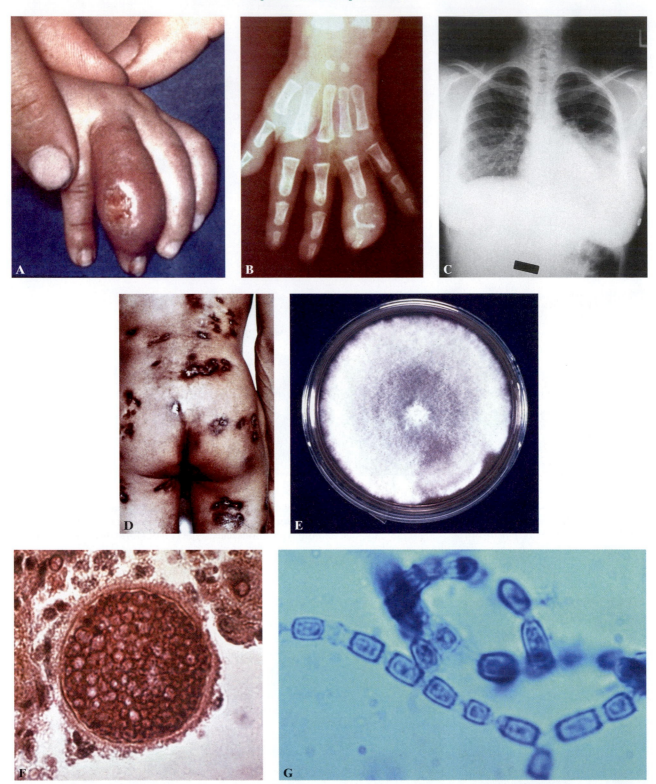

Figure 274

Coccidioides immitis. (A) Finger infection. (B) X-ray of infected finger. (C) A chest x-ray showing an infiltrate in the lower left lobe of a patient with the disease. (From S.A. Westphal and G.A. Sarosi, *CID* 18(1994):974–78.) (D) Surface lesions in a case of disseminated disease. (E) Mycelium on Sabouraud's dextrose agar. (F) A spherule containing endospores in tissue. (G) Stained arthroconidia from a culture grown at 25°C.

Systemic Mycosis *(Continued)*

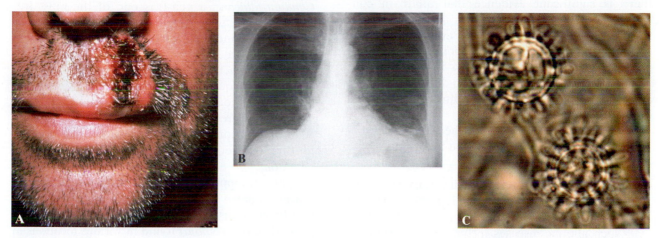

Figure 275
Disseminated *Histoplasma capsulatum.* (A) Skin infection. (B) Lung involvement. (C) Microscopic view from a 25°C culture grown on Sabouraud's dextrose agar. Note the large, thick-walled round tuberculated macroconidia. (From P.R. Cohen, J.C. Held, M.E. Grossman, M.J. Ross, and D.N. Silvers. *Internat. J. Dermatol.* 30(1991):104.)

system has been compromised or weakened by a burn or instrumentation, opportunists can cause serious and often fatal disease. Opportunists take advantage of the opportunity provided by an immunodeficient host's weakened state. Opportunistic infections (OIs) cause disease with higher frequency or greater severity, or both, among HIV-infected persons than among the general population, presumably because of immunosuppression. More than 100 microorganisms, including bacteria, fungi, protozoa, and viruses, cause OIs. Such infections are associated with considerable morbidity and mortality. Three examples of fungal opportunists will be described.

Aspergillus species

Infections with fungus pathogens have emerged as an increasing risk faced by immunocompromised individuals and patients under sustained immunosuppression. *Aspergillus* species account for most of these infections, and in particular *A. fumigatus* (Figure 276).

Aspergillosis is a fungus disease that occurs worldwide in distribution and can affect all ages and sexes. The disease can be broadly defined to include any infection caused by a species of *Aspergillus.* Of the hundreds of recognized species of aspergilli, only a little over 20 have been verified to cause human infections, and only five of these consistently and regularly are encountered as pathogens.

Transmission. Aspergilli generally do not have elaborate mechanisms for releasing their conidia

(spores) into the air. Spreading of the conidia relies on disturbances and strong air currents. Once in the air, the small size of conidia enables them to float, and to be airborne both indoors and outdoors.

Morphology and Cultural Properties. Aspergilli are rapidly growing molds producing fluffy, variously colored colonies known as mycelia (singular, mycelium) within 2 days on laboratory media or foodstuffs (Figure 259A). *Aspergillus* species are identified on the basis of differences in the structure of conidiophores and the arrangement of conidia (Figures 259B and C).

Pathology and Clinical Features. The most common *Aspergillus* species causing invasive disease include *A. fumigatus, A. niger, A. terreus,* and *A. nidulans.* In addition, the most common sources of **allergens** (allergy-causing antigens) include *A. fumigatus,* and *A. clavatus.* Some species of *Aspergillus* and of other genera produce **mycotoxins** (toxic substances).

One of the most important of these poisonous products are the **aflatoxins.** *Aspergillus flavus,* which often grows on stored cereal products, nuts (particularly peanuts), and animal feed, produces aflatoxins. (The first part of the name for the mycotoxin is derived by using the *A* from the genus, and the *fla* from the species designations, respectively.) The real and potential dangers of **aflatoxin poisoning** were dramatically shown in 1960 by the occurrence of large-scale trout poisoning in commercial fish hatcheries, and concurrent outbreaks of turkey X disease in

England. Although the risk to humans is unknown, there is significant evidence that aflatoxin contributes to liver cirrhosis and liver cancer in parts of the world, such as India and portions of Africa. It should be noted that aflatoxin poisoning does not produce immediate effects upon ingestion. These poisons are stored in the liver, and depending on the amounts ingested may not produce disease for a long time.

The specific diseases caused by the aspergilli, and in particular *A. fumigatus,* include the following:

1. allergic reactions and complications caused by the presence of conidia or temporary growth of the fungi in the mouth, nose and other orifices; such allergic diseases include allergic sinusitis (Figure 276),
2. colonization without extension in existing cavities and debilitated tissues within the respiratory tract, often called an **aspergilloma** or **fungus ball**;
3. superficial infection of the skin, nasal sinuses, external ear canal, and more rarely, burn areas after sloughing of tissue, nails, and other body sites;
4. invasive, granulomatous, necrotizing infection of the lungs; and
5. systemic and fatal disseminated disease.

Diagnosis. Establishing a laboratory diagnosis of invasive infection due to *Aspergillus* species can pose difficulties. Convincing evidence includes the isolation and growth of *Aspergillus* species from specimens obtained from normally sterile sites; also, histopathological findings showing the presence of fungal parts, such as typical branching hyphae and spores, clearly invading host tissue. Specimens of value for the isolation of aspergilli include bronchoalveolar lavage fluid (BAL), sputum, and nasal swabs. Bronchoscopy may also provide a suitable specimen for culture.

Pathogenic aspergilli generally grow easily and relatively rapidly on routine media preparations used for fungal isolations. Only pathogens are capable of growing at 35° to 37° C. In addition, *A. fumigatus* has the unique property of being able to grow and reproduce at temperatures of 50° C or higher.

Serological testing for the detection of circulating *Aspergillus* antigens or antibodies can be very helpful in the diagnosis. There are over 20 diagnostic procedures to detect anti-*Aspergillus* antibodies. Enzyme-linked immunosorbent assays (ELISAs) are used for the detection of circulating *Aspergillus* antigens.

Candida albicans

Candida albicans, a yeast, is one of the most frequently found fungal opportunists as well as one of the most common causes of several serious fungal diseases (Figure 277).

Although bacteria are the predominant source of hospital-acquired (*nosocomial*) infections, fungal infections continue to increase dramatically. *Candida* species are the agents responsible for almost 80% of nosocomial fungal infections and are the fourth-most-frequent cause of all such disseminated infections.

Among the approximately 150 fungal species known to cause human infections, *C. albicans* is one of the rare fungi that is a member of the normal microbial population (*microbiota*) found in various locations in the human body. The yeast is carried by about half of the human population.

Transmission. *Candida albicans* is a normal inhabitant of the oral cavity, the lower gastrointestinal tract, and the female genitalia. Most infections are caused by an organism in an individual's own microbiota. However, infections can also result from direct mucosal contact with lesions in others, as in cases of sexual contact or by the introduction of *C. albicans* with invasive procedures.

Morphology and Cultural Properties. *Candida albicans* grows in either of two basic forms. One is a yeast form with blastoconidia (Figure 244D) and hyphal elements, which include pseudohyphae and true hyphae. The second form grows on specialized media. This form develops thick-walled **chlamydoconidia**, or chlamydospores (Figure 278B), which distinguish it from other *Candida* species. The organism grows well on most standard media (Figure 278A).

All morphological forms, specifically yeasts, pseudohyphae and true hyphae, have been found in infected tissues. However, hyphal cells are thought to be the more invasive.

Pathology and Clinical Features. *Candida* infections involve the skin or mucous membranes, or spread to body organs (Figure 277C). In skin infections, *Candida* invades the skin and nails, producing effects similar to those of dermatophytes (Figure 277D). In mucous membrane infections of the mouth, the yeast may produce a condition known as **thrush**. Involvement of vaginal mucosal surfaces, a frequent site of *Candida* infection, is known as **vulvovaginitis**. Individuals with immunologic defects, such as persons with AIDS, may experience various mucocutaneous types of candidiasis (Figure 277B).

Opportunistic Mycoses

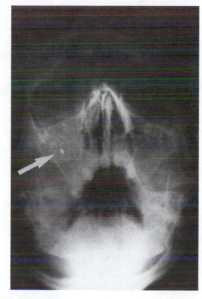

Figure 276
Aspergillus sinus infection. An x-ray showing the location
of a fungal mass.

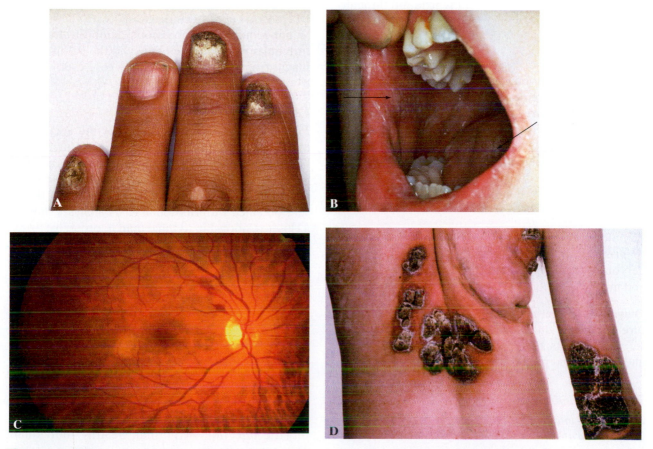

Figure 277
Candida albicans infections. (A) Candidiasis of the nails and surrounding tissues. (B) Mucocutaneous *Candida* infection of
the mouth. (Courtesy of C.J. Kirkpatrick, M.D., President, Innovative Therapeutics, Inc.) (C) Retinal lesions found in a
patient with candidal retinitis. (D) Extensive mucocutaneous candidiasis involving a large surface on the back (From
Dr. P. Phillips. *Revs. Inf. Dis.* 9:587, 1987.)

Death from an associated *Candida* bloodstream infections, or candidemia, is one of the most common of all bloodstream infections. Despite the emergence of other *Candida* species, *C. albicans* remains the leading cause of life-threatening, disseminated *Candida* infections.

Diagnosis. Superficial *C. albicans* infections provide adequate material with which to diagnose. Microscopic examination of KOH preparations or Gram stains will show numerous budding yeast cells (Figure 278C). Cultures of various types of specimens, including direct biopsy and bronchoalveolar lavage (BAL), are of value. Biochemical tests also are used to distinguish among *Candida* species.

Filobasidiella (Cryptococcus) neoformans

Filobasidiella (Cryptococcus) neoformans is the causative agent of cryptococcosis. The incidence of this disease has increased substantially in recent years because of the large number of AIDS cases in the world. One-half of the patients with cryptococcosis are infected with HIV.

Transmission. The disease agent is acquired through aerosols. Association with infected pigeons and bird-related habitats is a major factor.

Morphology. In the tissues of an infected host, the yeast occurs as a blastoconidium with a capsule (Figure 279A). In culture, the fungus produces mucoid colonies (Figure 280A) that contain encapsulated yeast cells, usually revealed by India ink preparations (Figures 251 and 280B). Special selective and differential media also are used for isolation purposes. Bird seed agar causes the yeast to produce the pigment melanin, a distinguishing feature (Figures 279B and 280A).

Pathology. Six general clinical types of cryptococcosis are known. They can involve the lungs, central nervous system, gastrointestinal system, bone, skin, and mucous membranes.

Diagnosis. Direct microscopic examination and culture of appropriate clinical specimens (Figure 241a), serological tests to demonstrate capsular material, and nucleic acid probes are used for diagnosis.

Penicillium marneffei

Penicillium marneffei is an emerging pathogenic fungus known to cause a fatal systemic mycosis in individuals infected with human immunodeficiency virus (Figure 280A). The fungus is a found in tropical Asian countries.

Transmission. Infection is acquired by inhalation of the fungus. It is likely that rodents, such as bamboo rats, and soil are sources of *P. marneffei*.

Morphology. *P. marneffei* has a unique feature relative to other penicillia. It is the only known *Penicillium* species that exhibits temperature-dependent dimorphic growth. At temperatures below 37° C., the fungus grows as mycelia consisting of sepatate hyphae, bearing conidiophores (Figure 280B). At 37° C, on artificial media or in human tissue, the fungus grows in a yeastlike form. Yeast cells represent the parasitic stage of the pathogen (Figure 280C).

Pathology. *P. marneffei* appears to be a primary pulmonary pathogen that spreads to other internal organs via the circulatory system. The severity of the disease depends upon the immunological condition of the infected individual.

Diagnosis. Laboratory diagnosis of *P. marneffei* infection requires the microscopic demonstration of intracellular yeast cells in tissue and the culture of the fungus from clinical specimens (Figure 280B).

Pneumocystis carinii

Pneumocystis infection is caused by the unclassified microorganism, *Pneumocystis carinii*. Currently, it has not been established as to whether the organism is a fungus or a protozoan. *Pneumocystis carinii* is included here as a probable fungus because of the results obtained with ribosomal RNA analyses showing a closer relationship to fungi than to protozoa.

Transmission. Infection is usually acquired by inhalation of droplets containing *P. carinii*. The organism is found worldwide. It is unlikely that lower animals are significant sources of infection.

Morphology. *Pneumocystis carinii* develops and reproduces going from cyst to trophozoite (active metabolizing form) to cyst. Both forms can be found in the lungs (Figure 282A).

Pathology. Pneumocystis does not produce symptomatic infection in normal, healthy individuals. The organism can exist in an inactive or latent

Opportunistic Mycoses *(Continued)*

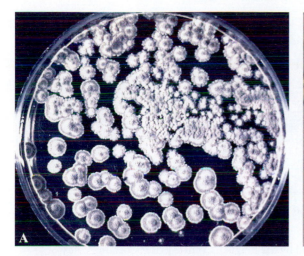
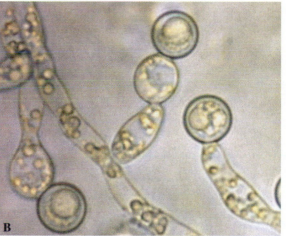
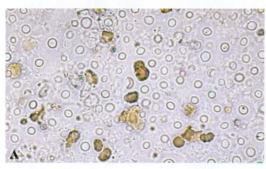

Figure 278

Microscopic and cultural features of *Candida albicans.* (A) *C. albicans* on Sabouraud's dextrose agar. (B) In the body and in culture *C. albicans* may develop threadlike hyphae and the spherical chlamydoconidia (chlamydospores). (C) A Gram stain of a vaginal thrush specimen showing the oval shape of the yeast form of *C. albicans.*

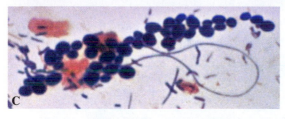

Figure 279

Filobasidiella (Cryptococcus) neoformans. (A) A microscopic view showing numerous, round, cryptococci obtained from a patient with meningitis. (B) Orange to brown colonies of *F. neoformans* on *Guizotia abyssinica* (bird seed) creatinine agar. (Courtesy of Dr. F. Staib, Former Chief, Mycology Unit, Robert Koch Institute.)

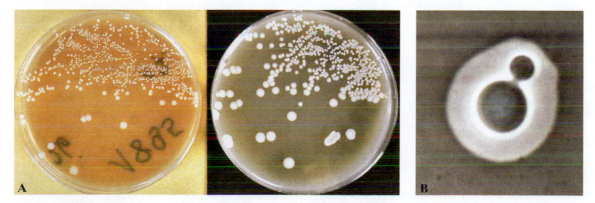

Figure 280

(A) *Filobasidiella neoformans* (small orange to brown) colonies on *Guizotia abyssinica* creatinine agar (left). The nonpigmented colonies are *Candida albicans.* The plate on the right primarily contains *C. albicans.* (B) A microscopic view of *F. neoformans* showing a bud and a surrounding halo-like capsule. (From F. Staib and M. Seibold. *Mycoses.* 31(1988):175–86.)

Opportunistic Mycoses *(Continued)*

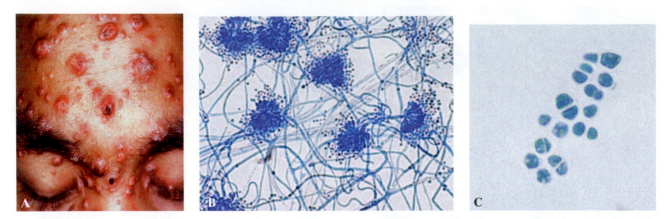

Figure 281

Penicillium marneffei, an emerging pathogen. (A) Skin lesions caused by the fungus. (Courtesy of Associate Professor Dr. Siri Chiewchanvit.) (B) Conidiophores and conidia. Note the wavy nature of the hyphae (original magnification 100×.). (C) The yeast form of *P. marneffei*. (Figures 280B and Courtesy of Prof. Dr. Nongnuch Vanittanakon.)

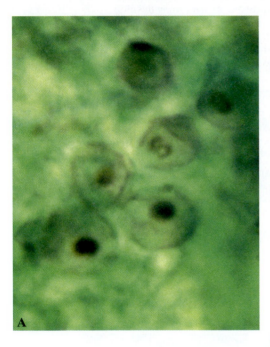

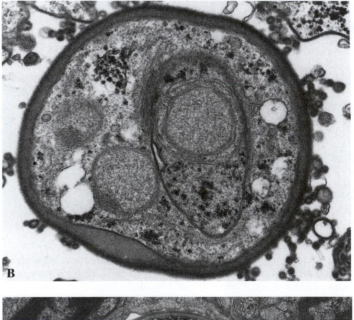

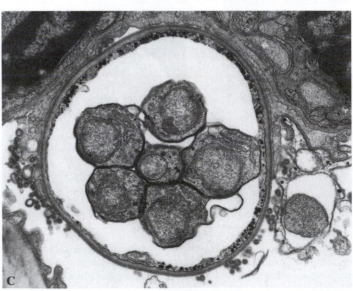

Figure 282

Microscopic views of *Pneumocystis carinii*. (A) A bronchial wash stained by the Gomori-methenamine-silver nitrate method and submicroscopic (Courtesy of Dr. L. Alpert, Pathology Department, The Sir Mortimer B. David Jewish General Hospital.) (B) A transmission electron micrograph showing a large active feeding form of *Pneumocystis* within lung tissue (original magnification, 52,000×). (C) A thin section revealing the interior of a mature cyst containing potential infective units (Original magnification, 38,750×). (Courtesy of Dr. M. Goheen.)

state unless the host becomes immunocompromised, as could be the case with premature infants, AIDS patients, and individuals receiving immunosuppressive treatment for tissue transplants, cancer, and other related conditions.

Symptoms of infection include fever, dry cough, difficulty in breathing, and progressive respiratory failure.

Diagnosis. *Pneumocystis carinii* infection diagnosis is generally made by radiography and demonstration of the organism in sputum or lung tissue (Figure 282A).

Algology (Protists—The Algae)

All we know is still infinitely less than all that still remains unknown.

—William Harvey, De Mota Cordis, dedication

In the early twentieth century, most aquatic plant life was lumped together and collectively referred to as algae. By the 1920s it became clear that algae contained several related yet distinct groups of microscopic and massive forms of life (Figures 283A through 291). The organisms once grouped under aquatic plants, or algae, are classified as shown in Table 16.

The five-kingdom classification system established the kingdom of Protista. According to this system the golden algae, diatoms, euglenoids, and dinoflagellates are placed into the Protista, and all others are placed into the plant kingdom. The general features of algae are compared in Table 16.

▲ Properties and Distribution

Algae are photosynthetic forms of life. Most algal forms are free-floating and free-living. Many of the single-celled species are suspended in vast numbers in various bodies of water. Occasional spurts of growth known as *algal blooms* (Figure 283A, and Figure 283B) may disrupt aquatic communities by increased accumulations of their waste products (Figure 283B).

Collectively, the photosynthetic types are *phytoplankton*, the food producers which form the basis of nearly all of the food webs in aquatic environments.

Table 16 Properties of Major Groups of Algae

Algal Group	Common Name	Morphology	Pigments	Major Habitats
Chlorophyta	Green algae	Unicellular to leafy	Chlorophylls *a* and *b*	Freshwater, soils, a few marine
Euglenophyta	Euglenoids	Unicellular, flagellated	Chlorophylls *a* and *b*	Fresh water, a few marine
Chrysophyta	Golden brown algae, diatoms	Unicellular	Chlorophylls *a, c,* and *e*	Fresh water, marine, soil
Phaeophyta	Brown algae	Filamentous to leafy, occasionally massive and plantlike	Chlorophylls *a* and *c,* xanthophylls	Marine
Pyrrophyta	Dinoflagellates	Unicellular, flagellated	Chlorophylls *a* and *c*	Fresh water, marine
Rhodophyta	Red algae	Unicellular, filamentous to leafy	Chlorophylls *a* and *d,* phycocyanin, phycoerythrin	Marine

Algology (Protists–The Algae)

Figure 283

Distribution of algae. (A) The light green appearance of an algal bloom on a waterway near a small town in Holland (arrow). (B) The result of an algal bloom. Note the dead fish in the midst of the accumulated algae. (C) Green algae occupying (colonizing) the microscopic air space system a few millimeters below the surface of a porous desert rock. (D) The presence of green algae on the surfaces of several stones at Stonehenge, a prehistoric ruin on the plain north of Salisbury, England. (E) Certain sea anemones (green) in an endosymbiotic relationship with green algae. Nonsymbiotic anemones are white.

Figure 284

A 2,000 year-old medusa head covered with algae. This sculpture, which is normally upside-down, was uncovered at the base of one pillar in the oldest underground reservoir in Istanbul, Turkey. (Courtesy of Reneé. Wisztreich.)

▲ Chlorophyta (Green Algae)

The green algae bear the greatest structural and biochemical resemblance of all the algae to plants and may be their nearest relatives (Figure 285). This group shows more diversity than other algal groups. They include microscopic colonial forms (Figure 285D), filamentous forms (Figure 285B), desmids (Figure 285E), and the larger sea lettuce (Figure 285A). Green algae have both asexual and sexual types of reproduction. Some, such as the filamentous green alga *Spirogyra,* form conjugation tubes to allow the cellular contents from one cell to pass into another (Figure 285C).

▲ Euglenophyta (Euglenoids)

Euglenoids can be found in fresh water, stagnant ponds, and lakes. These microorganisms contain a large number of organelles, including chloroplasts, contractile vacuoles, and light-sensitive eyespots (Figure 286A). *Euglena* also have a type of firm but flexible outer layer, the pellicle and a flagellum. *Euglena* species also can be grown in artificial media and in the presence of a light source (Figure 286B).

▲ Chrysophyta (Golden Brown Algae and Diatoms)

Many of the known species of golden algae are single-celled, are photosynthetic, and have scales or skeletal elements of silica. Many are important in freshwater habitats.

Diatoms are plentiful in aquatic habitats, and many are photosynthetic. The structure and organization of these algae are quite distinctive (Figure 287A). Diatoms are usually classified on the basis of the shape, symmetry, and structure of their cell walls, called frustules. The walls of diatoms consist of two halves, referred to as **valves**, which fit together much like the parts of a pillbox. The larger halve is the *epivalve,* and the lower one is the *hypovalve.* Organisms with circular valves are called **centric diatoms**, while those having a boat-shaped structure are known as **pinnate diatoms** (Figure 288). Various-sized chambers (*loculi*) are located internally on the margins of the valves. The structural details of the cell walls of diatoms help to distinguish them from certain green algae. These include holes (*puncta*) arranged in rows (*striae*) or ridges. Various substances move in and out from the plasma membrane through the puncta. In addition, the valves of some diatoms, which are usually supported by riblike supports, also have a long, narrow opening called a *raphe.*

Diatoms are chemically unique in that their cell walls contain large concentrations of silicates. These substances are the basic components of glass, granite, and sand. Diatoms continually absorb silicates and deposit them in their cell walls. When a diatom dies, the silica in the cell wall begins to dissolve rapidly. However, under favorable conditions, these glassy structures accumulate and form deposits of fossil diatoms called **diatomaceous earth**. Such deposits have been gathering for thousands of years, and in some parts of the world deposits 900 meters thick have been found. Diatom remains have commercial value and are used as abrasives, filters, and insulating materials.

▲ Phaeophyta (Brown Algae)

Brown algae are found in temperate or cool marine waters. Depending on their photosynthetic pigments, they may appear as olive-green, golden, or dark brown (Figure 289). Brown algae range in size from microscopic filamentous forms to giant kelps. Some of these organisms also exhibit tissue differentiation, including anchoring structures (**holdfasts**), stemlike parts (**stipes**), and photosynthetic, leaf-shaped parts (**blades**).

▲ Pyrrophyta (Dinoflagellates)

The dinoflagellates are marine, or sometimes freshwater, phytoplankton. Some have flagella that fit in grooves between stiff cellulose plates at the cell surface (Figure 290). Depending on their photosynthetic pigments dinoflagellates may appear yellow-green, green, brown, blue, or red. The red forms occasionally undergo population explosions and color seas red or brown. The well-known red tide is an example.

▲ Rhodophyta (Red Algae)

The red algae are especially abundant in warm currents or tropical seas and at great depths (Figure 291). Some are single-celled; others are colonial. Most species are multicellular with a filamentous, often branching organization. The cell walls of red algae incorporate a mucus-type material, which gives them a flexible, slippery texture. Agar, the polysaccharide

Algology (Protists–The Algae) *(Continued)*

Figure 285

Representative green algae. (A) Sea lettuces (*Ulva lactuca*) growing in a tidal pool. (B) *Spirogyra,* an unbranched green alga, showing the typical helical (springlike) arrangements of chloroplasts (original magnification 1,000×). (C) Conjugation (connecting) bridges of *Spirogyra* formed during sexual reproduction. (D) A *Volvox* colony consisting of smaller interdependent cells. (E) Desmids. *Micrasterias crux-melitensis* after cell division (1,100×).

Figure 286

Euglenoids. (A) Several *Euglena,* showing numerous chloroplasts. These organisms have a flexible outer plasma membrane (pellicle) that allows them to easily change shape (450×). (B) The dark green growth of *Euglena* in *Euglena* broth. An uninocluated preparation is shown for comparison.

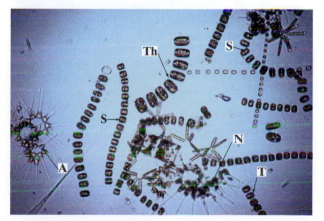

Figure 287

A light micrograph of a marine phytoplankton community including of various diatom genera such as *Asterionella* (**A**), *Navicula* (**N**), *Rhizosolenia* (**R**) *Skeletonema* (**S**), *Thalassionema* (**T**), and *Thalassiosira* (**Th**).

Algology (Protists–The Algae) *(Continued)*

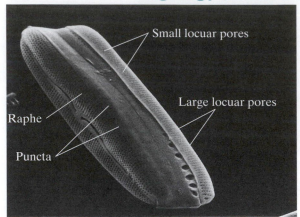

Figure 288
A scanning electron micrograph showing the external girdle view of the entire frustule (cell wall). Distinct puncta (P), the raphe (R), and large locular pores (L) also are shown.

Figure 289
Representatives of brown algae. (A) Rockweed, *Fucus,* an alga that grows on rocks in cold ocean waters. (B) A closer view, showing the branching, the typical air (round) bladders, the long stemlike stipes, and holdfast cells (H) anchoring the alga to a rock.

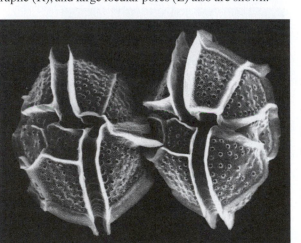

Figure 290
A scanning electron micrograph of the dinoflagellate *Pyrodinium bahamense.* (Courtesy of Dr. M. Ellengard, Botanical Institute, University of Copenhagen.)

Figure 291
Red algae are so named because of their red pigment, phycoerythrin. These organisms mainly are found in marine environments.

Figure 292
An example of monument stone decay as a consequence of lichen encrustation.

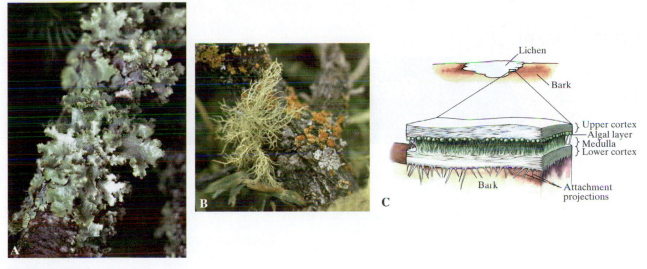

Figure 293
Lichens (A) fruticose (leaflike) lichen. (B) Several lichen species on a tree branch: the foliose form called "old man's beard" and orange and white crustose types. (C) Crustose lichen organization. This lichen, as well as similar crustose forms, consists of four layers: (1) the upper, protective cortex formed by a thick layer of fungal hyphae; (2) a layer in which algal cells are distributed among thin-walled hyphae; (3) the medulla, a thick hyphal layer; and (4) a lower cortex containing hyphae with fine projections that attach the lichen to the rock surface (substrate)

used for solidifying bacteriological and other media, is extracted from certain species of red alga.

▲ Lichens

Certain algal species can enter into a beneficial relationship with fungi to form a structure known as a **lichen** (Figures 292 and 293). The fungal hypha produces a tightly woven mycelium, which houses and protects its partner the alga. The fungi absorb environmental water and minerals, which the algae use for photosynthesis. Many different arrangements and colors are displayed by lichens. However, three major types are recognized: **crustose** (crustlike) lichens, found on rocks or on bark as irregular flat patches (Figures 292–293); **foliose** (leaflike or fibrous) lichens, curled and frequently having rootlike structures for attachment; and **fruticose** (shrublike) lichens, which are highly branched and either hang from different tree parts or originate in the soil (Figure 292A). Figure 293C shows the organization of crustose lichens.

Since lichens are capable of living in environments that cannot support most other forms of life (Figure 292), it may be that the association between the involved fungi and algae is formed as a matter of utmost necessity for their survival. Lichens survive in these environments partly by being able to dry or freeze and thereby reach a condition that may be considered to be a form of suspended animation. Once conditions improve, lichens recover and continue with their normal activities.

Although lichens are remarkably resistant to drying, cold, heat, and other environmental conditions, they are extremely sensitive to air pollutants and quickly disappear from heavily polluted urban regions. Apparently, lichens absorb pollutants from rain water. Because they have no means of excreting them, lethal concentrations of toxic air pollutants gradually build up.

Lichens are ecologically and economically important. They serve as food for arctic caribou and reindeer and are used in the preparation of litmus paper, a well-known indicator in chemistry and related scientific areas. Certain lichens also have been found capable of producing antibiotics, which may have potential in the treatment of certain bacterial and fungal diseases. Another role of lichens that is sometimes overlooked is their contribution to the organic content of soil. Quite often lichen activity aids in the decomposition of rocks. The decaying remains of dead lichens become intermixed with small rock particles, providing nutrients and a foundation for plant development.

Protozoology (Protists—The Protozoa)

Do there exist many worlds or is there but a single world? This is one of the most noble and exalted questions in the study of nature.

—*St. Albertus Magnus*

The 25,000 or so species of protozoa were formerly placed into the animal kingdom. Today, these animal-like microorganisms are considered to be members of the kingdom Protista and are grouped into phyla that differ from one another in several respects, including their means of movement (See Table 18).

Protozoa (singular, *protozoon*) are found in many different environments. Some are present in bodies of water, where they play an important role in the food webs of natural communities. Others have mutually beneficial (**symbiotic**) relationships with higher forms of animal life or with other microorganisms.

Protozoa are also known for their harmful activities. African sleeping sickness, amebic dysentery, malaria, and toxoplasmosis are but a few of the human diseases associated with these microorganisms. Several protozoa also infect wild and domestically important animals. In severe cases, infected hosts are crippled, disfigured, and eventually die. Protozoan protists of the same species are similar to one another, and none is specialized solely for feeding. Most of the individuals in a population of protists are produced by simple cell division of the parent, although sexual reproduction by the mating of two individuals also does occur.

▲ Structures and Functions

Protozoa are not functionally simple microorganisms. Essential functions require a division of labor of tasks involved with movement, obtaining and using nutrients, excretion and osmoregulation, reproduction, and protection. Table 17 lists and briefly describes the organelles involved in these tasks (Figures 294, 295, 298–300, 302, and 303 show several of them.)

Table 17 Protozoan Structures

Structure	Function(s)
Cilia	Movement
Contractile vacuole	Excretion and osmoregulation
Cytopage	Elimination of indigestible material
Cytosome (mouth)	Food gathering
Flagella	Movement, sense reception
Food vacuoles	Digestion
Macronucleus	Regulation of metabolism and development
Micronucleus	Overall cellular control and regulation of reproductive process
Pellicle (strong cellular covering)	Protection against chemicals, drying, and mechanical injury
Pseudopodium	Movement
Tentacle	Protection and trapping of food
Trichocyst	Defense and food capture
Undulating membrane (formed with plasma membrane and flagellum)	Movement

▲ Trophozoites and Cysts

Several pathogenic protozoan species as well as free-living ones found in bodies of water appear in a normal, active feeding form known as the **trophozoite** (Figures 294A, 297A, and 299A). Some protozoa form **cysts** to overcome the effects of various chemicals, food deficiencies, temperature or pH changes, and other harsh environmental factors. This stage also serves as a means of reproduction and for spreading the protozoa (Figures 297C and 299C). Trophozoite and cysts stages also are important to the diagnosis of protozoan diseases.

▲ Classification

Various properties of protozoa are used in their classification. Among them are method of obtaining nutrients; method of reproduction; cellular organization, structure, and function; biochemical analyses of nucleic acids and proteins from specific cellular structures; and organelles of locomotion (Table 18).

All parasitic and either medically or agriculturally important protozoa are placed into specific phyla, including the Sarcomastigophora, Ciliophora, Apicomplexa, and Microsporidia (Table 18).

Table 18 Characteristics of the Major Groups of Protozoa

Phylum (Subphylum)	Common Name	Means of Movement	Habitats
Sarcodina	Amoebae	Pseudopodia	Fresh water and marine, some parasitic
Mastigophora	Flagellates	Flagella	Fresh water, some parasitic
Ciliophora	Ciliates	Cilia	Fresh water and marine, some parasitic
Apicomplexa	Sporozoa	Generally nonmotile except for certain sex cells	Mainly animal parasites
Microsporidia	Microspora	Nonmotile	Intracellular parasites

Representative Protozoa, Their Distinctive Properties, and Selected Disease States

▲ Sarcomastigophora

The phylum Sarcomastigophora includes the two subphyla, **Sarcodina** and **Mastigophora**, and contains many important disease-causing species.

Sarcodina

Members of this subphylum, unlike other protozoa, have no definite shape and move by forming pseudopodia (Figure 294A). Their simple cells change form as they move. The nucleus, contractile vacuole, and food-containing vacuoles shift about with the protozoon as it moves.

The many kinds of protozoa that constitute the Sarcodina are found in all bodies of water. One of the most interesting is the large subgroup called the **foraminifera** (Figure 294C). These microorganisms form shells made of lime or of substances such as sand from the surrounding waters. Approximately 1,800 species, found mainly in salt water, are known.

Other members of the Sarcodina, the parasitic amoebae, may be found in most kinds of animals. The most important forms to attack humans include *Entamoeba histolytica* (Figure 295), *Acanthamoeba* (Figure 297), and *Naegleria*. *Entamoeba coli* (Figure 297C), another amoeba, is nonpathogenic.

Entamoeba histolytica

Entamoeba histolytica is one of six parasitic amoebae of the genus *Entamoeba* known to infect humans. Infections occur worldwide but are most prevalent in the tropics. Humans are the major reservoir of infection. *Entamoeba histolytica* causes amebiasis, amebic liver abscess (Figure 296), and cutaneous amebiasis.

Transmission and Life Cycle. Ingestion of cyst-contaminated food and drink and fecal-oral contact are the most common means of infection. The use of human feces for fertilizer is also an important source of infection. Flies and cockroaches can also spread the pathogen.

Morphology. *Entamoeba histolytica* is an enteric pathogen that exists in either a trophozoite or cyst

Protozoology (Protists–The Protozoa)
Sarcomastigophora
Sarcodina

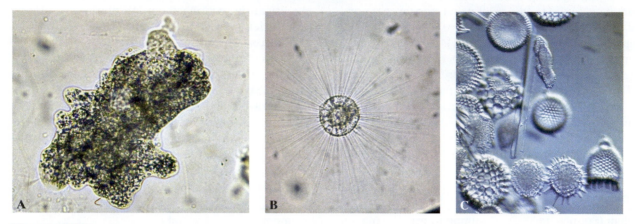

Figure 294

Examples of members of Sarcodina. (A) A temporary wet-mount of an *Amoeba* species trophozoite (Original magnification 100×). Note the presence of the extended pseudopodia. (B) *Actinophrys*. (C) Examples of foraminifera.

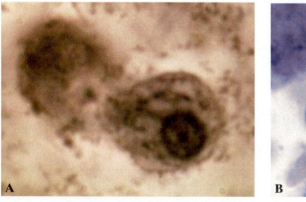

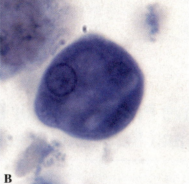

Figure 295

Microscopic views of *Entamoeba histolytica*. (A) A trophozoite emerging from a cyst in a stool specimen. (B) A stained preparation showing the typical wagon wheel nucleus.

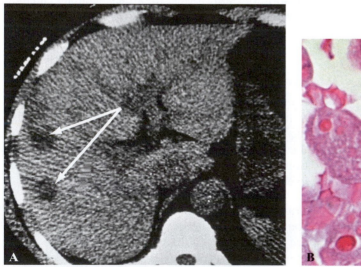

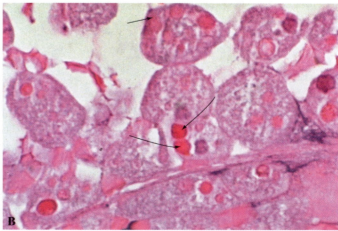

Figure 296

Clinical and diagnostic features of amebiasis. (A) A CAT scan showing liver destruction (arrows). (B) A stained preparation of the large intestine showing oval to round amoebae in the tissue. (Courtesy of Dr. L. Alpert, Pathology Department, The Sir Mortimer B. David Jewish General Hospital.)

stage. The trophozoite contains a single nucleus with a small, distinctive central body, known as a **karyosome**, and nuclear material distributed evenly in the nuclear envelope (Figure 295B). The resistant cyst can contain several similar nuclei and some deeply staining bundles of RNA called **chromatoidal bodies**.

Diagnosis. Diagnosis of infections depends primarily on examinations for *E. histolytica* cysts in stools, tissue biopsy for trophozoites, and serological tests.

Acanthamoeba

Acanthamoeba species (Figures 297A and B) cause several diseases, including a form of amebic encephalitis, keratitis (corneal ulcers), and skin ulcers. The organism is found worldwide.

Transmission. Infections may be acquired by inhalation or by direct contact with contaminated soil, water, or solutions contaminated with the organism. Keratitis usually is seen in contact lens users who have used nonsterile saline solutions for cleaning purposes.

Morphology. *Acanthamoeba* species occur in trophozoite and cyst stages.

Diagnosis. *Acanthamoeba* infections are diagnosed by finding the organism in tissue biopsies (Figure 297A), in scrapings, or in freshly drawn cerebrospinal fluid.

Mastigophora

The Mastigophora, or flagellates, are mostly unicellular and usually possess at least one flagellum at some stage of their life cycle (Figure 299A). These flagella are used for locomotion, for obtaining food, and as sense receptors.

The flagellates include more than half of the living species of protozoa and are an extremely variable group. They are believed to be the oldest of the eukaryotic organisms and the ancestors of the other major forms of life.

Free-living Mastigophora are common in both fresh and salt water. Many others inhabit the soil or the intestinal tracts of some animals. Some Mastigophora are free-living, commensal, mutualistic, or parasitic. Several flagellates are parasitic and medically important. These include *Leishmania* species, *Giardia lamblia*, *Trichomonas*, and trypanosomes.

Leishmania

Leishmaniasis is disease complex caused by 17 different species of protozoa belonging to the genus *Leishmania*. The disease has a worldwide distribution, with significant numbers of cases occurring in Central and South America, southern Europe, North and East Africa, the Middle East, and the Indian subcontinent.

Leishmaniasis is a general term for diseases caused by *Leishmania* species. The clinical disease depends on the species involved and the immunologic status of the host. The spectrum of clinical conditions includes: local infections of the skin and bone (Figure 298), subcutaneous tissue, and regional lymph nodes; **mucocutaneous leishmaniasis**, or **espundia**, spreading infections of the oronasal mucosa; and **visceral leishmaniasis**, or **kala azar**, disseminated infection involving visceral organs of the mononuclear phagocyte system.

Resistance to some of the drugs used in treatment is threatening the successful control and potential elimination of leishmaniasis from various world regions.

Transmission and Life Cylce. *Leishmania* species are transmitted by sand flies of the genus *Lutzomyia* in the New World and *Phlebotomus* in the Old World. Animal reservoirs include wild rodents, sloths, and various meat-eating mammals. In vertebrates, *Leishmania* are obligate, intracellular parasites. They invade macrophages and reproduce in membranous structures that surround the protozoon during the invasion process.

Morphology. The first invading form is the **promastigote** (Figure 299A), which is a long flagellated cell that develops extracellularly inside the digestive tracts of sand flies. Once inside the membranous structure, the promastigote changes into an **amastigote** (Figure 299B), which is a round, nonmotile form that multiplies rapidly. Amastigotes are released when parasitized cells burst; they then spread to neighboring host cells. All *Leishmania* species in humans are morphologically similar.

Diagnosis. Diagnosis involves the demonstration of amastigotes in tissues such as bone marrow or white blood cells. Cultures of blood and bone marrow also are of value. Such cultures may require 4 weeks for the isolation of organisms. Serological tests, such as immunofluorescence, and nucleic acid probes are satisfactory for indirect diagnosis.

Giardia

Giardia is a binucleate flagellated protozoon that causes intestinal infection in mammals, birds, reptiles, and amphibians. *Giardia lamblia* (Figures 300) is the cause of giardiasis in humans.

Transmission and Life Cycle. Giardiasis is spread by contaminated water and food. *Giardia* has a simple life cycle, consisting of an infectious cyst (Figure 300B) and a vegetative trophozoite (Figures 300A and B). After a cyst is ingested, it excysts in the small intestine to form two trophozoites. Each trophozoite divides into two new cells in the small intestine, and

Protozoology (Protists–The Protozoa)
Sarcomastigophora
Sarcodina *(Continued)*

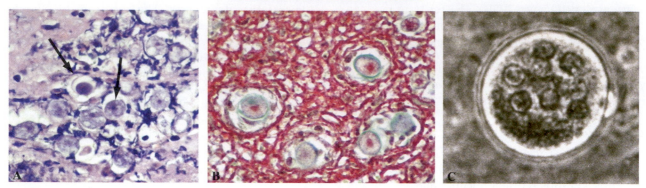

Figure 297

Other pathogenic and nonpathogenic amoebae. (A) Stained *Acanthamoeba* trophozoites at the base of a peptic ulcer. (B) Clusters of *Acanthamoeba* (40×). (Courtesy of Drs. K.L. Thamrasert, S. Khunamornpong, and N. Morakote.) (C) An *Entamoeba coli* multinucleated cyst.

Protozoology (Protists–The Protozoa)
Mastigophora

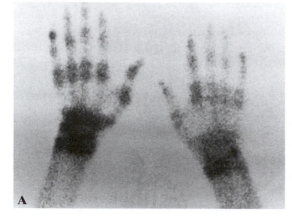

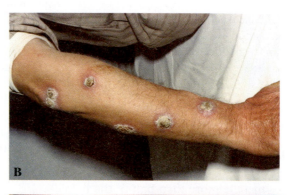

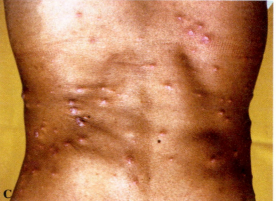

Figure 298

Leishmaniasis. (A) A bone scan of a patient with *Leishmania* infection. The darkened areas, which are the sites of infection, show increased radioactively labeled (radiopharmaceutical) material. (From A. Perell'o Roso. *CID.* 22(1996):1113–14). (B) A case of cutaenous leishmaniasis. Six ulcers covered with white crusts are shown. (Courtesy of Dr. M. Al-Taqi, University of Kawait.) (C) The appearance of disseminated leishmaniasis, a special form of cutaneous disease. (Courtesy of Professor Achilea L. Bittencourt, Bahia, Brazil.)

these forms are responsible for symptoms of the infection. Some of the trophozoites transform into cysts. The cycle is completed when such cysts are passed in the feces and ingested by another host.

Diagnosis. Microscopic examination of stools for cysts and trophozoites is usually the first diagnostic test. Cysts are found in most patients with giardiasis. Removal of intestinal fluid and tissue biopsies also are of value.

Trichomonas vaginalis

Trichomonas vaginalis is the cause of a common urogenital disease, trichomoniasis in women. Infected women usually experience a vaginitis, with a foul-smelling discharge and small, bleeding lesions.

Transmission and Life Cycle. Most cases of trichomoniasis are acquired by sexual intercourse. Fomites, such as shared towels, also are possible sources of the protozoon. Neonatal infections also have been reported.

Morphology. *Trichomonas vaginalis* is a pear-shaped organism (Figure 301). It has four anterior flagella and a fifth flagellum that forms the outer edge of an undulating membrane. The organism has only a trophozoite stage and divides in the urogenital tract by splitting in half.

Diagnosis. Demonstrating the organism in vaginal secretions, washings, or scrapings is generally used for diagnosis. Culture and immunofluorescence are also of value.

Trypanosoma brucei rhodesiense and T. b. gambiense

Human African trypanosomiasis (sleeping sickness) is a systemic and central nervous system infection caused by two geographically distinct forms of trypanosomes (Figure 302A). *Trypanosoma brucei gambiense* (West Africa) and *T. brucei rhodesiense* (East Africa). Infected individuals exhibit fever, headache, and enlarged lymph glands early in the disease. Menigoencephalitis develops later.

Transmission and Life Cycle. The trypanosomes are spread through the bites of tsetse flies, which are members of the genus *Glossina* (Figure 302B). Both domestic and game animals are major reservoirs.

Morphology. The form of the protozoon found in the human is the **trypomastigote**. It is slender and spindle-shaped measuring about 15 μm in length, and has an undulating membrane extending the full length of the protozoon and ending in a flagellum (Figure 302A). The protozoon does not have a cyst stage.

Diagnosis. In the early stages of the disease, demonstrating trypanosomes in blood and lymph node

specimens is most successful. Spinal fluid also is of value in cases of early central nervous system invasion. Culture, laboratory animal inoculations, and serological tests such as immunofluorescence are usually successful for diagnosis.

Trypanosoma cruzi

American trypanosomiasis, also known as Chagas' disease, is caused by *T. cruzi* (Figure 303A), a protozoon with a large animal reservoir in South and Central America.

Clinical conditions resulting from infection include local inflammation at the site of inoculation, severe inflammation of the heart tissue, encephalitis, and multi-organ involvement.

Transmission. *Trypanosoma cruzi* is usually transmitted to humans by blood-sucking triatomid (reduviid) bugs (Figure 303B) that deposit infectious feces at the site of a bite or on the mucosal surface. Transmission may also occur with the infection of contaminated human blood products or transplacentally from infected mother to fetus.

Morphology. In the blood *Trypanosoma cruzi* appears as short, stumpy cells with undulating membranes and free flagella. These forms are able to infect a variety of cells in which they become small, round protozoa lacking flagella.

Diagnosis. *Trypanosoma cruzi* can be easily demonstrated microscopically in stained blood smears in the early stages of the disease. Xenodiagnosis, a procedure in which laboratory-raised reduviid bugs feed on a suspected patient, also is used. The gastrointestinal contents of these arthropods are examined for organisms 1 to 2 months later. Polymerase chain reactions (PCR) and nucleic acid probes are expected to replace xenodiagnosis.

▲ Ciliophora

Members of the phylum Ciliophora are characterized by the presence of numerous cilia on their surfaces (Figure 304). These cilia function both in moving the organism and in obtaining food. The beating strokes of the cilia, as in the case of the *Paramecium* and others, cause the cell to revolve as it swims (Figure 304B). Ciliates also have a definite shape due to the presence of a sturdy but flexible outer covering, the pellicle (Figure 304C).

Of all the protozoa, the ciliates are the most specialized because they have organelles that carry out particular vital processes. These organelles include trichocysts, the macronucleus, the micronucleus, and contractile vacuoles. (see Table 17).

Protozoology (Protists–The Protozoa)
Mastigophora *(Continued)*

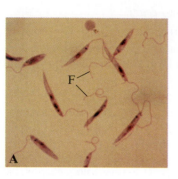

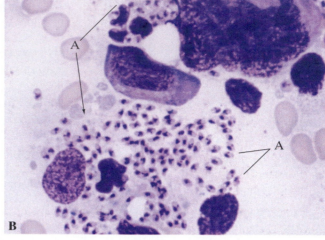

Figure 299

Leishmaniasis (A) *L. tropica* promastigotes flagella (F). (Courtesy of Dr. M. Al-Taqi, University of Kuwait.) (B) A microscopic view of amastigotes located within a mononuclear cell and externally The specimen was obtained from a case of visceral leishmaniasis. (Original magnification approximately 430×.) (From M.J. Garcia and J. Herraz, *NEJM*. 335:1034, 1996.)

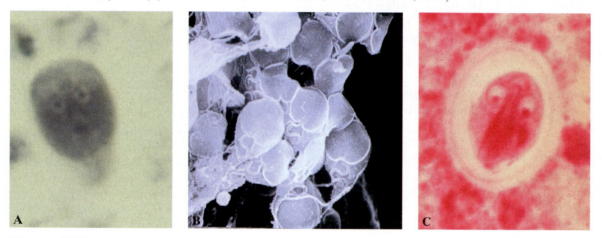

Figure 300

Stained preparations of *Giardia lamblia* in fecal smears. (A) A trophozoite. (B) A scanning micrograph showing *Giardia* trophozoites attached to tissue. (C) A cyst.

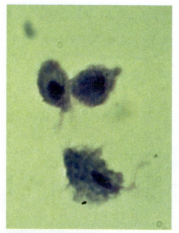

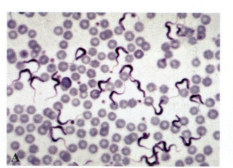

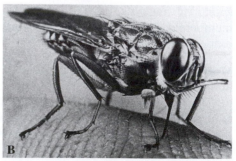

Figure 301

Trichomonas trophozoites in a stained smear.

Figure 302

African sleeping sickness. (A) The causative agent, *Trypanosoma brucei gambiense* in a blood smear. (B) The tsetse fly (*Glossina* species) on the tip of a finger. (Courtesy of the World Health Organization.)

**Protozoology (Protists–The Protozoa)
Mastigophora *(Continued)***

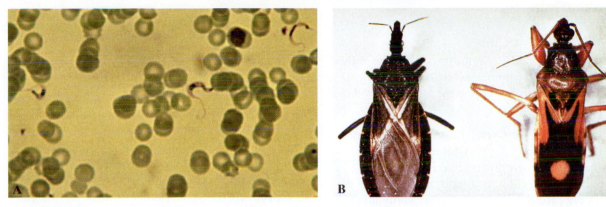

Figure 303
Chagas' disease. (A) The small causative agent, *Trypanosoma cruzi*, surrounded by numerous red blood cells. (B) Cone-nosed bugs of the family Reduviidae.

**Protozoology (Protists–The Protozoa)
Apicomplexa**

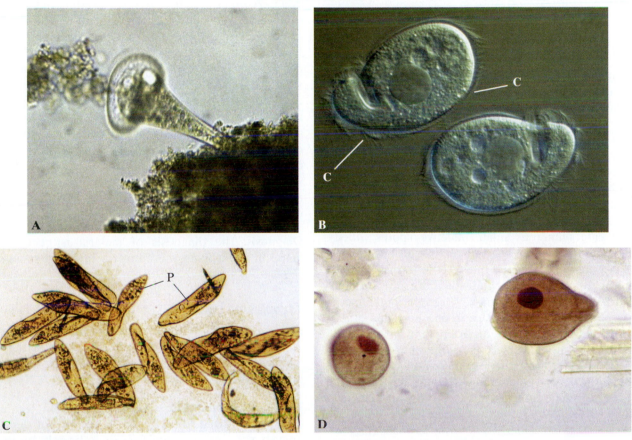

Figure 304
Ciliates. (A) *Vorticella.* (B) The marine anaerobe *Plagiopyla frontata*, with an obvious circular to oval nucleus. The Cilia (C) also are apparent. (C) *Paramecium.* A swarm of paramecia. Note the pellicle (P). (Original magnification 100×.) (D) The pathogenic *Balantidium coli* (trophozoites).

Ciliates are found in both fresh and salt water. Some are free-living; others, such as *Balantidium coli*, are parasitic (Figure 304D).

Balantidium coli

Balantidium coli causes balantidiasis. It is the only ciliate known to cause human infections. This organism produces ulcerations in the large intestine similar to those found in amebiasis.

Transmission and Life Cycle. Infection is acquired by the ingestion of cyst-contaminated food or water. Human-to-human transmission also is common.

Although the infection is found in a variety of mammals, pigs are the major reservoir. Many infected persons have a history of contact with pigs.

Morphology. *Balantidium coli* has both a trophozoite and a cyst stage.

Diagnosis. Finding trophozoites (Figure 304D) or cysts in stool specimens is diagnostic.

▲ Apicomplexa

Members of the phylum Apicomplexa, also called sporozoa, are intracellular parasites that contain several organelles organized into an **apical complex.** This structure helps the parasite penetrate into host cells.

All sporozoa are parasitic, absorbing nutrients from their hosts. Some are intracellular; others are found in body fluids or in various body organs. Adult sporozoa have no organelles for movement.

Both asexual and sexual reproduction occur among sporozoa. A number of sporozoa undergo sporulation, producing numerous small, infective spores called **oocysts.** Such spores reach a susceptible host by way of food, water, or arthropod bites. Spores typically contain one or more smaller individual organisms called **sporozoites.** Many sporozoa have complicated life cycles; certain stages take place in one host, and other stages in a different host. An example of such a cycle involves mosquitoes and the different species of *Plasmodium*, which cause malaria not only in humans but also in several other animals (Figure 305).

Several other pathogens in the Apicomplexa include *Toxoplasma gondii* (Figure 309), *Babesia* species (Figure 310A), *Theileria* species (Figure 309B), *Isospora* (Figure 311), and *Cryptosporidium* (Figure 312).

Plasmodium

Human malaria is caused by four *Plasmodium* species: *P. falciparum* (Figure 306), *P. vivax* (Figure 308 and 307), *P. ovale*, and *P. malariae. Plasmodium falciparum* accounts for the majority of deaths.

Malaria is endemic throughout the major tropical areas of the world.

Transmission and Life Cycle. Malaria is transmitted through the bites of female *Anopheles* mosquitoes. Blood transfusion and sharing of contaminated hypodermic syringes and needles are other modes of transmission.

Infective forms of *Plasmodium* (sporozoites) are introduced into the host by the mosquito while it is biting. These sporozoites are motile, and enter the host's circulatory system. In about one hour, after leaving the circulation the sporozoites enter liver cells, where they initiate the first phase of the asexual reproductive cycle, *exoerythrocytic schizogony.* Schizogony refers to the formation of a *schizont*, a stage resulting from the repeated division of the nucleus. The parasite enlarges and eventually fragments into a large number of daughter cells called *merozoites.* Released merozoites from the liver then begin the second phase of the asexual reproductive cycle, *erythrocytic schizogony*, by invading red blood cells (RBCs) in the circulatory system and continuing to reproduce asexually and forming schizonts. Upon maturing, the erythrocytic schizonts rupture and release merozoites. These forms then, in turn, invade other uninfected cells and continue the reproductive cycle. Reproductive cycles occur with typical regularity for each of the *Plasmodium* species and result in schizont stages with characteristic numbers of merozoites. The cycles occurring in both the mosquito and the human are shown in Figure 305. The various (RBC) stages of two species, *P. falciparum* and *P. vivax*, are shown in Figures 306 and 308, respectively.

Diagnosis. The diagnosis of malaria is generally made by identifying the parasites in blood smears (Figure 307).

Toxoplasma gondii

Toxoplasmosis is a worldwide zoonosis caused by *Toxoplasma gondii* (Figure 309). The cat is the definite host, with a broad range of mammals, including humans, serving as secondary hosts.

Transmission and Life Cycle. *Toxoplasma gondii* can be acquired directly by ingestion of fecal oocysts shed by infected cats or indirectly by ingestion of poorly cooked cyst-containing meats. Transmission of the protozoon also can occur by congenital passage of *T. gondii*, organ transplantation, transfusion with contaminated blood, and laboratory accidents.

Cats acquire *T. gondii* by eating infected rodents and birds that have tissue cysts or by ingesting oocysts from feces of other cats. The parasite reproduces sexually, resulting in the production of infectious oocysts

Protozoology (Protists–The Protozoa)
Apicomplexa

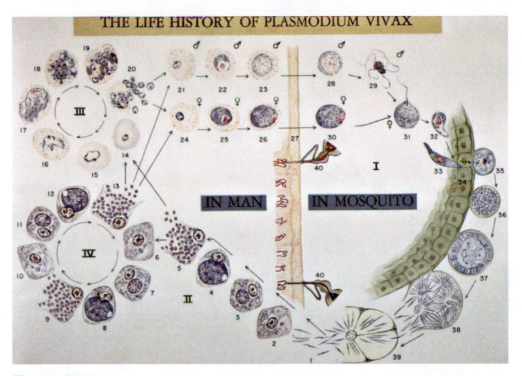

Figure 305

The life cycle of *Plasmodium* in a mosquito (right) and in the human (left). The human is infected by small elongated cells (no. 1), the sporozoites, which are produced in the *Anopheles* mosquito and invade the salivary gland of the insect (nos. in part I). The female mosquito introduces sporozoites into the human bloodstream when she takes a blood meal. The newly introduced parasites are removed from the blood by organs such as the liver and spleen (nos. 2–13, in II and IV). Here they multiply and produce other infective forms that are released into the bloodstream to attack and carry out the asexual cycle in red blood cells (nos. 14–26, in III). The outcome of this cycle is the formation of sex cells (gametocytes) and other infective units for red blood cells (nos. 24–26). If these sex cells are ingested by another female mosquito, they mature and participate in the sexual reproductive phase by forming a zygote (nos. 28–31). This fertilized cell undergoes development within the mosquito's stomach lining (I), where it enlarges and forms a large number of sporozoites. The cycle then continues.

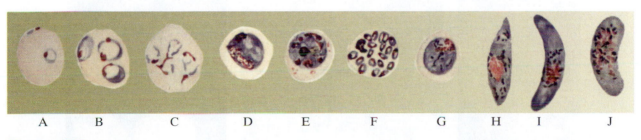

| A | B | C | D | E | F | G | H | I | J |

Figure 306

(The descriptions given here are adapted from the 1960 U.S. Department of Health, Education and Welfare publication *Manual for the Microscopical Diagnosis of Malaria in Man.*) *Plasmodium falciparum.* (A) A single erythrocyte showing a double infection with young trophozoites. The parasite close to the center of the red cell is a "signet ring" form, and the organism located at the periphery is referred to as a "marginal form." (B) One red blood cell with three somewhat more developed trophozoites. (C) The parasites shown are called estivo-automnal forms. (D) The parasite is undergoing initial chomatin (red area) division. (E, F) Mature schizonts with merozoites. Note the number of small merozoites. (G, H). These stages are representative of the successive events that take place in gametocyte (sex cell) development. Such forms generally are not found in the peripheral circulation. (I) A mature macrogametocyte (female sex cell). (J) A mature microgametocyte (male sex cell).

Protozoology (Protists-The Protozoa)
Apicomplexa *(Continued)*

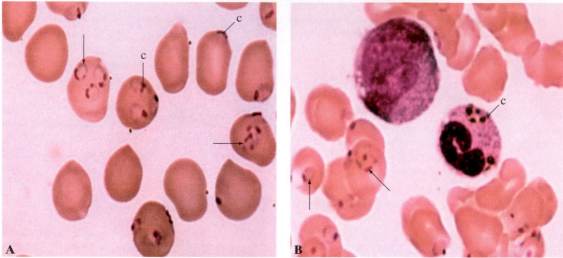

Figure 307

Two microscopic views of a blood smear from an individual with a severe infection with *Plasmodium falciparum*. (A) The ring forms (arrows) and chromatin dots (C) which are typical of this *Plasmodium* species are shown. (B) The extent of infection even involves white blood cells. The original magnification is approximately 1,000×.) (Courtesy of Drs. Y. Chalandon and A. Kocher.)

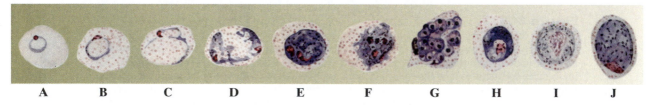

Figure 308

Plasmodium vivax. (A) A typical signet ring shape. (B) An enlarged erythrocyte with a ring form of trophozoite. The cell also contains Schüffner's stippling. Such stippling may not always be present in infected red blood cells. (C) Another trophozoite form. (D) The presence of two amoeba-shaped trophozoites. (E, F) Erythrocytes showing progressive stages in schizont division. (G) A mature schizont. Note the number of merozoites. (H) A developing gametocyte. (I, J) Mature microgametocyte, and macrogametocyte, male and female sex cells, respectively.

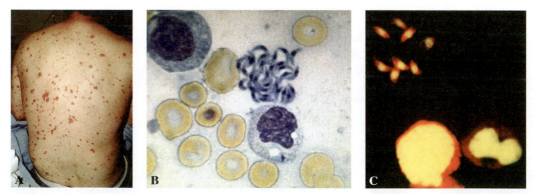

Figure 309

Toxoplasmosis. (A) The appearance of skin involvement with toxoplasmosis. (From P.J. Spagnuolo *et al. CID.* 14(1992):1084–88.) (B) The differently staining *Toxoplasma gondii* in a bone marrow smear. White and red blood cells are also shown. (C) Extracellular *T. gondii* in a blood specimen shown by immunofluorescence. The protozoon can develop intracellularly as well as extracellularly; note the crescent shape of this microorganism (tachyzoites).

that are excreted in the cat's feces. When humans or other animals ingest such oocysts, organisms escape and develop into trophozoites called **tachyzoites**. These tachyzoites are elongated, crescent-shaped cells that actively penetrate body organs and tissues (Figure 309C). Eventually, some tachyzoites form cysts (**bradyzoites**) in brain, skeletal muscle, and heart tissues.

Diagnosis. Serological tests are the principal means of diagnosis.

Babesia

Babesiosis is a malaria-like illness caused by *Babesia* species, which are intraerythrocytic protozoa (Figure 310A). *Babesia*, together with members of the genus *Theileria* (Figure 310B), are referred to as **piroplasmas** because of their pear-shaped red blood cell stages. Both genera infect a wide variety of wild and domestic animals. *Babesia microti* (in the United States) and *B. divergens* (in Europe) cause infection in humans.

Transmission. Young forms (nymphal) of *Ixodes dammini* (ticks) transmit the *Babesia* species. Infections also can be acquired by transfusions with infected blood. Many human infections caused by *B. microti* are either asymptomatic or subclinical. Severe and even life-threatening infections with high fevers and hemolytic anemia occur in immunocompromised and elderly individuals.

Diagnosis. Diagnosis is made by identifying the parasites in stained blood smears. The presence of four pear-shaped infective units, a tetrad, called the *Maltese cross* is diagnostic (Figure 310).

Isospora belli

Isosporiasis is caused by *Isospora belli*. The disease, which takes the form of a long-lasting diarrhea, is uncommon in the United States, but it occurs frequently in South America and Southeast Asia.

Transmission and Life Cycle. Infection is acquired by ingestion of oocysts (Figure 311). Oocysts are excreted in stools and contain sporocysts consisting of four sporozoites. After they have been ingested, the sporozoites escape from the oocyst and penetrate the intestinal lining, where they reproduce and eventually cause the chronic, disabling diarrhea. AIDS patients

and others with some type of immunosuppression are quite susceptible to infection.

Diagnosis. Diagnosis is dependent on finding oocysts in stool specimens. Modified acid-fast stains of materials are of value.

Cryptosporidium parvum

Cryptosporidium parvum causes **cryptosporidiosis**, an infection of the epithelial cells lining the human digestive tract resulting in diarrhea. This organism is recognized as a major cause of diarrhea worldwide.

Transmission and Life Cycle. *Cryptosporidium parvum* is acquired by ingestion of infectious forms, oocysts, in contaminated water; by contact with infected lower animals; or by person-to-person transmission.
Cryptosporidium parvum has a life cycle similar to that of *Isospora belli*. In the immunocompetent host, illness is self-limited. However, in individuals with immunosuppression, infection may result in prolonged life-threatening cholera-like diarrhea.

Diagnosis. Modified acid-fast and other staining techniques are used with stool specimens for diagnosis (Figure 312). Immunofluorescent techniques also are of value.

▲ Microsporidia

Microsporidia are obligate intracellular protozoa that parasitize numerous species of vertebrates and invertebrates and may coexist as commensals with their animal hosts. The epidemiology and pathogenesis of microsporidial disease in humans are unclear. Since 1985, increasing numbers of cases of human microsporidiosis, mostly in HIV-infected or other immunocompromised patients, have been reported. Most infections reported in AIDS patients are caused by *Enterocytozoon* species (Figure 313).

Transmission. Intestinal and systemic infections are probably acquired through the ingestion of spores.

Diagnosis. Diagnosis usually is made by demonstrating the organisms in biopsies or in staining fecal preparations. Currently, electron microscopy is required to identify microsporidia species.

Protozoology (Protists-The Protozoa)
Apicomplexa *(Continued)*

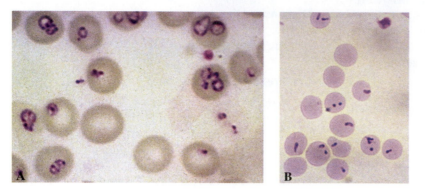

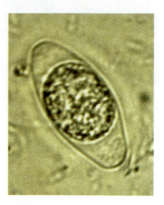

Figure 310

Piroplasms in blood smears. (A) *Babesia microti*. (B) *Theileria parva*. Note that these organisms are located within several of the circulating red blood cells.

Figure 311

Isospora sporulated oocyst.

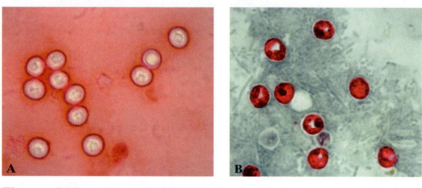

Figure 312

Cryptosporidium parvum oocysts in human stool. (A) The results of a rapid negative stain procedure with methyl green. The methyl green acts as a background for the oocysts, which do not stain. (B) Red oocysts stained by a modified acid-fast technique (1,000×). (Courtesy of Drs. M. Scaglia and G. Chichino, Laboratory of Clinical Parasitology. Institute of Infectious Diseases, University-IRCCS, San Matteo, Pavia, Italy.)

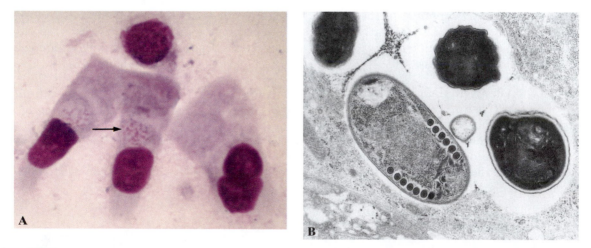

Figure 313

(A) A Giemsa-stained intestinal biopsy specimen showing darker microsporidia, of *Enterocytozoon bieneusi* surrounding the nuclear area (1,000×). (From J.-M. Molina, et al., *J. Inf. Dis*. 167(1993):217–21.) (B) A transmission electron micrograph of *E. bieneusi*. (Courtesy of Dr. J.M. Orenstein, Department of Pathology, George Washington University.)

They can strike anywhere, anytime. On a cruise ship, in the corner restaurant, in the grass just outside the back door. And anyone can be a carrier.

—Michael D. Lemonick, Time Magazine

▲ Viruses and Their Effects

Viruses are very different from the microbial groups mentioned thus far. They are so small that most can be seen only with an electron microscope, and they are acellular (not cellular). Structurally, a virus particle (**virion**) contains a core made of only one type of nucleic acid, either DNA or RNA. This core may be surrounded by a protein coat. Sometimes the coat is encased by an additional layer, a lipid-containing membrane called an **envelope**. All living cells have RNA and DNA, can carry out chemical reactions, and can reproduce as self-sufficient units. Viruses can reproduce only by using the cellular machinery of other organisms. Thus, all viruses are parasites of other forms of life.

Virtually every kind of life can be infected by viruses—vertebrate and invertebrate animals, plants, prokaryotes, and eukaryotic microorganisms, such as fungi, protozoa, and certain algae (Figures 314–318). There are even some "satellite" viruses, which are, in a sense, parasites of other viruses. Viruses are unlike any other microorganism. This difference is obvious not only from their submicroscopic size, but also from other differences related to the way they function (Table 19).

Viroids and Prions

Until about 1972, it was generally believed that the smallest infectious disease agents were viruses. This view changed in the 1980s with the discovery of smaller and less complex agents called **viroids**. A number of plant diseases are known to be caused by viroids, which consist of small uncovered (naked) molecules of ribonucleic acids (Figure 319).

Table 19 Properties of Viruses and Other Microorganisms

Microbial Group	Cell Wall	Microbial Components		Growth Requirements	
		Internal membrane parts	Nucleic acid content	Cultivation in or on artificial media	Require living cells
Algae	Present	Present	DNA, RNA	Yes	No
Bacteria	Present	Absent	RNA, RNA	Yes	Some
Fungi	Present	Present	DNA, RNA	Yes	No
Protozoa	Absent	Present	DNA, RNA	Yes	Some
Viruses	Absent	Absent	DNA or RNA[a]	No	Yes
Viroids	Absent	Absent	RNA	No	Yes
Prions	Absent	Absent	None	No	?

[a]Individual virus particles contain either DNA or RNA, never both.

157

VIROLOGY
Viruses and their effects

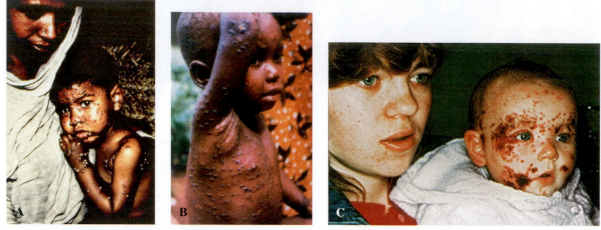

Figure 314
The effects of viruses. (A) Smallpox. A victim of a human disease eradicated by immunization. (Courtesy of the World Health Organization). (B) A case of monkey pox, a disease that resembles smallpox. (Courtesy World Health Organization.) (C) The recurrences of fever blisters (caused by herpes simplex virus) on the lips of a mother, as a result of close contact probably caused the widespread infection on the child's face.

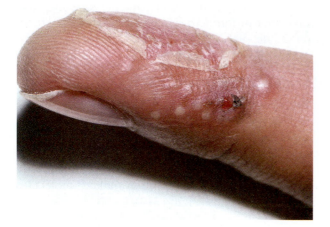

Figure 315
Herpetic whitlow, a herpes simplex virus type 1 infection of the finger.

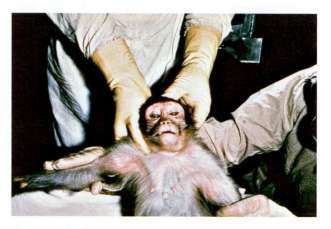

Figure 316
A simian version of chickenpox.

Figure 317
A virus-infected plant. The yellow mosaic areas contain virus.

Figure 318
Colonies of the bacterium *Streptomyces azureus* with plaques (surface holes) caused by bacterial viruses.

The discovery of subviral agents of diseases apparently did not end with viroids. Another smaller and quite different disease agent of mammals, the **prion** (Figure 321A), causes fatal neurodegenerative diseases of the central nervous system. Most of these diseases are characterized by a long incubation period and a neuropathologic feature resulting in the production of numerous sponge-like lesions in various parts of the central nervous system, loss of neurons, destruction of star-shaped neurological cells called *astrocytes*, and an absence of inflammation. Infectious prions damage brain tissue through a process resulting in the formation of vacuoles, making neurons appear like sponges with pitted holes, hence the term **spongiform** (Figure 321B). These transmissible spongiform encephalopathies (**TSEs**) include Creutzfeldt-Jacob disease in humans, kuru, a disease originally found among cannibals in the South Pacific (Figure 320), bovine spongiform encephalopathy, or "mad cow disease," in cattle, scrapie in sheep, and chronic wasting disease in deer and elk of North America. These diseases are similar to non-transmissible protein deposition conditions such as Alzheimer's disease, where a host-source of protein is misfolded and may damage nearby cells. **Amyloid plaques**, which are accumulations of a protein-polysaccharide complex with starch-like properties, may occur in these diseases.

As indicated in Section 1, prions occur in two different forms, namely the *normal cellular prion protein* (PrP^C) and the *disease-causing (scrapie) form* (PrP^{Sc}). The normal host-derived protein involved in misfolding and amyloid plaque formation is required for susceptibility to disease and replication of infectious prions. Although the mechanism responsible for the neurological deterioration or neurodegeneration in TSEs is unclear, it apparently involves the accumulation in the brains of afflicted individuals of the disease-associated protein, resistant to digestion by an enzyme known as proteinase K. This form is also designated as PrP^{RES}. After infection, the enzyme-sensitive, normal cellular prion protein is converted to the enzyme-resistant form associated with brain pathology. Enzyme-sensitive normal cellular prion protein (PrP^C) can be found in the neurons and glia of the brain and spiral cord, as well as in several peripheral tissues and in leukocytes. The protein is localized on cell surfaces. Its normal physiologic function, however, currently is unknown. Determining the role of this normal protein (PrP^C) may be relevant to understanding prion-caused diseases, since the protein may fail to perform its normal function when it is converted to the disease-causing form (PrP^{Sc}).

▲ The Introduction of Electron Microscopy—The History of a Great Invention

In 1878, Ernst Abbé proved that the ability to see detail with an optical microscope is limited by the wavelength of light. This meant that even when immersion optics and ultraviolet light were used, the smallest detail that could ever be obtained optically is of the order of 100 micrometers, or 10 Angstroms (see Figure 6).

No means were conceived of obtaining finer detail until two discoveries were made. These were L. de Broglie's defining the wave properties of the moving electron in 1924, and the discovery by H. Busch in 1926 of the similarity between the effect of a magnetic coil and a focusing coil on an electron beam. Since the wavelength of the moving electron is significantly shorter than the shortest wavelength of light, these discoveries made it possible to see extremely small objects with the use of an electron beam and magnetic lenses.

In 1931, leading a skilled research team, Max Knoll and Ernst Ruska incorporated the findings of de Broglie and Busch in building the first electron microscope (Figure 322). The development of this instrument at the High Tension Laboratory of the Technical University in Berlin was one of the greatest and most far-reaching achievements of the twentieth century, especially in the fields of virology and related sciences.

▲ Basic Structure of Viruses

Each mature virus particle or virion has a characteristic morphology—basic structure and organization. This feature of viruses is best studied by electron microscopy and such techniques as **negative staining** (Figure 323) and **thin sectioning** of infected tissues (Figure 324).

The viral DNA or RNA forms a central core of a virion, which, depending on the virus, may or may not be encased in a protein shell known as the **capsid** (Figure 323A). Each capsid is constructed from a definite number of subunits called **capsomeres**, which are grouped together to form a characteristic shape, or morphology (Figures 323A and C). Some virions may in turn be wrapped in a membranous envelope derived from the membranes of host-infected cells (Figures 323B and 324). Viral envelopes may have additional structures such as spikes associated with them (Figure 323B).

Several basic viral morphological types are recognized: (1) icosahedral (20 triangular faces) capsid, (2) icosahedral capsid and an outer envelope, (3) helical

Viroids and Prions

Figure 319

The effects of viroids. The smaller plant's growth has been stunted due to a viroid infection.

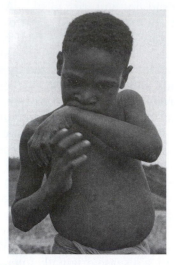

Figure 320

A young victim of kuru. This child has lost the use of his limbs, and is unable to eat. This prion disease also interferes with swallowing. (Courtesy of the World Health Organization.)

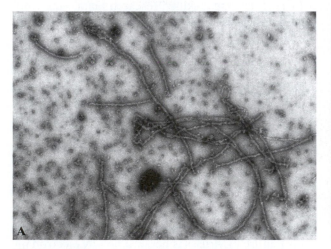
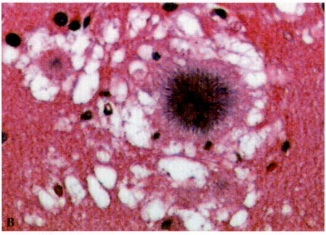

Figure 321

(A) A transmission electron micrograph showing negatively stained long particles from a scrapie-infected animal. (Courtesy of Dr. H.K. Narang, The London Hospital Medical College.) (B) A light microscopic view of a stained brain tissue specimen from a patient with Creutzfeldt-Jakob disease (CJD). The spongiform appearance of an amyloid plaque is shown. (Courtesy of Dr. Paul Brown.)

Figure 322

Members of the research team involved with the development of the first electron microscope in 1931. M. Knoll (1) and E. Ruska (2) are shown in the photo. (Courtesy of Martin M. Freundlich.)

VIROLOGY
Basic Structure of Viruses

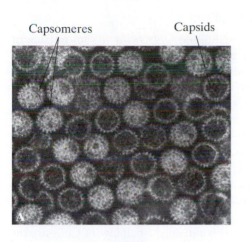

Capsomeres   Capsids

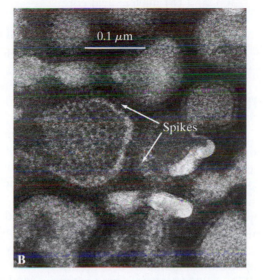

0.1 μm

Spikes

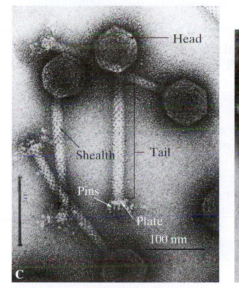

Head

Shealth   Tail

Pins

Plate

100 nm

C

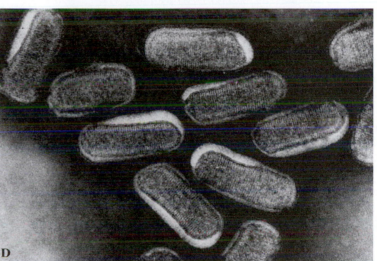

D

Figure 323
Electron micrographs of negatively stained viruses. (A) The capsids and capsomeres of rotavirses. (B) The virions of mouse mammary tumor viruses showing an enveloped surface covered with projections known as spikes. (C) Bacterial viruses (bacteriophages) with typical parts. (Courtesy of J. Lembke and M. Tuber, Federal Dairy Research Center.) Sowthistle yellow vein virus of plants. (Original magnification 160,000×.)

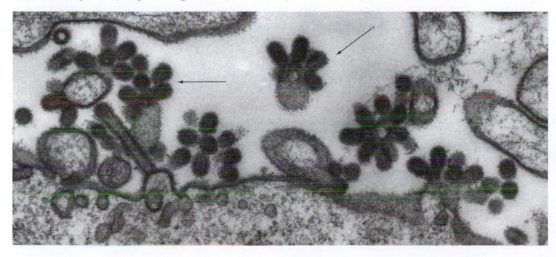

Figure 324
An ultrathin section through cells from which influenza viruses are budding (emerging). (Original magnification 85,500×.)

(springlike) capsid, (4) helical capsid and outer envelope, (5) binal, a combination of icosahedral and helical capsids (Figure 323C), and (6) nucleic acid core surrounded by an envelope (Figure 323B). Many virus particles lacking a basic shape or containing accessory structures are called **complex**.

▲ Cultivation

The replication (multiplication) of viruses differs significantly from the reproduction mechanisms of other microorganisms. Because viruses do not have the cellular structures of eukaryotic and prokaryotic cells, they are incapable of independent activity outside of a host cell. Viral replication can occur only within a host cell. The essential nature of most viruses is to infect a host cell, replicate, package its nucleic acid, and exit the host cell. Replication of animal viruses occurs within animal cells (Figure 325A), plant viruses within plant cells, and bacterial viruses (bacteriophages) within bacteria (Figure 325B). Viruses exhibit a high degree of specificity for host cells.

Viral replication begins with the **attachment** of the virus to a susceptible and compatible host cell (Figure 325). This step is followed by a series of events, including (1) the entry of the viral **genome** (genetic material) into the host cell, (2) the synthesis of viral proteins and nucleic acid molecules under the direction of the viral genome, (3) the assembly of new viral particles within specific cellular sites (Figure 326A), and (4) the release of mature viruses from the host cell.

Certain viruses acquire a covering of host cell membranes (envelopes) as they emerge (Figures 324A and 326B). Virus release usually causes the death of host cells.

Animal Virus Cultivation

A number of laboratory animals can be used for animal virus cultivation. These include rats, guinea pigs, hamsters, and mice. Chicken, duck, and turkey embryonated eggs also are of value and provide an excellent experimental system for isolating viruses, for obtaining large quantities of viruses for vaccines and diagnostic reagents, and for studying viral replication mechanisms and the effects of drugs on such processes. The chicken embryo (Figure 327A) provides several tissues and cells that readily support viral replication. These include the yolk sac, amnion, and the chorioallantoic membrane (Figure 327B). Viruses replicating in chicken embryos may cause a variety of visible destructive effects. These include death, embryonic growth defects, and localized areas of membrane damage resulting in bleeding or the formation of discrete opaque areas known as **pocks**

(Figure 327C). A pock contains large concentrations of viruses.

Cell (Tissue) Culture

Although the technique of growing tissues outside of an animal or plant (*in vitro* cell or tissue culture) is almost as old as the study of viruses, early virologists could not use this technique because of problems of contamination by bacteria and fungi. It was only after three advances in virology that tissue culture gained routine acceptance for the cultivation of viruses. These were the introduction of antibiotics, which greatly reduced the contamination problem; the development of an excellent, defined growth medium for cells; and the introduction of the enzyme trypsin to free cells from fragments of tissue so that they could be grown in single-cell layers. Cells are introduced into growth containers (tubes, Petri plates, and so on) so that they can form a monolayer, or a single sheet of cells (Figure 328A). Monolayers can be examined macroscopically and microscopically for signs of viral infection. Only the cultivation of animal tissues will be considered.

Two types of cell cultures are in common use: primary and continuous. **Primary cultures** generally consist of cells taken from normal embryonic, fetal, and adult animal tissues. Primary cultures have several characteristics of the original tissue from which they were obtained but will be limited in the number of times they can be subcultured. Eventually, primary cultures die or mutate into different cell strains. **Continuous cultures** are derived from mutant cell lines or cancerous tissues. Such cell lines can undergo an unlimited number of divisions.

Virus-infected cells often develop abnormally and show visible changes in appearance. These changes are known as **cytopathic effects** (CPEs) and include (1) abnormal cellular rounding and detachment (Figure 328B); (2) cell destruction (Figure 328B); (3) syncytium, a joining of several cells or giant cell formation (Figures 328C and 329B), and (4) inclusion body formation consisting of masses of viruses or damaged host cell organelles (Figure 329C). Viruses also can be detected by overlaying infected cells with a soft agar preparation to hold cells in place. Areas of virus destruction appear as clear, well-defined holes in the monolayer. The clear patches are called **plaques**. Similar techniques can be used to detect or determine the numbers of bacteriophages in specimens (Figure 330A).

Another procedure known as **bacteriophage (phage) typing** can be used to identify or to distinguish among bacterial strains. A set of specific bacteriophages is tested against cultures of different strains of a bacterial species. Because different strains of bacterial species are generally killed by different bacteriophages in the

Cultivation

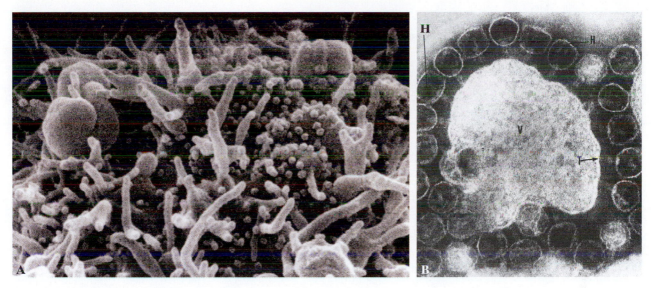

Figure 325

Virus replication. (A) A scanning electron micrograph showing an infected cell studded with simian immunodeficiency virus particles (arrows). (Courtesy of Dr. William N. Norton.) (B) A transmission electron micrograph showing bacteriophages absorbing onto the surfaces of a bacterial cell. Contracted viral sheaths (T) can be seen with some viruses. Viruses that have injected their nucleic acid cores appear as empty heads (H).

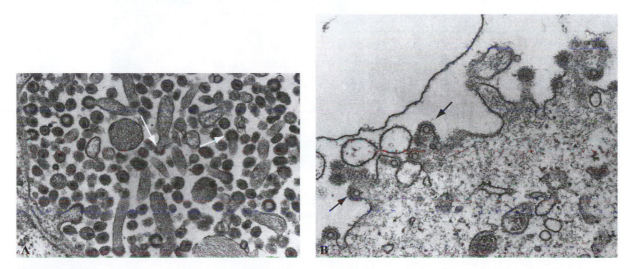

Figure 326

Electron microscopic views of viral replication. (A) Numerous cytoplasmic virus particles in different stages of development are shown. (B) Viruses budding from cell surfaces (arrows).

Cultivation

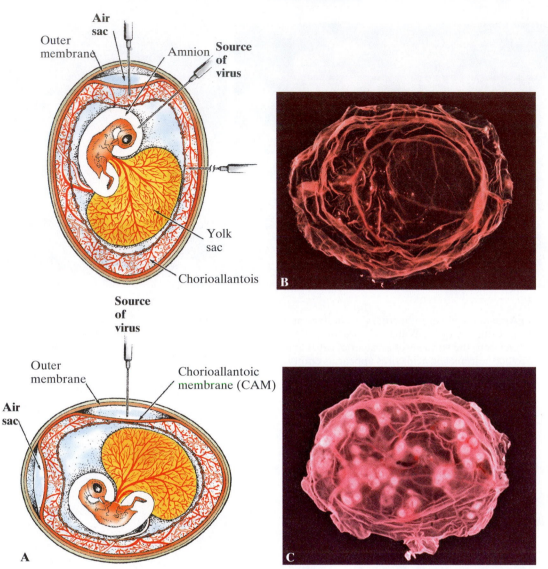

Figure 327

Embryonated chicken egg inoculation. (A) Cutaway diagram of the different parts of the animal and the chorioallantoic membrane (CAM) site of inoculation are shown. (B) The appearance of a normal CAM. (C) An infected CAM showing pocks, localized dense areas, and clouding.

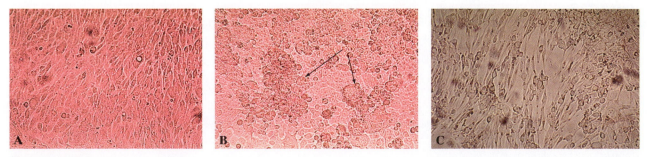

Figure 328

Light microscopy of cytopathic effects. (A) Unstained normal rabbit corneal cells growing in a continuous layer. (B) Giant cell (multinuclear) formation caused by rubella virus. (C) Destruction of cell layer by adenoviruses.

set used, it is often possible to distinguish one strain from another. The identification is based on the appearance of plaques caused by a specific phage (Figures 330B, and 330C).

▲ Classification

The taxonomy of viruses undergoes changes as more is learned about their properties. Only the taxonomic principles and selected features of animal viruses are briefly considered here.

The most widely used taxonomic criteria are based on four physical viral characteristics: (1) the nature of the nucleic acid in the virion, (2) virus morphology, (3) the presence or absence of a viral envelope, and (4) virion size. Other characteristics, such as the effects on host cells and immunologic properties, provide a basis for placing viruses into taxonomic families, the names of which end in "viridae." There are 12 families of RNA-containing viruses and six families of DNA-containing viruses. Selected features of representatives from these two groups are briefly considered here.

▲ Diagnosis

Proving the specific viral cause of an infection is generally expensive, time-consuming, and difficult to do in most clinical settings. Therefore, laboratory virological tests are usually reserved and used for situations in which the results will significantly contribute to the management of infections.

Laboratory Diagnosis

A wide variety of laboratory tests are available for viral disease diagnosis. They can be grouped into four categories: (1) morphology, (2) serology, (3) virus isolation, and (4) nucleic acid technology.

Morphological tests involve direct examination of appropriate clinical specimens for the presence of viruses or their parts. They include electron microscopy and immunofluorescence techniques (Figure 331A).

Serological tests include procedures used to show the presence of viral antigens, increases in antibody concentration during recovery, or the appearance of specific antibody (IgM) during the early (acute) phase of the disease.

Virus isolation techniques involve inoculation of appropriate clinical specimens into laboratory animals or cell cultures and the subsequent demonstration of specific virus replication (Figures 324 through 328).

Nucleic acid techniques are used to detect the presence of virus genetic material (genome) in cells or secretions. Nucleic acid probes are quite effective (Figure 331B). A probe is defined as a fragment of DNA or RNA, typically labeled with radioactive material, that can be applied to show the presence of a complementary sequence of DNA or RNA in cells or secretions. This reaction is generally referred to as **nucleic acid hydridization.**

▲ Features of Generalized Infections Accompanied with a Rash

Individual skin lesions may occur as generalized rashes resulting from or occurring during certain stages of virus- and other types of microbial-caused infections.

Such lesions include the following:

1. **macule**: a small discolored area or patch on the skin that is neither depressed nor elevated; may appear in various colors and sizes (Figure 332A);
2. **papule**: a red elevated area ; usually firm and circumscribed (Figure 332B);
3. **vesicle**: a small elevated area filled with fluid (Figures 332C and 332D);
4. **pustule**: a small elevated area filled with lymph and pus (Figure 332B); such lesions may be flat, circumscribed, rounded, or centrally depressed.

The primary involvement of the outer skin layer (epidermis) usually results in vesicle formation, ulceration, and scab formation (Figure 332E). Vesicles can convert to pustules if there is no large amount of cellular secretion. Prolonged dilation and severe damage of local skin capillaries may lead to the formation of small, purplish, bleeding spots called *petechiae*.

Viruses as Biological Weapons

Biological weapons are not new. Biological agents have been used as instruments of warfare and terror for thousands of years to produce fear and harm to humans, and significant losses of lower animals and plants. Because they are not visible to the naked eye, and are odorless, tasteless, and silent, biological agents such as viruses may be used as an ultimate weapon. *Biologic terrorism* can be defined as the intended use or even the threatened use of biologic agents (which include microorganisms and/or their toxins) against individuals, a group or groups, or larger populations to create fear or cause human deaths or disease.

The United States Public Health System has prioritized the potential biologic agents that could be used by bioterrorists on the basis of the risk to national security and has placed them into three specific categories,

Cultivation

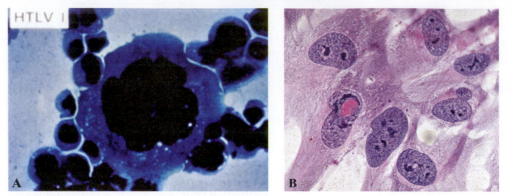

Figure 329
(A) Stained giant cell formation caused by human T-lymphotrophic virus, type 1. (From R.C. Gallo. *J. Inf. Dis.* 164(1991):235–43) (B) Eosinophilic nuclear inclusion body formation (arrow).

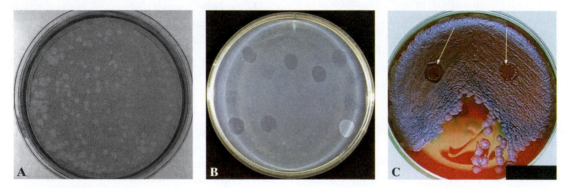

Figure 330
Bacteriophage applications (A) Result of virus number determination by the plaque assay procedure. (B) Plaque formation. The effects of bacteriophages on a lawn of susceptible bacteria (clear areas) lead to their identification. (C) The clear areas represent a positive identification of a strain of the anthrax bacillus. The Petri plate contains *Bacillus anthracis* colonies. (From T. Abshire and J. Ezzell. *Clin. Microbiol.*, 4780–4788, 2005.)

RNA Viruses

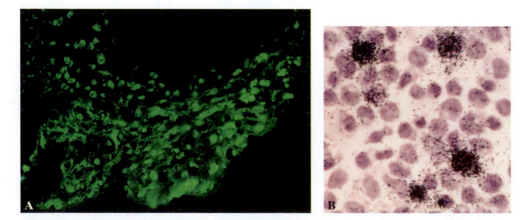

Figure 331
Laboratory diagnosis. (A) The results of applying an immunofluorescence technique to a specimen containing human papillomavirus, the cause of genital warts. The presence of viral antigen is indicated by the green fluorescence. (Approximate magnification 100X.) (B) The results of applying a nucleic acid probe. The clusters of black dots indicate the location of virus-infected cells. A DNA probe was used to identify viral messenger RNA within the cells.

Skin Lesions

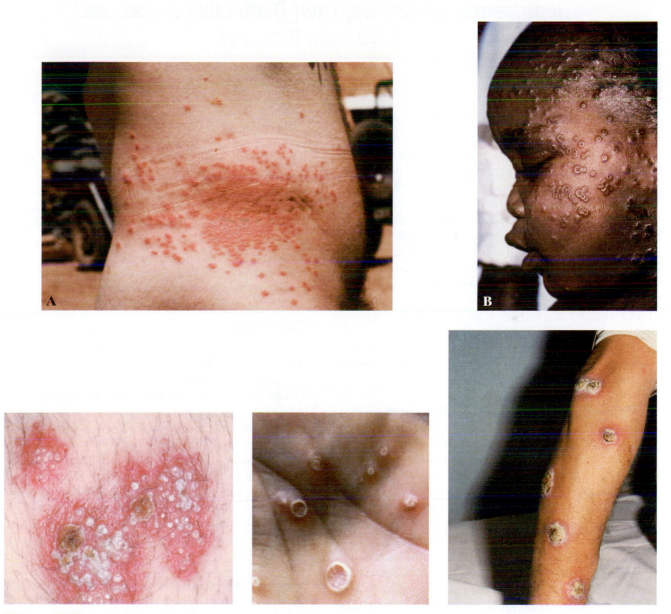

Figure 332
Skin lesions associated with viral and other microbial diseases. (A) The appearance of macules. (B) The presence of papules and pustules on a child with smallpox. (C) Various stages of vesicle formation. (D) Complete and broken vesicles on the palm. (E) Crust formations along the length of the arm.

namely **A**, **B**, and **C**. These categories were established by the Centers for Disease Control and Prevention (CDC) on December 30, 2003, and include pathogens and/or their products that are not commonly seen in the United States. A number of viruses together with such bacteria as *Bacillus anthracis* and *Clostridium botulinum* are considered likely candidates for use as biological weapons. The viruses include smallpox (Figure 314A), monkeypox (Figure 314B), and Ebola and Marburg viruses (Figure 338). Most of these pathogenic viruses are in Category A, which contains the biological agents posing the greatest risk and threat.

Representative Viruses, Their Distinctive Properties, and Selected Diseases

▲ RNA Viruses

Corona Viruses

A multi-country outbreak of an unexplained atypical pneumonia referred to as severe acute respiratory syndrome (SARS) came to the attention of the world beginning on or about March 11, 2003. Investigations by researchers in Hong Kong and reports from the Centers for Disease Control and Prevention identified the major cause of SARS to be a coronavirus. Originally SARS was believed to have been caused by human metapneumovirus (Figure 333). This virus is responsible for a fair number of respiratory infections in early infancy and childhood, and also in the elderly and immunocompromised individuals.

Coronavirus infections occur mainly during winter months, and primarily affect adults. The common cold is the usual result from such infections.

Morphology and Genome Properties. Coronaviruses are spherical, and contain a single plus-strand of ribonucleic acid (RNA), which serves as the **genome** (genetic material). These viruses have the largest viral RNA genome known to date. The RNA together with one of the virus' three proteins forms a **nucleocapsid** component that is surrounded by a viral **envelope**, or covering (Figure 334A). The overall diameters of **virions** range from 80 to 160 nanometers (nm). Coronaviruses, as their name indicates, are distinguished from other viral agents by possessing distinctly separated petal-shaped projections or spikes (*peplomers*) attached to the surfaces of their envelopes.

In addition to the coronavirus protein forming the helicocapsid, coronaviruses contain a membrane glycoprotein (**E1**) embedded within the viral envelope, and a glycoprotein (**E2**), which forms the smaller petal-shaped projections on the viral envelope. The E2 glycoprotein is used by the coronaviruses for attachment to susceptible cells, and is associated with laboratory diagnostic tests for virus identification.

Transmission. The transmission of corona viruses and especially the SARS agent has been well documented among health-care workers administering to victims of the associated diseases. Close person-to-person or, more specifically, face-to-face contact was found to be responsible for most infections. The virus is transmitted via the inhalation of infectious respiratory droplets. There is also the possibility that infection can be spread by contaminated fingers and/or fomites.

The results of various studies point to bats as likely natural reservoirs for the SARS virus (Figure 334B). Since bats are among the live animals sold in markets in southern China, it is fairly reasonable to assume that they could serve as a plausible source of infection for civets, one of the animals originally suspected as source of virus. The weasel-like civets are delicacies sold in Chinese markets. Three species of horseshoe bats (*Rhinolophus* species) are now officially recorded as natural reservoir hosts of the coronavirus that causes SARS in humans.

Rotaviruses

Rotaviruses (Figure 335) are the only members of the *Reoviridae* that produce significant human disease in the United States.

Rotavirus is found worldwide, infecting children in industrialized and developing countries in the first few years of life. Infections cause diarrhea and vomiting that is often mild, but can be severe enough to require hospitalization. In the developing world, where access to lifesaving rehydration treatment is either not adequate or unavailable, rotavirus is responsible for more fatal cases of diarrhea than any other single pathogen. Developments with rotavirus vaccines are expected to provide great promise for the prevention of severe dehydrating diarrhea in the next generations.

Morphology and Genome Properties. The virion consists of double capsid layers that are icosahedral. There is no envelope. The capsomeres of the virus particle give it a striking wheel-like appearance (Figure 335). The rotavirus genome is a double-stranded RNA molecule.

Transmission. Rotaviruses are transmitted by the fecal-oral route.

Rabiesvirus

The rabiesvirus is the most important member of the Rhabdoviridae. The disease rabies is an acute encephalitis.

Morphology and Genome Properties. All rhabdoviruses are enveloped and have a helical capsid (Figure 336). These viruses exhibit a bullet-shaped morphology. The genome consists of a single-stranded RNA molecule.

Transmission. Rabiesviruses are transmitted by the bite of infected animals.

Few viruses have played a more central role in the historical development of virology than that of influenza. The pandemic that swept the world in 1918, just as the Great War (World War I) ended, killed more people than the war itself. In some of the more recent severe epidemics in the United States, influenza viruses attacked over 25 percent of the population and killed over 40,000 individuals.

Influenzaviruses

Three species of *Orthomyxoviridae* that cause the human respiratory disease influenza are influenza viruses A, B, and C, so classified on the basis of the antigenic differences of their nucleoprotein antigens. Influenza A viruses are further subdivided in a number of "subtypes," such as A0, A1, and A2, on the basis of antigenic differences in the surface proteins, *hemagglutinin* and *neuraminidase*. The subtypes in turn are subdivided into "strains." All the strains of these viruses within each type share common internal proteins, namely the nucleoprotein, and membrane protein, but differ in their surface proteins (**antigens**) briefly described in the following section.

The avian flu virus (H5N1) is a member of the type A group. This virus subtype may cause infection of the lower respiratory tract and severe pneumonia in humans.

Morphology and Genome Properties. Typical influenza virus particles, or **virions**, are irregularly shaped spherical forms measuring approximately 100 to 120 nanometers (nm) in diameter. Variations in shape and size including elongated particles are commonly seen in specimens (Figure 337). Influenza viruses contain a single-stranded, segmented genetic component (**genome**) that is enclosed with a covering known as an **envelope**. The genome of both influenza A and B viruses is composed of ribonucleic acid (**RNA**), which forms a helical (spring-like) **nucleocapsid** and consists of eight distinctive segments. The genome of influenza virus C also is made up of RNA, but it is distributed among only seven fragments. The RNA genome segments contain the information for the *internal protein antigens* and the surface or envelope-associated protein antigens *hemagglutinin*(**H**) and *neuraminidase*(**N**), in the form of **projections** or **spikes**. The influenza virion also contains internal proteins which include nucleoprotein and three polymerase molecules. The polymerase molecules are important to the formation of viral nucleic acid. The

envelope of influenza C virus does not have neuraminidase.

The **hemagglutinin protein antigen (H)** derives its name from its role in attaching to and causing the clumping (agglutination) of red blood cells in a laboratory test used to identify viruses, namely the hemagglutination test. The H antigen is responsible for the attachment of influenza viruses to susceptible host cells at the onset of infection. Blocking the virion's hemagglutinins with specific antibodies against them prevents virus particle attachment to susceptible cells.

The second envelope (surface) protein antigen, **neuraminidase (N)**, has the enzymatic activity to liquefy host mucus and to facilitate the local spread of influenza virions. Host antibody to this viral part does not prevent infection, but does restrict multiple cycles of viral replication and reduces the severity of the illness.

The **nucleoprotein** is the major protein of the helical internal ribonucleoprotein complex associated with the virion's RNA segments and polymerase proteins. It also contains the antigenic determining factors that permit influenza A, B, and C viruses to be distinguished from one another.

A naming or nomenclature system recommended by the World Health Organization (WHO) in 1980 makes use of the influenza virus antigenic composition for identification and related purposes. The system designates the virus type, geographic origin, strain number, the year of isolation, and the subtype. The nomenclature of the system follows this definite order. Thus the influenza A virus responsible for the pandemic of 1968 which arose in Hong Kong is designated as **A/Hong Kong/1/68/(H3N2)** to specify specific properties of the isolated virus as follows:

> **the type**: A
> **the geographic origin**: Hong Kong
> **the isolate (strain number)**: 1 (first)
> **the year**: 1968.
> **the subtype**: (H3N2)

Transmission. Influenzaviruses are spread by aerosols and contaminated objects. Human-to-human transmission is common.

Influenza viruses are known to undergo permanent genetic changes (**mutations**) rather rapidly. The fact that they also have segmented genomes means that their genes do not lie along a continuous strand of nucleic acid, as is the case with most forms of life. Instead, influenza virus genes are carried on unconnected RNA segments. Thus, if two different influenza virus strains infect the same cell, mixing or **reassortment** of their genes becomes possible. Such

RNA Viruses

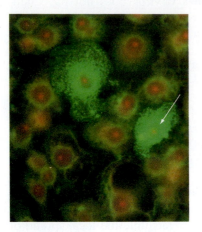

Figure 333
The presence of human metapneumovirus in a nasopharyngeal specimen detected with the application of direct fluorescent antibody. (From E. Percivalle, A. Saraini, L. Visai, M. G. Revello, and G. Gerna. *J. Clin. Microbio.* (2005) 43. 3443–3446, and with permission of G. Gerna.)

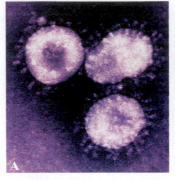

Figure 334
Coronavirus. (A) A transmission electron micrograph showing the general features of coronaviruses. (B) A wall of bats. These mammals are currently considered to be the source of infections.

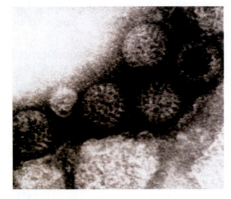

Figure 335
An electron micrograph of negatively stained rotavirus capsids (160,000×).

Figure 336
A tissue section showing a rabies virus (*Lyssavirus*) budding from the plasma membrane of an infected cell (150,000×).

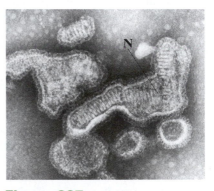

Figure 337
A negative stain of influenza virus showing its nucleic acid (NA) core (260,000×).

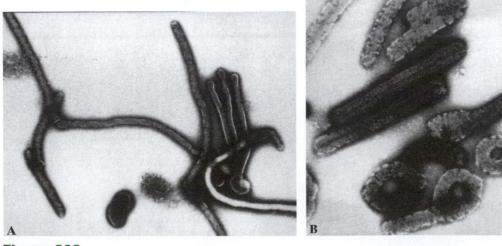

Figure 338
Transmission electron micrographs of negatively stained filoviruses. (A) Ebola virions. (B) Marburg virions. (Courtesy of Centers for Disease Control and Prevention.)

gene reassortment can create a new hybrid influenza virus, which increases the chances of a virus jumping from one host species to another.

Ebola Virus

Among the viruses known to cause hemorrhagic fever, Ebola virus stands out for its extreme lethality. After infection, humans and nonhuman primates develop severe, rapidly progressing illnesses, generally accompanied by high fever, bleeding in multiple organs, and shock. In large outbreaks, human morality rates have reached approximately 90%. Four species of Ebola virus are currently recognized.

Morphology and Genome Properties. Ebola virus belongs to the enveloped RNA virus family, *Filoviridae*. The designation filovirus comes from the Latin meaning "thread." Filovirus virions are described as being long and threadlike and often take the shape of a fishhook or a U (Figure 339A). They range in length from 790 to 970 nanometers (nm) and are consistently 80 nm in diameter. Each infectious virus particle contains a nucleocapsid surrounded by a host-cell-derived covering or envelope bearing small spike-like projections in a regular pattern. The genetic material (genome) of filoviruses is contained within a single-stranded RNA molecule.

Transmission. The most likely means of transmission of Ebola virus appears to be either by close contact with infected persons or through the use of virus-contaminated syringes and needles in the course of patient care.

Direct contact with infectious organs and semen also is a potential source of the virus. The period of highest risk is during the late stages of illness, when patients are vomiting, experiencing diarrhea, and hemorrhaging. Health-care-facility infections have been frequent.

Marburg Virus

Marburg disease is named after Marburg, Germany, where an outbreak occurred in 1967. The causative virus was identified in tissues obtained from green vervet monkeys imported from Africa.

Marburg disease has been recognized on at least six different occasions. Most of the cases have occurred in Kenya, Democratic Republic of the Congo, and Zimbabwe.

Morphology and Genome Properties. Marburg virus belongs is a single negative-stranded enveloped RNA member of the *Filoviridae* family (Figure 339B). As is the case with Ebola virus, each infectious virus particle consists of a nucleocapsid surrounded a host-cell-derived envelope bearing small spike-like projections.

Transmission. Marburg virus infection results from direct contact with infected organs, or body fluids, such as blood, semen, and other secretions. Under natural conditions, aerosol transmission among humans has not been documented.

Hantavirus

Hantaviruses take their name from the Hantaan River in Korea. These members of the *Bungaviridae* cause hemorrhagic fever often complicated by varying degrees of acute kidney failure. A newly isolated virus is known to cause a respiratory illness that can be difficult to distinguish from influenza.

Morphology and Genome Properties. Hantaviruses are enveloped RNA viruses (Figure 339A). The genome consists of a segmented single-stranded RNA molecule.

Transmission. Hantaviruses are believed to be acquired by humans most often by inhalation of infected rodent excreta (urine, feces, and saliva), by contamination of the eye, or by direct contact with open cuts. Deer mice are believed to be the main reservoirs for these viruses (Figure 339B).

Hepatitis A Virus

Well before the viral cause of hepatitis was known, the disease was defined on the basis of its point of attack, namely the liver. The disease was subdivided on the means of transmission and the particular signs and symptoms. As diagnostic techniques improved with time, several different pathogens, both RNA- and DNA-containing viruses, were shown to cause hepatitis. The list of known major pathogens includes hepatitis viruses A, B, C, D, and E. As this list continues to grow, additional viruses have been added. Two examples of recent additions are hepatitis F and G viruses that are associated with minor hepatitis-associated diseases.

Hepatitis A (Figure 340), a member of the Picornaviridae and also known as *Hepatovirus*, is the cause of viral hepatitis.

Morphology and Genome Properties. Picornaviruses are nonenveloped and have an icosahedral capsid. The genome consists of a nonsegmented single-stranded RNA molecule.

Transmission. The major means of transmission for hepatitis virus A is the fecal-oral route.

Retroviruses

Retroviruses represent a unique family of viruses. Members of the *Retroviridae* use DNA in their replication. Genetic information flows from the viral RNA genome to DNA and then back to viral RNA. This reverse flow of genetic information depends on the specific viral enzyme reverse transcriptase and is the basis for the family designation. The viruses in this group include several cancer-causing agents and the human immunodeficiency viruses (HIVs). HIV is the major focus here.

Infection with HIV type 1 and 2 disables the human immune system and thus predisposes individuals to a wide range of opportunistic infections (Figure 341A) and certain types of cancers.

The sequence of events involved with the slow destruction of an infected person's immune system is as follows: 1) **HIV infection**: *HIV invasion of the body, and subsequent virus reproduction*; 2) **HIV disease**: *HIV reproduction resulting in disease signs and symptoms*; *and/or demonstrable body changes and injury*; and 3) **AIDS**: *the end stage of HIV disease.*

Morphology and Genome Properties. HIV, a lentivirus, is an enveloped virus with a nucleocapsid of varying morphology (Figure 341B). This virus has a centrally located "nucleoid" (nuclear acid core). The genome consists of two identical single-stranded RNA molecules.

HIV-1 viruses can be grouped into subtypes, or **clades**, based on their nucleic acid composition and arrangement. Members of each *clade* are closely related genetically.

Transmission. HIV infection generally can be acquired sexually, by transmission from an infected mother to her fetus, through breast feeding, and by exposure to contaminated blood or body fluids.

▲ DNA Viruses

Papillomaviruses

Cutaneous virus-caused warts are common, usually harmless, self-limited tumors (growths), with a preference for the hands and feet (Figure 342). Such warts or papillomas are caused by papillomaviruses, members of the *Papovaviridae*. The main targets in papillomavirus infections are the skin and mucosal epithelial linings (Figure 342B).

More than 50 subtypes of papillomaviruses are known to cause a variety of differently appearing warts. Among these forms are genital warts, known as **condyloma acuminata**. The greatest danger associated with genital warts is an association with cancer. Although this danger is more prevalent with women and cervical cancer, men can develop penile cancer. The presence of papillomaviruses, specifically subtype 16, shows a significant correlation with the causation of cervical and anogenital cancers.

The introduction of a recently developed vaccine is expected to significantly reduce the number of cases of cervical cancer cases in the future.

Morphology and Genome Properties. Virions of human papillomavirus are nonenveloped icoshedral nucleocapsids measuring 55 nanometers in diameter. The genome consists of a single double-stranded circular DNA molecule (Figure 342C).

Transmission. Genital warts are transmitted through sexual activities and direct contact with infected individuals. The source of the virus is the wart.

Hepatitis B Virus

Hepatitis B virus (HBV) infections are common and worldwide in distribution. This virus, a member of the *Hepadnaviridae*, is one of several hepatitis viruses pathogenic for humans. The virus causes severe liver damage. Some infections may become chronic. The virus also is strongly associated with liver cancer.

Morphology and Genome Properties. Hepatitis B virus exhibits several morphologies, including spherical and filamentous forms (Figure 343). It is an enveloped virion with an icosahedral capsid. The genome consists of a partially double-stranded DNA molecule.

A virion can contain outer-shell surface antigens (HB_sAg), capsid proteins in the form of a core antigen (HB_cAg), and a derivative of the core antigen (HB_gAg). Figure 343 shows the relationship of certain parts.

Transmission. Hepatitis B virus is transmitted through contact with contaminated blood. Transfusions with contaminated blood or blood products, sexual activity with infected persons, and transplacental infection (from infected mother to fetus) are the major means of transmission.

Hepatitis C Virus

Hepatitis C Virus (HCV) is the nation's most common blood-borne infection, and is viewed as one of the most serious of the hepatitis viruses. The highest prevalence of HCV infection, which amounts to about 70–90%, is reported among persons with substantial or repeated direct percutaneous exposures to blood. These include injection drug users (IDUs), recipients of transfusions from HCV-positive donors, and hemophiliacs treated with clotting factor concentrates that did not undergo viral inactivation. Infection can result in cirrhosis, liver cancer, and liver failure, and is the major reason for liver transplants in the United States, accounting for 1,000 of the procedures annually.

RNA Viruses (Continued)

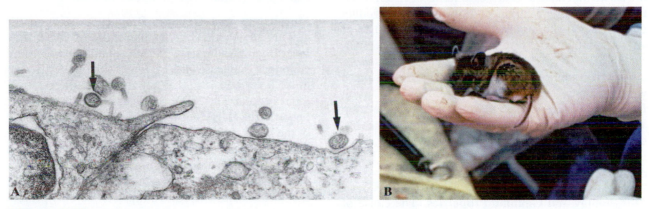

Figure 339
Hantavirus. (A) A section of cultured cells showing budding *Hantavirus* particles. (Courtesy of Drs. J.M. Hughes, C.J. Peters, B.W. Cohen, and J. Mahy, Centers for Disease Control and Prevention.) (B) A typical deer mouse.

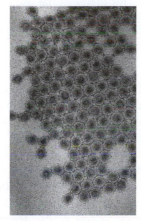

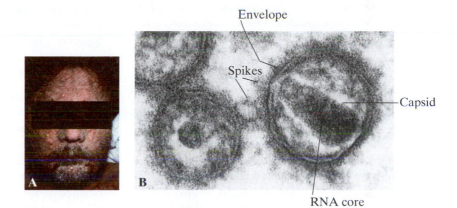

Figure 340
A transmission electron micrograph of Hepatitis A virus, *Hepatovirus*. (100 nm)

Figure 341
Retroviruses. (A) A case of cystic acne with increased severity in an HIV-infected patient. (From A.G. Martin et al. *Brit. J. Dermatol.* 126(1992):617–20.) (B) A greatly enlarged tissue section of human immunodeficiency virus particles to show viral parts.

DNA Viruses

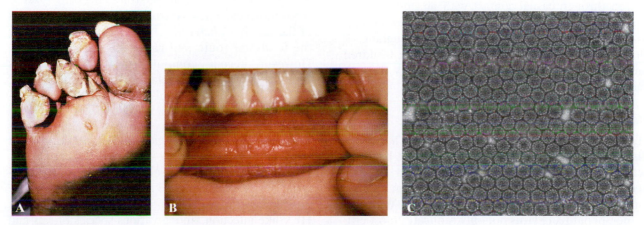

Figure 342
Papillomaviruses. (A) A case of plantar warts. (B) Warts on the lower lip. (C) The nonenveloped virions of human papillomavirus (100,000×). (From T. Iwasaki et al. *J. Path.* 168(1992):293–300.)

HCV is a member of the family *Flaviviridae,* a group of viruses that have similarities in terms of virion structure, genome organization, and replication (reproductive) strategy. The family includes the viral causes of yellow fever, dengue fever, and Japanese B encephalitis. HCV has been assigned to the genus *Hepacivirus.*

Aside from blood donor screening at present there is no vaccine or other related means of preventing HCV infections. Thus the ultimate goal is to develop an effective vaccine with which to prevent new cases, especially in underdeveloped countries, where HCV infection is more prevalent and treatment is financially beyond the reach of most persons.

Morphology and Genome Properties. The cause of hepatitis C infection is a small, 30 to 34 nm, enveloped virus. HCV has a single-stranded, positive-sense RNA genome, and its envelope contains lipid.

Various studies indicate that the virus circulates in the body in two forms, interactive virus particles or virions, and nucleocapsids without a lipid envelope.

Transmission. In the recent past, persons receiving blood transfusions were a main risk group for HCV infection. This was mainly because prior to 1990, there was no reliable test with which to screen the blood supply for HCV, so many persons were unknowingly infected. Because of improvements in blood screening, the odds of contracting HCV infection from donated blood are 1 in 100,000 units of blood.

Transmission from an infected mother to her fetus is rare, and results in infant infections in only 5% of cases. Currently, there is no evidence that breastfeeding spreads HCV.

Sporadic infections are possibly caused by acquiring the virus through contamination of cuts, wounds, or medical injections with the blood or body fluids of infected persons. Further, some health experts say there may be an as-yet-unknown transmission pathway for the virus. Most experts agree that HCV cannot be acquired through casual contact with an infected person, such as shaking hands, hugging, or even kissing. The virus also is not spread by sneezing, coughing, or sharing eating utensils or drinking glasses. Children have become infected with HCV primarily through the transfusion of blood or blood products.

Hepatitis D Virus

Hepatitis D (delta) virus (HDV) is a novel infectious disease agent quite unlike any form that has been described among human pathogens. It is a subviral agent that can only replicate with help from another virus. In nature, that helper virus is HBV. With this helper, hepatitis D virus not only replicates most efficiently in liver cells, but greatly increases the severity of liver damage caused by HBV. The main functions or roles of HBV are to assist in HDV entry into susceptible cells and subsequent virus assembly.

Morphology and Genome Properties. HDV particles are somewhat smaller than those of HBV. They measure approximately 35 to 37 nm in diameter and contain a single strand of RNA surrounded by an outer layer consisting of hepatitis B surface (envelope) protein. The HDV particles are similar to their helper virus (HBV) at least in terms of the presence of the same envelope (surface) proteins.

The RNA genome of HDV is combined with an HDV-specific protein (delta antigen) without forming any specific structure such as a nucleocapsid with the nucleic acid. The delta antigen is the only protein made by HDV.

Transmission. Natural HDV infections seem to be transmitted in the same manner as those caused by HBV. The virus in the blood of an infected individual is spread parenterally. This includes the use of contaminated needles, transfusions with contaminated blood or blood products, and sexual transmission.

Hepatitis E Virus

Hepatitis E virus (HEV) infection is a major cause of epidemic hepatitis, and acute sporadic hepatitis in adults in several developing countries. The existence of the virus was suspected on epidemiological grounds for many years. It was isolated and genetically investigated in the early 1990s. With the exception of a few isolated cases, individuals from developed (industrialized) countries who have had acute HEV infection have traveled to countries where the virus is endemic. HEV is considered to be less infectious than HAV.

Morphology and Genome Properties. Hepatitis E virus (HEV) is an icosahedral (20-sided), nonenveloped particle measuring 32 to 34 nm in diameter. The HEV genome is a single-stranded, positive-sense RNA molecule. HEV is structurally similar to viruses of the family *Caliciviridae,* which are transmitted by the fecal-oral route and cause diarrhea in humans. However, recent computer analysis of HEV's genome shows a closer relationship to rubella virus. Eventually, with additional studies, HEV may be reclassified and placed into a separate virus family.

Transmission. Epidemiologic studies indicate that HEV is transmitted by the fecal-oral (enteric) route. Contaminated water has been the primary source of infection in most investigated cases. Recent studies have provided convincing evidence of widespread HEV- or HEV-like infection among various species of rats and mice in the United States. Such findings suggest the existence of reservoirs for HEV other than humans.

Herpes Viruses

Eight members of the family *Herpesviridae* are known to infect humans: herpes simplex virus types 1 and 2 (HSV-1 and HSV-2), varicella-zoster virus, cytomegalovirus, Epstein-Barr virus, and human herpesviruses 6, 7 (Figures 344 through 352). All of these viruses persist for life after the primary infection and are reactivated at intervals during the infected individual's lifetime.

Herpesvirus-6 has been associated with disorders involving white blood cell increases, febrile illness, hepatitis, and non-Epstein-Barr virus-caused infectious mononucleosis. This virus also has been shown to be the cause of the skin infection exanthum subitum or roseloa infantum.

Herpesvirus-7 has been isolated from patients with chronic fatigue syndrome, roseola infantum, enlarged livers and spleens, and decreased blood platelets

Herpesvirus-8 mainly has been isolated from patients with Kaposi's sarcoma and certain other forms of cancer.

Morphology and Genome Properties. Herpesviruses share a similar morphology. The virus particles consist of a central core containing linear double stranded DNA; an icosahedral capsid of 162 capsomeres surrounding the core with an envelope that, on the outside, displays glycoprotein spikes; and material called the **tegument** filling the space between the capsid and the envelope (Figure 347B). The envelope is derived from the host cell's nuclear membrane.

Herpes Simplex Viruses (Simplexvirus)

Herpes simplex viruses (HSV) types 1 and 2 are the most frequent causes of disease. Infection involves the body surface, where, in addition to replicating and causing cellular destruction (Figure 344A and B, F), the virus enters sensory nerve endings. Common sites for herpetic lesions are the skin and mucous membranes around the mouth (Figure 344A) and genital openings, the oral cavity itself, the cornea, the cervix, the anus, and the urethra.

Important secondary targets include the brain, the meninges, and the cornea. Herpetic whitlow (see Figure 315) is a special type of herpetic skin infection. It involves the finger, may be quite painful, and is commonly found among health care workers. Recovery from the first infection results in latency.

Transmission. Herpes simplex virus infection is acquired through direct contact with active herpetic lesions. Neonatal infections also can occur through contact with HSV in the birth canal of infected mothers.

HHV-2 infection is acquired from persons shedding the virus at the time of sexual intercourse or direct contact with genital secretions. HHV-2 shed on the genital mucosal surfaces is the principal source of virus in transmitted infections. In addition, the practice of anal intercourse has resulted in an increasing number of cases of primary *herpes proctitis*, an inflammation of the rectum and anus. Direct contact with infectious material can also result in eye infections.

The presence of HHV-2 infection during pregnancy carries with it an increased risk of spontaneous abortion, premature birth, and the possibility of a life-threatening neonatal infection.

Varicella-zoster Virus (Varicellovirus)

Chickenpox and herpes zoster (shingles) are different clinical effects of infection by the same virus, varicella-zoster virus (VZV; Figure 348F). Chickenpox is generally regarded as a childhood disease. Infections, however, can be severe and even fatal in high-risk groups, such as immunocompromised persons and newborns. Recovery from primary infections leads to latency. Herpes zoster results from reactivation of the virus in sensory nerve ganglia (Figures 348A, B). Its incidence and severity increase with age, and it seldom occurs in the young. Once activated, the virus moves down the sensory nerves until it reaches the skin and produces lesions similar to those of chickenpox. The lesions are confined to the areas affected in chickenpox, including thoracic, lumbar, and facial regions (Figures 348A–D). In the pre-vaccine era, chickenpox was primarily a disease of children under 10 years of age, while shingles was an illness mainly involving individuals over 50 years of age. Today, shingles remains a disease of persons over the age of 50, and has extended its attack zone to include immunocompromised persons. Among the latter group, bone-marrow-transplant recipients and children with human immunodeficiency virus are at a particular risk for VZV infection.

Shingles occurs as a localized, unilateral, and painful vesicular rash. The lesions appear in crops and, although often larger than those seen with chickenpox, progress through pustular and crusting stages, and heal as in cases of chickenpox, but at a slower pace. Scabs are noninfectious but may persist for two or more weeks.

Infected cells may show the presence of prominent red (eosinophilic) intranuclear inclusions called Cowdry A bodies (Figure 348E).

Transmission. Infection with VZV is acquired by aerosol.

Cytomegalovirus

Cytomegalovirus (CMV) is considered to be the most important cause of congenital infections in the United States and a major opportunistic disease

DNA Viruses (Continued)

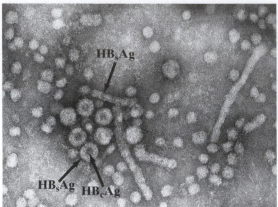

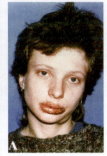

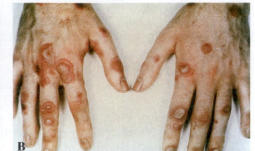

Figure 344

Herpes simplex virus (HSV) type 1 (*Simplexvirus*). (A) A severe crop of blisters (vesicles) in a primary herpes infection. (B) Typical target lesions associated with a form of recurrent HSV infection.

Figure 343

Hepatitis B virus, *Hepadnavirus*. Virions appear as circular and filamentous forms.

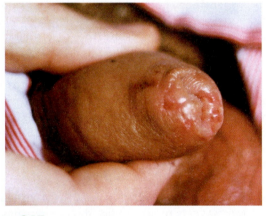

Figure 345

A case of herpes simplex type 2 infection. Several lesions are located on the prepuce or foreskin or this individual. (Courtesy of Dr. Jyoti Dhar, *et al.*, University Department of Genitourinary Medicine, Royal Liverpool University Hospital, Prescot St., Liverpool.)

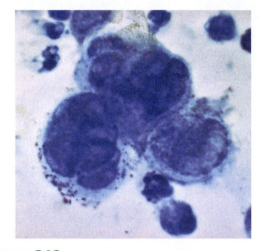

Figure 346

The results of the Tzanck test with a tissue smear showing multinucleated giant cells typical of herpes infections.

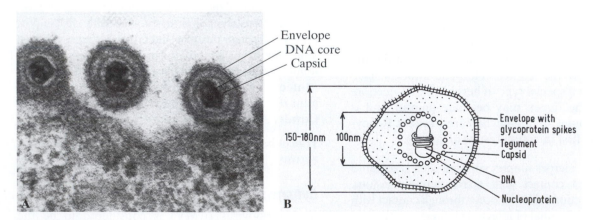

Figure 347

Herpesvirus structure. (A) A HSV particles in tissue shown by transmission electron microscopy (120,000×). (B) A diagrammatic view of herpes simplex type 1 virus. (M. Norval. *Clin. & Exp. Dermatol.* 17(1992):221–37.)

agent in organ transplant recipients and in immuno-suppressed individuals (Figure 349). The most fre-quent manifestation of CMV in patients with AIDS is a sight-impairing condition, known as chorioretini-tis, that can result in blindness. In addition, CMV is an occasional cause of infectious monocucleosis in apparently normal individuals.

Cytomegalovirus causes a systemic primary infec-tion and becomes widely distributed in the body. Most of such infections are without signs and symp-toms. Severe fetal infections known as **Cytomegalo-inclusion disease** occur mainly in pregnant women experiencing a primary infection.

The development of immunity results in latency, but the virus can be reactivated by the use of immuno-suppressive agents during pregnancy.

Transmission. Cytomegalovirus is spread by direct contact with infectious body fluids, such as saliva, urine, semen, and cervical secretions.

Epstein-Barr Virus (Lymphocryptovirus)

Epstein-Barr (EBV) is the major cause of infectious mononucleosis. It is also associated with other dis-eases, including hairy leukoplakia (Figure 350A), na-sopharyngeal cancer, African Burkitt's lymphoma (Figure 351), Hodgkin's disease, and salivary gland cancer. The primary targets of the virus are the sali-vary glands, where susceptible B lymphocytes be-come infected. These infected cells spread the virus (Figure 350B) by way of the lymphatics and blood to distant lymphoid and related organs.

Infectious mononucleosis is generally a self-limiting disease.

Transmission. The major route of infection for infec-tious mononucleosis (IM) in young adults is proba-bly through intimate oral contact, as in kissing and the exchange of saliva. The mechanism for transmis-sion of EBV in nurseries and in young children in low socioeconomic environments presumably occurs by contact with infectious saliva on fingers, toys, and other inanimate objects. The virus replicates in cells in the oropharynx, and nearly all seropositive per-sons (individuals having antibodies against the virus) actively shed EBV in their saliva.

Kaposi's Sarcoma

Kaposi's sarcoma (KS) has received immense atten-tion since 1981, when it was first considered as a major sign of AIDS. Four forms of KS have been described: 1) an inactive endemic condition which appears as nu-merous pigmented growths on the skin of elderly men of Eastern European or Mediterranean descent; 2) a form primarily found among African children; 3) a

disorder occurring in persons receiving immunosupres-sive therapy, as in cases of organ transplantation; and 4) a modern form associated with HIV infection. Vari-ous studies have demonstrated the presence of a virus designated as Kaposi's sarcoma–associated her-pesvirus or human herpesvirus-8 in all forms of KS. Currently, HHV-8 is recognized as a gamma her-pesvirus that is an important cause of conditions not only of KS, but primary effusion lymphoma and Castleman's disease, a condition in which benign giant lymph node enlargement occurs.

Kaposi's sarcoma and HHV-8 have a worldwide distribution.

Transmission. HHV-8 has been found in certain body fluids of HIV-infected persons, such as nasal se-cretions and semen. While the mode of transmission is not known in all cases, a number of factors and sources have been implicated. These include sexual intercourse, intravenous drug use with infectious ma-terial, and contaminated blood-clotting factor prepa-rations used for hemophiliacs.

Poxviruses

The poxviruses are classified into the family of *Poxviridiae*, which consists of the largest group of ani-mal viruses known. The family contains a number of medically important members, such as smallpox, mon-keypox, vaccinia, molluscum contagiosum, camelpox, buffalopox, and catpox viruses.

Morphology and Genome Properties. Individual virus particles observed by electron microscopy appear somewhat rounded, brick-shaped, or ovoid (Figure 353A). They range in length from 220 to 450 nanome-ters (nm), and in width from 140 to 260 nm. All poxviruses contain a nucleoprotein core, which in spe-cially prepared specimens appears dumbbell-shaped and surrounded by a number of membranes (Figure 353B). The genome (genetic contents), which is in the nucleoprotein core, consists of a single molecule of double-stranded deoxyribonucleic acid (DNA).

Smallpox

From the earliest days of recorded history smallpox virus has left it indelible mark on the medical, political, and cultural affairs of humankind. Records emphasize the horrendous nature of epidemics from these earliest of times. By the eighteenth century, smallpox had be-come endemic in major European cities. Two basic forms of smallpox are recognized: **variola major**, a highly virulent form, and **variola minor** or **alastrim**, a less dangerous version of the disease (Figure 354).

In January 1967, the World Health Organization (WHO) launched an immunization program for the

DNA Viruses *(Continued)*

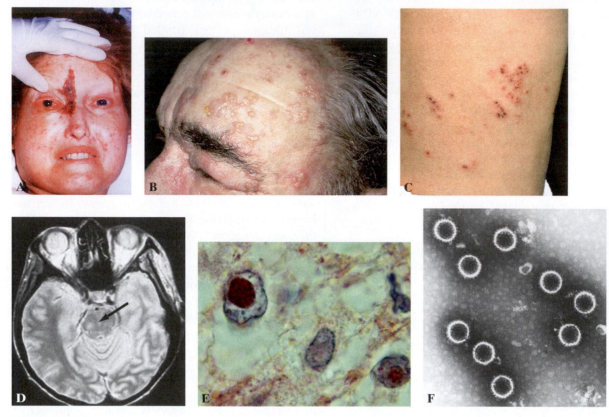

Figure 348

Varicella-zoster virus (*Varicellovirus*). (A) A case of herpes zoster ophthalmicus and optic nerve dysfunction. (B) and (C) Appearance of shingles on different parts of the body. (D) A magnetic resonance image showing central nervous system involvement (arrow). (From F. J. Lexa *et al. AJNR.* 14(1993):185–90.) (E) Cowdry type A intranuclear eosinophilic inclusion within a neuron (400×). (From A. Moulignier *et al. CID.* 20(1995):1378–80.) (F) A transmission electron micrograph of negatively stained varicella nucleocapsids. (Original magnification 99,200×.)

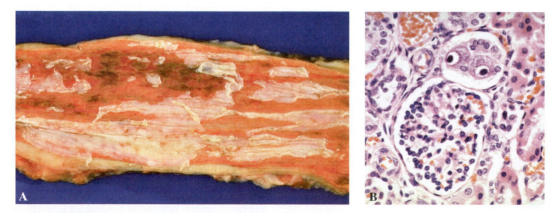

Figure 349

Cytomegalovirus (CMV) infection. (A) Infection of the esophagus, showing numerous ulcers and tissue damage. (From F. S. Buckner and C. Pomeroy, *CID* 17(1993):644–56.) (B) A kidney tissue section showing large intranuclear CMV inclusion bodies. The inclusions are referred to as owl's-eye cells. (From R. Herriot and E.S. Gray. *NEJM.* 331(1994):649.)

DNA Viruses *(Continued)*

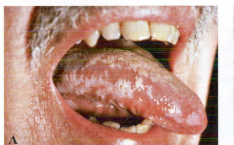 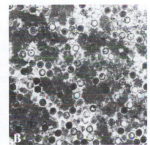

Figure 350
Epstein-Barr virus (*Lymphocryptovirus*). (A) A case of oral hairy leukoplakia (OHL). (Courtesy of Dr. P. Itin, Kantonsspital, Basel.) (B) Electron micrograph showing large numbers of Epstein-Barr virions (20,000×). (From J. Kantakis et al. *Brit. J. Dermatol.* 124(1991):483–86.)

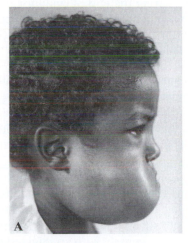

Figure 351
A child with the malignant tumor known as Burkitt's sarcoma. (Courtesy of Dr. S. R. Prabhu.)

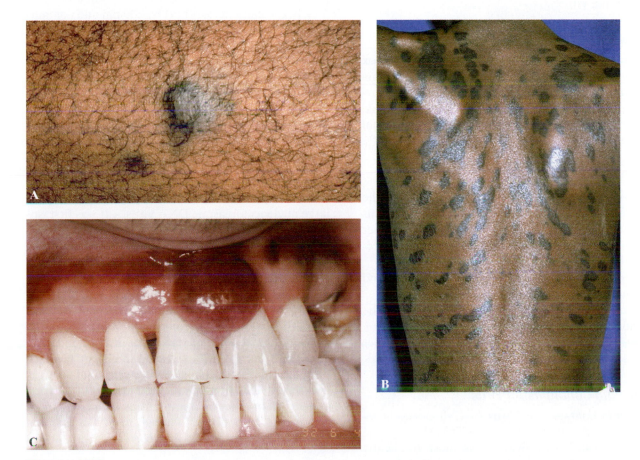

Figure 352
Kaposi's sarcoma (KS) a newly associated herpesvirus-8 infection. (A) A close-up of a lesion. (B) Kaposi's sarcoma in an HIV-infected person. (C) Intraoral KS. (From B.C. Muzyka and M. Glick. *NEJM* 328(1993):210.)

global eradication of smallpox. This goal was achieved through a conscientious and worldwide effort on May 8, 1980, when the WHO issued its historic finding declaring the end of smallpox on this planet. Unfortunately, soon after the announcement was made, smallpox virus was added to the list of viral and bacterial bioterrorist weapons.

Transmission. Transmission of smallpox occurs from person to person, primarily by respiratory tract secretion. Even though smallpox virus is considered to be highly contagious, its spread is slow, and the probability of infection from a single exposure is considered to be low.

Oral secretions are the main source of contaminating the face, the body in general, the clothes, and the bedding of infected individuals. Direct face-to-face contact with an infected person via infectious droplets or physical contact with an infected person's contaminated articles is usually responsible for the spread of smallpox. Later in the infection cycle, skin lesions, including vesicles, pustules, and crusts, may become a source of the virus.

Vaccinia Virus

Vaccinia virus is serologically related to the smallpox virus, although its exact origin is unknown. Vaccinia virus was used for smallpox immunization. It is currently used for immunization against monkeypox. The virus can cause potentially serious and lethal complications (Figure 239a).

Cowpox

Cowpox is closely related to vaccinia virus. It is the cause of a rare occupational human infection that is acquired through contact with infected cows or other animals. Infected cats, rats, and zoo and circus elephants have been sources of the virus.

Cowpox has been reported in parts of Europe and in adjoining regions of Asia.

In humans, lesions occur mainly on the fingers. The infection site first becomes reddened and swollen, and then becomes papular in 4 to 5 days. Vesicles forms soon thereafter. Healing takes about 3 weeks.

Molluscum Contagiosum Virus

Molluscum contagiosum virus (MCV) causes a benign form of skin tumors (Figure 355b). Although the disease is found worldwide, it is most frequently found as an easily treated disease of childhood.

Transmission. Transmission of MCV can occur by direct contact with contaminated articles, such as towels, and by sexual contact.

Monkeypox Virus

From about the early 1970s to the present, cases of human monkeypox (Figure 314B) have been observed only in the rain forest of Africa, mostly in Zaire, but also in Benin, Cameroon, Gabon, Ivory Coast, Liberia, Nigeria, and Sierra Leone.

The results of a number of virologic and related investigations of wild animals in the 1980s suggest that primates are sporadically infected with monkeypox virus, as are humans. The role of wild primates in sustaining virus transmission or as a source of human infections has been difficult to determine; however, human infection does correlate with contact with infected animals.

Morphology and Genome Properties. Monkeypox virus contains a single molecule of double-stranded DNA, and is genetically distinct from the other poxviruses. The virion of the virus measures approximately 250 nanometers (nm) by 200 nm and contains distinctive polypeptides located in its envelope.

Transmission. Normally, monkeypox is a rare zoonosis, with virus transmission between humans occurring extremely infrequently. Normally, monkeypox virus appears to enter the mucosa of the upper respiratory tract or through skin abrasions.

Mimivirus—An Unusual Find

Mimivirus, which is short for "mimicking microbe," is a relatively new virus related phylogenetically to large nucleocytoplasmic DNA viruses. It was first isolated in England by T.L. Rowbotham and his associates. The virus (Figure 357) contains a large genome consisting of a double-stranded DNA molecule which is larger than that of at least 25 bacteria. Mimivirus resembles intracellular microorganisms such as the rickettsia. However, it also has a number of viral properties, which include not undergoing binary fission, and the absence of ribosomal protein-encoding genes and any enzymes involved in energy metabolism. The mimivirus challenges conventional views as to where viruses fit along the boundary setting viruses apart from other microorganisms such as bacteria. The mimivirus is distinct enough to be the first isolate of a new virus family, *Mimiviridae*.

Pox Viruses

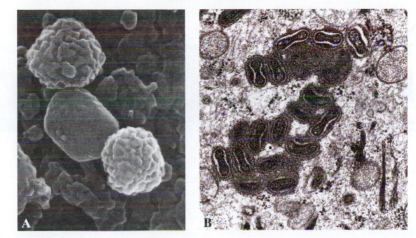

Figure 353
Two different views of poxviruses. (A) A scanning electron micrograph showing the brick-like appearance of virions. (B) A transmission micrograph showing the internal features of a cluster of poxviruses. Note the complex organization of individual particles.

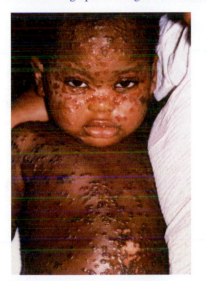

Figure 354
One of the last victims of the viral disease, smallpox. The skin surface shows various stages of the disease, including vesicles, papules, and pustules.

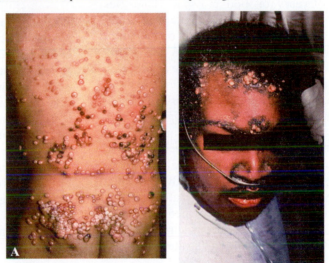

Figure 355
Molluscum contagiosum virus (*Molluscipoxvirus*). (A) Two examples of molluscum contagiosum infection. (A) Severe infection showing buttons. (B) An AIDS patient with an opportunistic molluscum infection. (Courtesy of Dr. Michael M. Lederman, Case Western Reserve University, Cleveland, Ohio.)

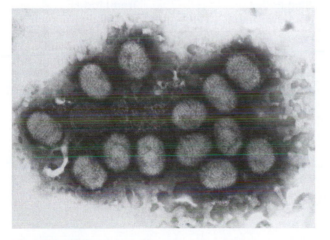

Figure 356
A transmission electron micrograph of a cluster of molluscum contagiosum virions shown by negative staining.

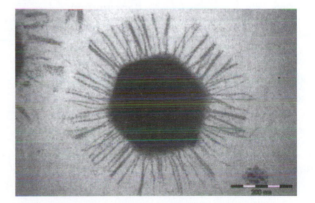

Figure 357
An electron micrograph of mimivirus. (From D. Raoult. ASM News. 71: 278–284, 2005. and Pr. D. Raoult, Unite des Rickettsies, Marseille, France.))

Lines on the antiquity of parasites: "Adam . . . had 'em."

—Gilian Strickland

Animals and plants have competed for millions of years for food and space. Over this time parasites have invaded almost every type of living host, exhibiting various degrees of dependence. The major groups of animal parasites are found among the helminths (Figure 358) and the arthropods.

Parasitic worms live in varying environments and must adapt to conditions in order to survive a host, body's cellular and chemical defenses (Figure 359). A parasitic helminth's (worm) existence and survival are in large part determined by the development of certain structural and metabolic modifications. These include an especially hard outer body covering; hooks (Figure 358C), spines, cutting plates, suckers, and other structures for purposes of penetration or attachment to body tissues; various enzymes, and elaborate reproductive systems and strategies.

▲ Classification

The common human helminthic parasites can be placed into one of three classes on the basis of body and digestive system properties, general body organization, nature of the reproductive system, and the need for more than one host species for the completion of the worm's life cycle. These classes are the **cestodes** (tapeworms); **nematodes** (roundworms), and **trematodes** (flukes). Table 20 lists the main features of these parasites.

The general laboratory diagnosis of helminthic diseases is based on the finding and identification of ova, or eggs (Figure 360A), larvae (young or developing forms), or adult worms (Figure 358D). Skin tests and various serological tests also are of value.

Table 20 Comparison of Cestode, Nematode, and Trematode General Features

Feature	Cestodes (tapeworms)	Nematodes (roundworms)	Trematodes (flukes)
Shape	Tapelike, in segments	Cylindrical, unsegmented	Flat, leaflike;[a] unsegmented
Suckers	Present	Absent	Present
Hooklets	Often present	Absent	Absent
Digestive canal	Absent	Present	Present, although not complete
Sex organs	Present in same worm	Separate	Generally present in same worm[a]

[a]Blood flukes (schistosomes) are exceptions.

Helminthology

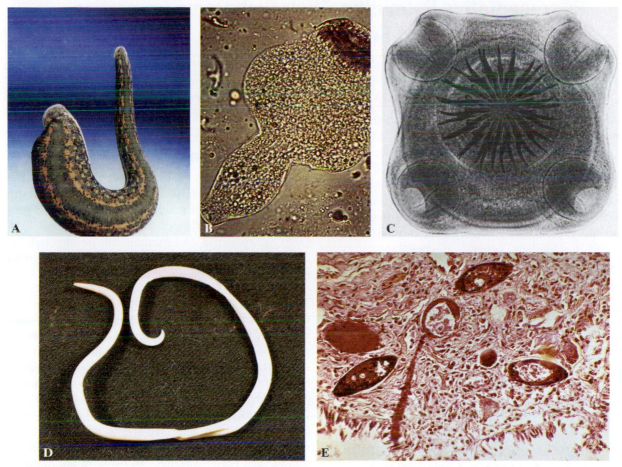

Figure 358

Examples of helminths. (A) Leeches such as this one, *Hirudo medicinalis,* because of their ability to repeatedly suck blood from a variety of hosts, may serve as a source of microbial pathogens (2.5×). (Courtesy of Dr. H. Mehlhorn.) (B) A microscopic view of the sheep tapeworm. Note the hooklets (dark top) on the scolex, or head. (C) Anterior view of a tapeworm scolex showing both suckers and sharp teeth-like hooklets. (D) An adult *Ascaris,* a roundworm, that is visible to the eye. (E) Flukes (trematodes) in a tissue specimen.

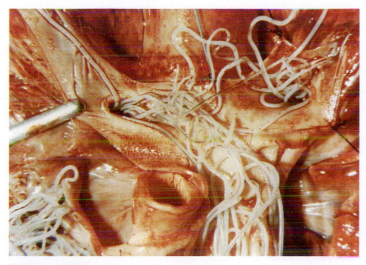

Figure 359

Opened heart, showing worms in the right portion of the organ and in the pulmonary artery (upper center and lower right side).

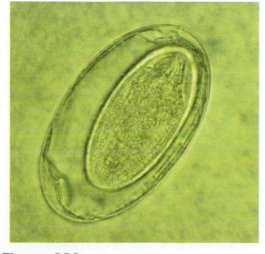

Figure 360

Diagnostic properties of helminths. (A) Unstained *Moniliformis moniliformis* ovum found in a stool specimen.

Representative Helminths, Their Distinctive Properties, and Selected Diseases

▲ Cestodes

Cestodes, or tapeworms, have flat, ribbon-shaped bodies. They have a **scolex,** or head, at the anterior, or front, end (Figure 358C). For purposes of attachment, scolexes have **suckers** (Figure 365) and at times, depending on the worm, **hooklets** (Figures 358B and C). Immediately behind the scolex is a neck region that produces reproductive body segments known as **proglottids** (Figure 361). Each segment contains both male and female sex organs, making the worm **hermaphroditic**. The proglottids located farther from the head are sexually mature and, by cross-fertilization of proglottids of either the same worm or another one, produce ova (Figure 360). Several examples of cestodes and their distinctive microscopic diagnostic properties will be described.

Diphyllobothrium latum

Diphyllobothriasis is caused by the fish tapeworm, *Diphyllobothrium latum*. Most infected persons are asymptomatic. Some, however, experience a variety of gastrointestinal effects. Infection with *D. latum* also can interfere with vitamin B_{12} absorption.

Transmission and Diagnosis. Infection is acquired through the ingestion of raw, uncooked, or pickled freshwater fish containing young forms (larvae) of the worm. Diagnosis includes finding ova (Figure 366B) or typical proglottids in stool specimens.

Diphylidium caninum

Diphylidium caninum (Figure 362) is the cause of dog tapeworm infection. Cats and humans also can be infected. Children are usually the ones infected, especially if they have been around infected dogs or cats. Diarrhea and general discomfort are typical signs and symptoms.

Transmission and Diagnosis. Human infection is acquired by the ingestion of infected adult fleas. Such fleas contain larval forms of the worm. Once inside the gastrointestinal tract of the host, the larval form develops into a mature tapeworm (Figure 362).

Diagnosis of *D. caninum* infection is based on finding adult worm proglottids in stools or released around the anus. The scolex of this worm contains a fleshy protrusion or swelling with one or more rows of hooks (Figure 362).

Echinococcus granulosis

Echinococcosis, or hydatid disease, is caused by the sheep tapeworm, *Echinococcus granulosis*. A hydatid is the larval stage of the adult worm. The adult is very small and has only three proglottids (Figure 363). Dogs, sheep, cattle, and humans can be infected. Infection can result in blockage and interference with the functioning of organs such as the liver.

Transmission and Diagnosis. Humans acquire hydatid disease as a consequence of ingesting ova, usually found in infected dog feces. Sheep ingest eggs on contaminated grass. The eggs hatch in the intestine and develop into embryos, which eventually make their way to the liver. Some embryos may be carried by the bloodstream to other body organs. Embryos develop into hydatid cysts (Figure 364A). Special structures called **brood capsules** bud from the cysts and give rise to several potential adult worms known **protoscolexes** (Figure 364B). Ingestion of such brood capsules can result in the development of adult worms.

Diagnosis is best made by showing cysts through the techniques of ultrasound, magnetic resonance imaging (MRI), or computed tomography (CT scans). Surgically removing cysts and finding potential tapeworms and hooklets (Figure 364B) are diagnostic.

Taenia

Taenia solium is the pork tapeworm (Figure 365) and *T. saginata* is the beef tapeworm. Infected individuals generally complain of slight pain, but most are asymptomatic.

Transmission and Diagnosis. *Taenia solium* infection results from the ingestion of raw or inadequately cooked pork containing larval forms (cysticerci) of the worm. Similarly, *T. saginata* infection occurs as a consequence of eating raw or undercooked beef containing larvae of the worm.

Diagnosis is based on finding the characteristic proglottid associated with each *Taenia* species. The ova of the two species are identical (Figure 366A).

Helminthology *(Continued)*

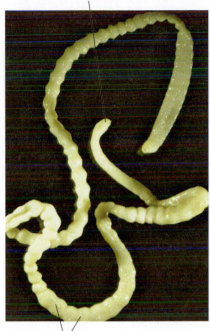

Scolex

Proglottids

Figure 361
A complete adult worm measuring 133 mm in length. (From R.C. Naafie and A.M. Marty, *Clin. Microbiol. Revs.* 6(1993):34–56.)

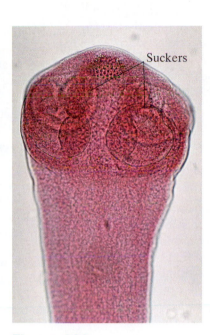

Suckers

Figure 362
Diphylidium caninum adult worm. Note the round suckers and rostellum (a fleshy swelling or protrusion) with its rows of hooks near the top of scolex.

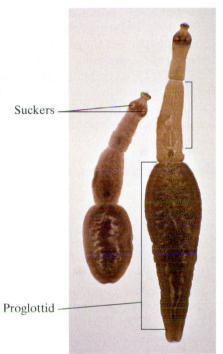

Suckers

Proglottid

Figure 363
Adult sheep tapeworms, *Echinococcus granulosus.*

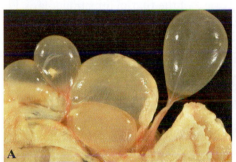

A

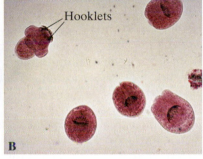

Hooklets

B

Figure 364
Echinococcus granulosis (sheep tapeworm). (A) Hydatid cysts *in situ.* (Courtesy of Dr. L. Alpert, pathology Department, The Sir Mortimer B. David Jewish General Hospital.) (B) Developing worms called protoscolexes. Note the hooklets on the scolexes.

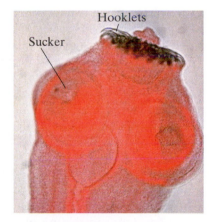

Hooklets

Sucker

Figure 365
Taenia solium (pork tapeworm). Adult worm.

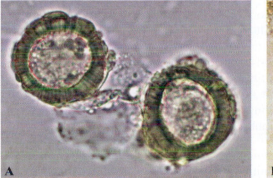

A

B

Lid

Figure 366
Cestode ova. (A) Taenia solium (pork tapeworm). (B) *Diphyllobothrium latum* (fish tapeworm) ovum. Showing its operculum, or lid.

▲ Nematodes

Nematodes, or roundworms, have a cylindrical tapered body (Figure 358D). Their tubular digestive tract extends from the mouth at the anterior end to the anus at the posterior end. The sexes are separate, with males typically being smaller than females. Nematodes can be divided into those that typically inhabit the gastrointestinal tract of hosts and those that are found in the blood and other tissues of hosts.

Ancylostoma duodenale and Necator americanus

Ancylostoma duodenale and *Necator americanus* are the causes of hookworm infection (Figures 367A–D). The signs and symptoms are generally proportional to the number of worms (**worm burden**) in the host. Abdominal pain, diarrhea, increased numbers of eosinophils (**eosinophilia**), and anemia can occur. Skin lesions also may be found in areas where hookworm larvae (Figure 367E) enter the host. The hookworm larvae of dogs and cats can also penetrate human skin and cause **cutaneous larva migrans** (creeping eruption). These larvae wander about in the skin, producing linear skin lesions (Figure 367E).

Transmission and Diagnosis. Infection is acquired by direct contact with soil that contains infective larvae. Larvae hatch from ova in the soil and often pass through two distinct stages. The first, known as *rhabditiform larvae* go on either to develop into free-living adults or to become infective *filariform larvae*. Worms having rhabditiform larvae include hookworms and *Strongyloides stercoralis* (Figure 372). Filariform larvae develop no further unless they make contact with exposed human skin and gain entrance into the body by penetration. The most common area of such invasion is the feet. After penetration, larvae are transported via the circulatory system to the lungs. Subsequently, the larvae escape from the lungs, are swallowed, and enter the intestinal tract, where they develop into mature male and female worms. Mating results in the formation and depositing of ova (Figure 367A). Identification of the ova of the two hookworm species is based on a comparison of the mouthparts; *A. duodenale* has pairs of teeth, and *N. americanus* has cutting plates.

Diagnosis depends on finding ova or larvae or both in stool specimens.

Ascaris lumbricoides

Ascaris lumbricoides is the cause of ascariasis (Figure 368A). Infected individuals may experience abdominal pain, diarrhea, and other gastrointestinal problems. Heavy intestinal infection may produce obstruction and liver, gallbladder, and pancreas involvement.

Transmission and Diagnosis. Infections are acquired by the ingestion of ova (Figure 368B) in contaminated food or water. Fomites and fecal-contaminated fingers also are possible sources of infection.

Diagnosis is based on finding ova in stool specimens (Figure 368B).

Enterobius vermicularis

Enterobius vermicularis is the cause of enterobiasis, or pinworm. Most cases are asymptomatic. However, depending on the migration activity of pinworms, infected persons may develop severe itching around the anus or the vagina (Figure 369A).

Transmission and Diagnosis. Pinworm infection is acquired by the ingestion of ova (Figure 369B). Contaminated fingers, toys, and other objects may be sources of ova. Ova also may be inhaled.

Diagnosis can be made by applying cellulose double sticky surface tape to the perianal area in the morning and before the infected individual bathes or has a bowel movement. Microscopic examination of the tape should show the presence of the ova (Figure 369B).

Loa loa

Loa loa is the cause of the filarial infection loiasis. Filaria are slender nematodes that have complex life cycles involving various insects as intermediate hosts. Adult worms migrate in subcutaneous tissues and at times move across the eye (Figure 370). Localized swellings are commonly found in infected persons. The disease is found in various parts of Africa.

Tranmissions and Diagnosis. *Loa loa* is transmitted by mango flies belonging to the genus *Chrysops*.

Diagnosis is based on finding young (embryonic) forms of the nematode, known as **microfilariae.**

Onchocerca volvulus

Onchocerca volvulus is the cause of another filarial infection, onchocerciasis (river blindness). Infected individuals develop painless subcutaneous swellings over bony body parts (Figure 371). Associated skin areas also itch and thicken, and lymph nodes enlarge. Microfilariae may also involve the eye and cause visual problems. The disease is found in tropical areas including Central and West Africa, Mexico, and various South American countries.

Transmission and Diagnosis. *Onchocerca volvulus* is transmitted by black flies from the genus *Simulium*.

Nematodes

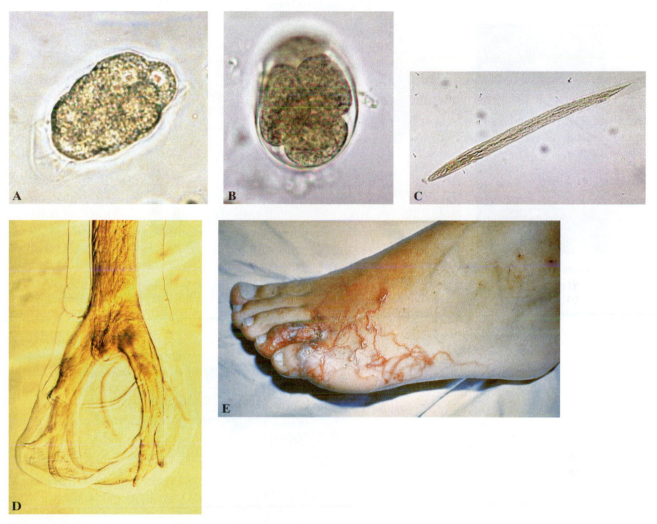

Figure 367

Ancylostoma duodenale and *Necator americanus* (hookworm). (A) *A. duodenale* ovum. (B) *N. americanus* ovum. (C) An infective (filariform) larva of *N. americanus.* (D) The bursa, or expanded posterior end, of the male worm. (E) A case of creeping eruption, also known as cutaneous larva migrans.

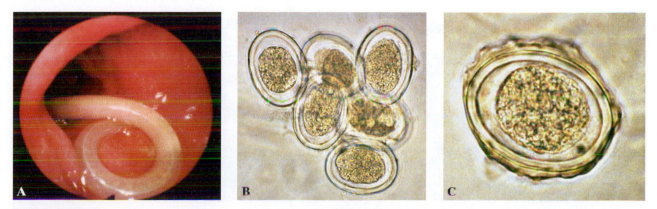

Figure 368

Ascaris. (A) A single adult *Ascaris* observed in the digestive tract. (From H.S. Fuessl, *NEJM* 331(1994):301.) (B) Unfertilized ova. (C) A fertilized ovum.

Nematodes *(Continued)*

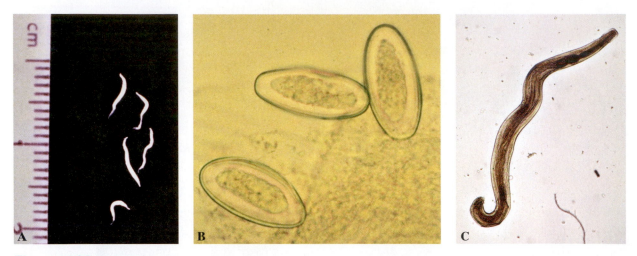

Figure 369
Enterobius vermicularis (pinworm). (A) Pinworms recovered from the perianal region of a young child. (Courtesy Dr. L. Alpert, Pathology Department, The Sir Mortimer B. David Jewish General Hospital.) (B) Ova on the surface of transparent tape. (C) An adult male worm stained.

Figure 370
Loa loa (eye worm). The worm (swollen area) can be seen near the surface of the eye.

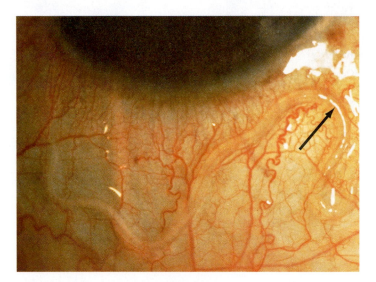

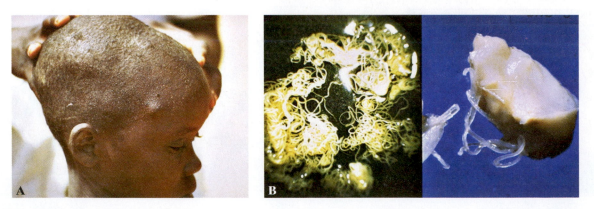

Figure 371
Onchocerca volvulus, the cause of onchocerciasis (river blindness). (A) Nodules on the head of an infected child. (B) A mass of live male and female *O. volvulus* (left) and a nodule with worms (right).

Laboratory diagnosis is based on finding microfilariae in skin specimens.

Strongyloides stercoralis

Strongyloides stercoralis causes strongyloidiasis. The cycle of this nematode is similar to that described for hookworms such as *Necator*. Infected individuals may experience respiratory problems such as coughing and difficulty in swallowing because of large numbers of larvae migrating through the lungs. Disseminated strongyloidiasis results in the involvement of other body organs. The infection is found worldwide.

Tranmission and Diagnosis. *Strongyloides stercoralis* infection is acquired by direct contact with soil that contains infective larvae. Such larvae can penetrate unbroken skin.

Diagnosis involves finding larvae in stool or other specimens (Figure 372).

Trichinella spiralis

Trichinella spiralis causes trichinosis. Early signs and symptoms of infection associated with the gastrointestinal tract and include nausea, vomiting, diarrhea, and abdominal pains. Later effects due to muscle invasion by larvae include fever, muscle pain, and tissue swelling of involved areas. Cardiovascular and central nervous system involvement also may occur.

Transmission and Diagnosis. *Trichinella spiralis* infection is acquired by the ingestion of undercooked pork or pork products and the tissues of wild animals containing the larvae of this nematode.

Diagnosis is usually based on clinical findings. Muscle biopsies are performed to confirm the infection (Figure 373).

Trichuris trichiura

Trichuris trichiura the whipworm, causes trichuriasis. Severe infections, especially those that occur in children, may result in bloody diarrhea and protrusion of the anus (**prolapse**). Light worm infections are generally asymptomatic.

Transmission and Diagnosis. *Trichuris* infection is acquired by ingestion of whipworm ova (Figure 374). Fomites and unsanitary conditions contribute to transmission.

Diagnosis is based on finding the typical ova in stool specimens.

Wuchereria bancrofti

Several nematodes cause lymphatic filariasis, also known as elephantiasis (Figure 375A). These include *Wuchereria bancrofti* (Figure 375B), *Brugia*

malayi, and *B. timori.* Early signs and symptoms of infection include inflammation of lymph nodes and channels fever and general discomfort. Males experience inflammation of the epididymis and the testes. Chronic infection leads to permanent enlargement of the legs (Figure 375A). Filariasis is widely distributed throughout tropical areas, including parts of Africa, Asia, Southeast Asia, Central and South America, Pacific islands, and some Caribbean islands.

Transmission and Diagnosis. *Wuchereria bancrofti* can be transmitted by a number of blood-sucking mosquitoes including *Culex tarsalis.*

Diagnosis is based on finding microfilariae in blood smears or other types of preparations (Figure 375B).

▲ Trematodes

Most trematodes, or flukes, are bilaterally symmetrical, flat, and leaf-shaped, and they generally have oral and ventral suckers for attachment (Figure 377). All species except schistosomes contain both sex organs in the same worm (hermaphroditic).

Flukes vary with respect to their life cycles. Many begin with the development of ciliated larval forms called **miracidia** (Figures 376A and 378 B) within ova. These larvae escape through hatching and penetrate the tissues of snails (first intermediate host). Further development results in the formation of other larvae, which escape from snails. Depending on the trematode species, other larval forms develop in other hosts. These include **sporocysts** and **rediae** (Figures 378C and D).

Cercariae, tail-bearing larvae (Figures 376B and 381H), represent a later stage in the life cycle of trematodes which are released from snail hosts and ready to attach to and attack susceptible hosts. With hermaphroditic flukes, cercariae round up and form **metacercariae** on aquatic plants or animals (Figure 376C).

Clonorchis (Opisthorchis) sinensis

Clonorchis sinensis (Figure 377), also known as *Opisthorchis,* is the Chinese liver fluke and is the cause of clonorchiasis. Most infections are asymptomatic. However, heavy fluke infections can result in bile duct blockage and gallbladder inflammation.

Signs and symptoms can include fever, diarrhea and abdominal pain.

Transmission and Diagnosis. Clonorchiasis is acquired by ingestion of raw, pickled, and dried freshwater fish containing metacercariae. Diagnosis is based on finding ova in stool specimens (Figure 377B).

Nematodes *(Continued)*

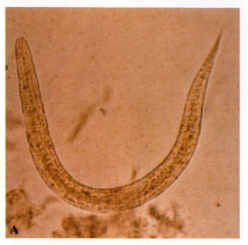

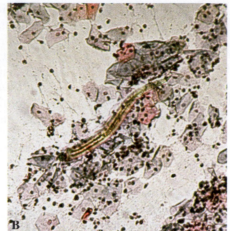

Figure 372
Strongyloides stercoralis.
(A) The rhabditiform larva.
(B) The presence of a long
larval form of the roundworm
in a vaginal smear. The worm
is covered by a number of
host cells (400×).

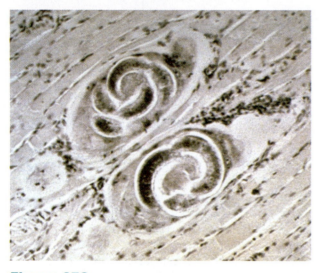

Figure 373
Trichinella spiralis (pork roundworm) in muscle.

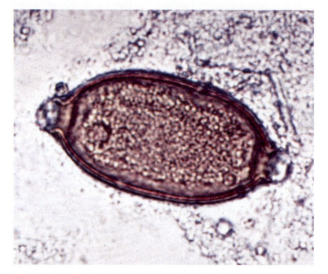

Figure 374
Trichuris trichiura (whipworm) ovum.

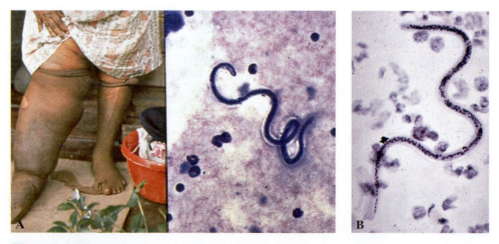

Figure 375
Wuchereria bancrofti, one cause of elephantiasis. (A) A case of the disease (left); nematode from the leg (right). (B) *W. bancrofti* (a microfilaria) in a tissue smear, showing its typical parts.

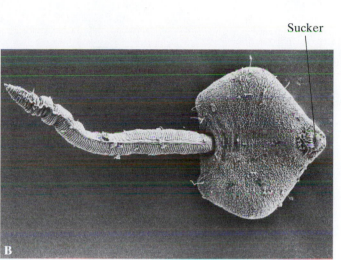

Sucker

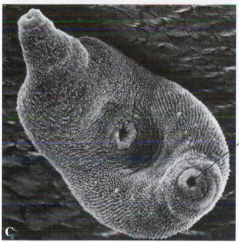

Sucker

Figure 376
Scanning electron micrographs showing stages in the life cycle of trematodes. (A) Highly ciliated miracidium. (B) Tail-bearing cercaria. (C) A metacercaria. (Figures B and C courtesy of Dr. Maianne Koie.)

Fasciola hepatica

Fasciola hepatica, the sheep liver fluke, causes fascioliasis. The infection is found in sheep-raising areas of the world. Adults flukes cause damage to and blockage of bile ducts and the gallbladder.

Transmission and Diagnosis. Infection is acquired by the ingestion of unwashed, raw aquatic vegetation containing *F. hepatica* metacercariae.

Diagnosis generally is based on finding ova in stool specimens (Figure 378A).

Fasciolopsis buski

Fasciolopsis buski causes fasciolpsiasis. Symptoms vary depending on the number of flukes. Abdominal pain and diarrhea are commonly experienced. The disease is found in various areas in Asia.

Transmission and Diagnosis. Infection is acquired by the ingestion of freshwater aquatic plants contaminated with metacercariae. Pigs, dogs, and rabbits are reservoirs of the fluke.

Diagnosis is made by finding ova in stool specimens (Figure 379).

Paragonimus westermani

Paragonimus westermani and other species of *Paragonimus,* the lung flukes, cause paragonimiasis. As is the case with most other fluke infections, the signs and symptoms depend on the number of flukes. Cough,

Trematodes

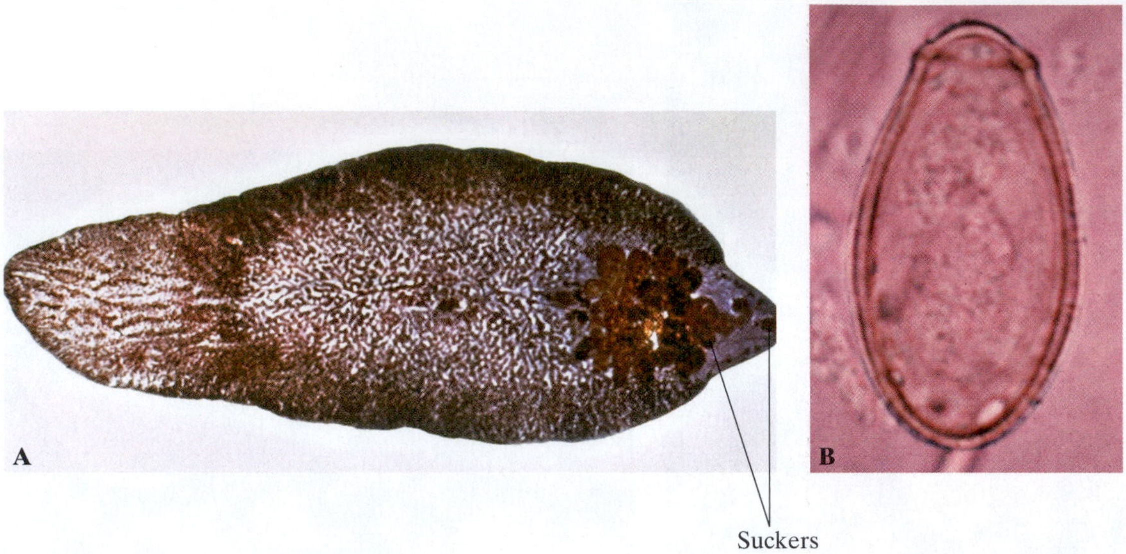

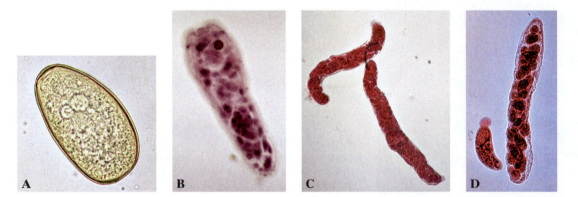

Suckers

Figure 377
Clonorchis (Opisthorchis) sinensis (Chinese liver fluke). (A) The adult fluke and its parts. (B) The ovum.

Figure 378
Fasciola hepatica (sheep liver fluke). (A) The ovum. (B) A stained miracidium (2.5×). (C) Stained sporocysts (2.5X).
(D) Stained rediae.

Figure 379
Fasciolopsis buski ovum.

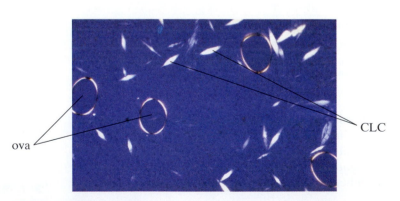

ova

CLC

Figure 380
Paragonimus westermani (lung fluke). Oval ova and Charcot-Leyden crystals (CLC) are shown under polarized light. (Courtesy of Dr. L. Alpert, Pathology Department, The Sir Mortimer B. David Jewish General Hospital.)

Trematodes (Continued)

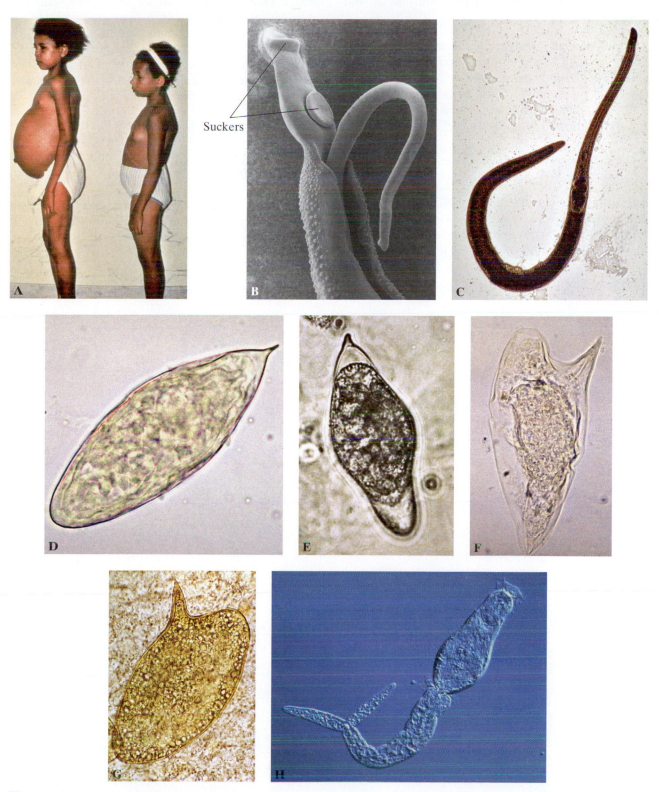

Figure 381

Schistosomes. (A) Victims of schistosomiasis. (Courtesy of the World Health Organization.) (B) Scanning electron micrograph of both the larger male and the smaller female schistomes (400×). (C) Adult female *Schistosoma mansoni* (2.5×). (D) *S. haematobium* ovum. (E) *S. intercalatum* ovum. (F) *S. mansoni* ovum. (G) *S. japonicum* ovum. (H) A live schistosome cercaria. Note the forked tail.

increased sputum production, and chest pain are typical. Paragonimiasis is found predominantly in Asia and Africa.

Transmission and Diagnosis. Infection is acquired by the ingestion of improperly cooked freshwater crabs or crayfish containing metacercariae.

Diagnosis generally is made by finding ova in sputum (Figure 380). Ova can be found in stools also because they may be swallowed.

Schistosoma

Various species of *Schistosoma*, blood flukes, cause schistosomiasis (Figure 381A). Unlike other flukes, schistosomes have separate sexes (Figures 381B and C). The effects of schistosomes result from the depositing of ova (Figures 381D–G) in the large intestine and urinary bladder and the reactions of blood vessels in the liver and lungs. Adult worms live in veins and venules of infected hosts. Schistosome species are found in different parts of the world: *S. haematobium* (Africa and the Middle East), *S. japonicum* (various regions of Asia), *S. mansoni* (tropical Africa, eastern part of South America and some Caribbean islands), and *S. intercalatum* (portions of Africa).

Transmission and Diagnosis. Schistosome infection is acquired when cercariae (Figure 381H) from infected snails penetrate intact skin upon contact in freshwater environments such as ponds and rivers.

Diagnosis includes demonstrating the characteristic ova of schistomes in stools. Serological tests such as enzyme immunoassays also are of value.

10

Arthropods and Disease

So, Naturalists observe, a flea hath Smaller fleas that on him prey; And these have smaller still to bite'em; And so proceed, ad infinitum

—Johnathan Swift

The largest group of species in the Animal Kingdom is found in the phylum Arthropoda. The importance of this group of animals is primarily due to the number of cases, deaths, and global distribution of the various diseases they transmit and cause.

Properties. Although arthropods vary greatly, they all share several properties. These include 1) a rigid or semi-rigid exoskeleton, 2) a complete digestive tract, 3) an open circulatory system (with or without a dorsally situated heart), and 4) excretory, nervous, and respiratory systems. Insects which include fleas, flies, cockroaches, and lice have segmented bodies and jointed legs.

Relationship to Disease. Arthropods can transmit diseases mechanically and biologically. In the mechanical means of transmission, the arthropod is not a part of the life cycle (reproduction) of the pathogen and just serves to transport it. The cockroach (*Blatta* species) and the ordinary house fly (*Musca domestica*, MUSS-ka, dŌ-MES-tik-ah) are good examples. (Figure 382). In the case of biological transmission, the arthropod has an integral role to play in the pathogen's development and/or reproduction. Examples of such vectors include fleas, a number of mosquito species (*Culex, Aedes,* etc.), ticks (*Dermacentor, Ornithodoras,* etc.), and the body louse (*Pediculus humanus corporis*) (Figures 383 A & B).

In addition to serving as transmitters of various pathogens, certain arthropod species are well known for the effects they cause by stinging or biting. Several arthropod species, such as the black widow spider (*Latrodectus mactans*, lat-RŌdek-tis, MAC-tans) and the brown recluse or violin spider (*Loxosceles* species),

are known for this property (Figure 384). Still other arthropods, such as pubic or crab lice, can cause infestations involving the hair and skin (Figure 385). Humans and other mammals also can be infested with mites. The mite is an obligate parasite that completes its entire life cycle on humans and other mammals. These arthropods, while able to serve as vectors for certain pathogens, such as rickettsia, also can infest humans, causing a noninfectious disease (Figure 386). Mites cannot fly or jump, but crawl at the rate of 2.5 centimeters per minute on warm skin. Only female mites burrow into the skin (Figure 387). The number of mites can reach hundreds on the skin of an infested individual and usually are spread by direct skin-to-skin contact. Transmission among family members and in institutional settings is fairly common. A well-know mite infestation is scabies SKĀ-bez. The skin-burrowing condition is called acariasis and is caused by *Sarcoptes scabies* (SAR-kop-tēz, SKĀ-bēz) (Figure 388). Scabies is particularly devastating with immunocompromised individuals such as persons with AIDS.

Myiasis (MĪ-a-sis), is another infestation of human and other vertebrates by the larvae of certain flies, which feed on the available host's dead or living tissue, liquid body secretions, or ingested food. Various types of myiasis are known and are grouped according to the affected tissue. The fly larvae can penetrate and cause varying degrees of damage in such body regions as the skin, eye, ear canal, vagina, and open wounds (Figure 389). Dozens of fly larvae cause myiasis.

Arthropods and Disease

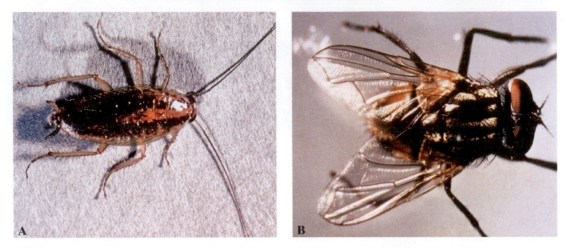

Figure 382

Mechanical transmitters of disease. (A) The oriental cockroach (*Blatta orientalis*). (B) The ordinary house fly (*Musca domestica*). The numerous body hairs of this arthropod can serve as individual inoculating needles.

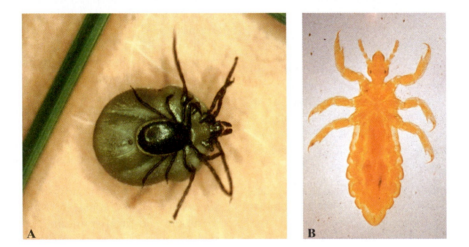

Figure 383

Ticks and lice. (A) The soft-shell tick *Ornithodoras*. A large female and young tick are shown. (B) *Pediculus humanus corporis* (the body louse).

Figure 384

Stinging or biting arthropods. (A) The black-widow spider (*Latrodectus mactans*). Note the characteristic red hour glass on the abdomen. (B) The brown recluse or violin spider (*Loxosceles* species). Note the outline of a violin (arrow).

Arthropods and Disease *(Continued)*

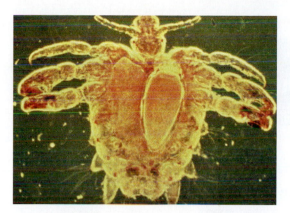

Figure 385
Phthirus pubis (the pubic or crab louse).

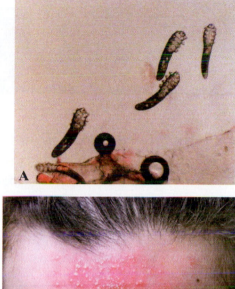

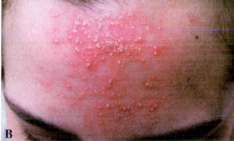

Figure 386
Skin manifestation of mites. (A) A light microscopic view of a skin scrapping showing *Demodex* (dē-MŌ-dex) mites. (B) Mite infestation of the forehead. (Courtesy of Dr. Joseph C. English III, University of Pittsburgh, department of Dermatology.)

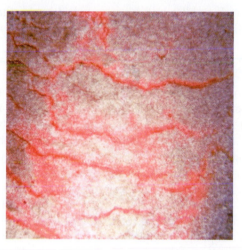

Figure 387
The effects of skin-burrowing by female mites in a case of scabies.

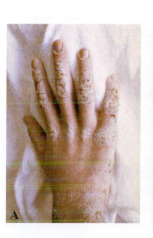

Figure 388
Sarcoptes scabiei (the itch mite), the cause of scabies. (A) A case of Norwegian scabies infestation in a patient with AIDS. (B) Live scabies mites in a skin scraping seen under dark field microscopy. (From T.L. Meinking *et al.* NEJM 333(1995):26–30)

Figure 389
An actual 25 millimeter-long larva covered with spines that emerged from the buttock of a patient with myiasis. (From E. Pijpers. *Clin. Infect. Dis.* 26 (1998):172–173.

Science advances through tentative answers to a series of more and more subtle questions which reach deeper and deeper into the essence of natural phenomena.

—*Louis Pasteur*

Humans and other vertebrates are protected in varying degrees from disease-causing microorganisms and cancer cells by a surveillance mechanism referred to as the **immune system.** Collectively, the various components of this system provide protection, or **immunity,** by imposing barriers to invasion by microorganisms or other disease agents or by selectively neutralizing or eliminating materials that they recognize as foreign. Immunologic responses, which may be either **nonspecific** or **specific,** serve three functions: defense against invasion by microorganisms, maintenance of a stable internal environment (homeostasis), and surveillance, or recognition of abnormal and foreign cell types.

▲ Immune System Components

The various components of the immune system are combined in an exquisitely complex communications network that functions as an effective defense against foreign microorganisms and against body cells that have become abnormal (cancerous). One laboratory procedure used together with other tests to determine the state of health of an individual is the differential white blood cell count (Figure 390). A stained blood smear is used to establish the percentage of **agranulocytic** and **granulocytic** leukocytes (white blood cells). *Lymphocytes* and *monocytes* represent the agranulocytes or cells without cytoplasmic granules, while the *neutorphils, eosinophils,* and *basophils* comprise the granulocytes or cells with cytoplasmic granules, (Figures 391A, B, and C). The shape and arrangement of the nuclei in both cell groups together with the stained appearance of the cytoplasmic granules help to distinguish these white cells. Monocytes, the largest of the leukocytes, usually exhibit kidney-shaped nuclei (Figure 392), while lymphocytes

are much smaller and have rounded nuclei that frequently occupy a major part of a cell's cytoplasm (Figure 391D).

Two major types of lymphocytes are known. These are *T* and *B lymphocytes.* Both cell types are important in the development of immunologic responses to **antigens**, which include cells and substances recognized by the immune system as foreign. Additional properties of T and B cells are presented in the following section. Monocytes are active phagocytic cells (Figure 393). Phagocytosis is the process by which cells engulf particles, microorganisms, and a variety of other cells. After attachment to the phagocytic cell surface, the cell extends a portion of itself in the form of pseudopodia around the particle or microorganism and engulfs it (Figure 393). Once inside the phagocyte, a digestive process ensues, with the end result being the destruction of the engulfed material, such as bacteria and the remains of dead cells resulting from infection. It should be noted that in certain situation pathogens are not destroyed by phagocytic cells. On the contrary, pathogens such as the meningococcus, gonococcus, tubercle bacillus, and even the human immunodeficiency virus can use phagocytes for protection against immune system defenses, reproduction, and even transport. Monocytes that migrate to areas of infection differentiate (transform) into wandering macrophages (Figure 394).

The nuclei of granulocytes are generally arranged in segments (lobes). In addition, eosinophils have red-staining granules, basophils contain dark blue to purple granules, and neutrophils are more or less neutral in color (391). Eosinophils are phagocytic in cases of worm infections and release substances that are involved in allergic reactions. The numbers of these granulocytes increase substantially in such situations. Basophils also are involved in allergic and inflammatory reactions. Basophils are in some ways

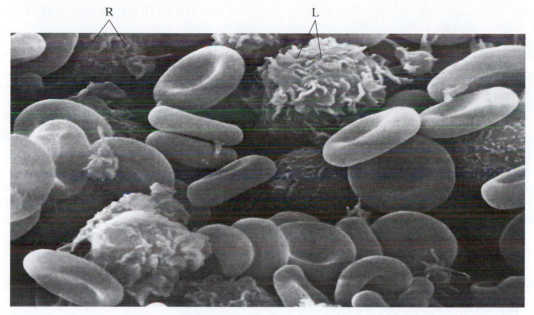

Figure 390
A scanning electron micrograph of a normal peripheral blood specimen showing red blood cells (R) and leukocytes (L). (Original magnification 4,900×.)

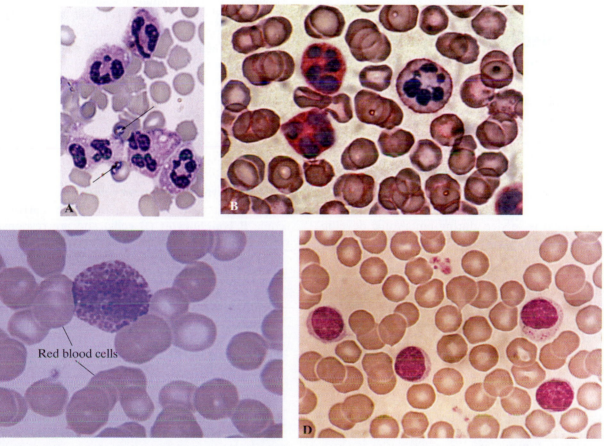

Figure 391
Representatives of leukocytes in stained blood smears. (A) A number of neutrophils. The ring stage of the malaria parasite (arrows) is evident in some red blood cells. (B) Two eosinphils and a larger neutrophil. (C) The features of a basophil. (D) Several lymphocytes. (Original magnification all micrographs 1,000×.)

similar to mast cells, another cell type that plays an important role in allergic reactions. Mast cells are present in most tissues adjoining blood vessels. They contain numerous granules containing such chemicals as heparin and histamine that are important participants in host cellular-defense mechanisms (Figure 395). Neutrophils are also phagocytic and respond quickly to tissue destruction caused by bacteria.

The respective numbers, increases, and decreases in certain cell types, and the appearance of abnormal cells are important clues to the diagnosis of a number of diseases. This procedure can be used to detect an infection (Figure 391A), identify certain infections such as malaria and meningococcal meningitis, uncover certain cancerous states, determine the presence and effects of possible poisoning by drugs and other chemicals, and detect allergic reactions. The normal values based on a count of 100 cells are as follows:

agranulocytes, lymphocytes 25–33,
monocytes, 4–8;

granulocytes, basophils 0–1,
eosinophils 1–4,
neutrophils 60–70.

▲ Selected Immune System Components

T and B Lymphocytes

As indicated earlier there are two basic types of lymphocytes that are involved in antigen- specific responses. They are known as T- and B- lymphocytes, respectively (Figure 396).

T Lymphocytes.
These cells develop in the thymus gland (Figure 398B), which is seeded during embryonic development by lymphocyte stem cells provided by bone marrow. Immature T lymphocytes occupy the outer cortex region of the thymus, while mature cells are found in the inner medulla area. During maturation, a variety of identifying protein molecules known as **Cluster of Differentiation (CD) markers** are expressed on T cell membranes and result in the formation of T lymphocyte subsets. These surface proteins are unique and serve as specific antigen receptors that make identification of subsets possible.

Two main subsets of T lymphocytes are recognized, **Th1** and **Th2.** These subsets are distinguished by the presence of cell surface markers, CD4 (CD4+) and CD8 (CD8+). T lymphocytes expressing the CD4 marker are also known as T helper cells, and are regarded as the most prolific producers of cytokines. Cytokines are the protein messenger molecules responsible for most of the biological effects in the

immune system. The general properties of the T lymphocyte subsets are as follows:

1. **Th1** cells produce protein messenger molecules known as cytokines, which carry signals between cells. The cytokines of this cell type include interferon--gamma, interleukin-2, and tumor necrosis factor-beta, all of which are important participants in phagocytosis and the destruction of microbial pathogens. They also promote the development of CD8 cytotoxic cells and activate another cell type, the macrophage. Macrophages prepare and present antigens to T helper cells to start an immune response.

2. **Th2** cells produce cytokines such as interleukin 4 that stimulate B cells to produce antibodies and are involved in related immune and allergic responses. This CD4 lymphocyte subset also functions as T helper cells in specific immune responses involving antibodies and antigen recognition.

Many CD8+ lymphocytes act as cytotoxic T (Tc) cells in the cellular immune response. Some other lymphocytes act as **suppressor (Ts) cells,** or **T-regulatory cells,** to suppress immune responses.

Certain T cells play an important role in **cell-mediated immunity,** the direct destruction of body cells that have been invaded by various infectious disease agents or that undergo degeneration. Other T cell types play a significant regulatory role in the development and activation of various types of immune responses by providing help to other cells capable of killing infected or defective cells.

B Lymphocytes and Plasma Cells.
Lymphocytes that mature in the bone marrow become **B lymphocytes (B cells)** and are responsible for the **humoral response** or antibody (immunoglobulin) production. B cells are identified by the presence of immunoglobulins on their surfaces. Each B cell expresses only the single specific antibody molecule it will eventually secrete.

Mature B cells carry surface immunoglobulins which function as antigen receptors. These cells move through the circulation to secondary lymphoid tissues, where they respond to antigenic stimuli by dividing and differentiating into immunoglobulin- producing *plasma cells* (Figure 397). **Immunoglobulins** (antibodies) are a class of blood proteins which are induced following contact with antigens. These proteins bind specifically to the antigens which caused their formation. Five different classes of immunoglobulins are recognized: *IgG, IgM, IgA, IgD,* and *IgE.*

Plasma cells have an expanded cytoplasm with parallel arrangements of rough endoplasmic reticulum.

▲ Diagnostic Tests

A large number of immunodiagnostic tests are used to detect and monitor infectious diseases and to follow the recovery from disease. Several of these techniques are applicable not only to the areas of infectious and immunologic diseases and disorders but also to the entire spectrum of clinical medicine.

Knowing whether or not an individual has antibody to a given antigen is extremely significant in establishing the identity of a disease agent, in monitoring a patient's recovery from infection, or in determining the effectiveness of treatment or immunizations. Because of the specificity of the antigen antibody reaction, if either antigen or antibody is known, it is possible to identify and determine the relative concentration of the other reactant by one of a number of diagnostic techniques. Several of these serologic or immunodiagnostic tests are applicable not only to the areas of infectious or immunologic diseases and disorders, but also to other specializations such as forensics, the broad spectrum of clinical medicine, and food technology. For example, specific procedures are used for measuring levels of certain drugs, hormones, determining the relative concentration of serum proteins alpha, beta, and gamma globulins (Figure 405), and albumin, detecting tumor and transplantation antigens, and identifying blood group differences (Figures 402B and C) and incompatibilities related to transfusion reactions.

The Question of Titer. It is important to note that the **titer**, or concentration of antibody, fluctuates as a consequence of immunizations and of subclinical (mild) as well as full-blown infectious states. To distinguish the antibody production associated with an actual ongoing infection from the effects of immunization or from antibodies formed during a past infection, at least two specimens, *acute* and *convalescent*, are needed. The acute specimen is obtained soon after the onset of the disease, and the convalescent specimen is obtained 12 to 14 days later. Both specimens are tested to determine if a rise in titer of the antibody to a suspected pathogen has occurred. If the titer has risen, as indicated by a greater antibody activity in the convalescent specimen, identification of the causative agent is possible. If little or no difference in titer is found between the two specimens, it can be assumed, barring any testing process abnormalities, that an organism other than the one being tested for is the cause of the disease.

Immunodiagnostic Procedures. Historically, the names of the antibodies involved in immunodiagnostic tests generally indicate how they interacted with an antigen. Examples of such procedures include agglutination,

precipitin, and complement fixation (the binding of the serum protein complement by an antigen-antibody combination). As refinements in testing develop, names are given to immunodiagnostic techniques that reflect the nature of the procedure. Examples of these techniques include gel diffusion, radial immunodiffusion, latex particle agglutination, and fluorescent antibody techniques. A representative number of the tests will be described. Table 21 briefly characterizes a number of immunologic tests, and Figures 401–412 show corresponding test results for several of them.

Agglutination Reaction. Agglutination is the term used to describe the clumping of *particulate* (particle-like) antigens such as bacterial and other cells in suspension by corresponding (homologous) antibodies (Figure 401A). In this process, the antibody is referred to as the *agglutinin*, and the particulate antigen as the *agglutinogen* (Figure 402A.) In blood-typing procedures, the red blood cell antigens are referred to as *hemagglutinogens* and the antibodies as *hemagglutinins*, respectively, and the procedure is called *hemagglutination* (Figures 401B and C). *Latex agglutination tests* are rapid commercial variations of the classic agglutination reaction. In latex agglutination, antibodies to bacterial antigens or other proteins are attached to latex particles. This reagent is then used to detect homologous antigen in various types of specimens by the clumping of the latex particles (Figure 412). Variations of this test are used for the identification of bacterial pathogens such as staphylococci, *Listeria* (Figure 412), and streptococci.

Identification of certain viral pathogens can be done by the prevention of a virus-caused hemagglutination reaction (Figure 403A). Several viruses, such as influenza, mumps, and Newcastle disease, are capable of agglutinating a variety of red blood cells. During the course of infections with these pathogens, hemagglutination-inhibiting (HI) antibodies are formed and can be used for diagnosis. In the HI procedure, known virus suspensions are mixed with and subsequently incubated with the blood serum specimen under investigation. If homologous HI antibodies are present, hemagglutination will not occur when a standard suspension of red bloods is added to the mixture (Figures 403B and C). In short, the virus hemagglutination activity will be neutralized. The absence of HI antibodies in the serum specimen will cause virus hemaglutination to take place.

Immunoprecipitin (Immunoprecipitation) Reactions. The immunoprecipitin reaction results from the mixing of a functional class of antibodies called *precipitins* with soluble (dissolved) antigens known as *precipitinogens*. In test procedures, a visible precipitate generally appears where optimal proportions of antibody and antigen exist (Figure 404A).

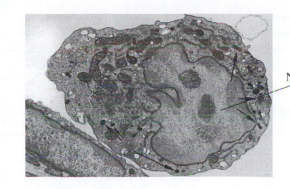

Figure 392
An ultrathin section through a monocyte showing several of its general features. (Original magnification 11,000×.)

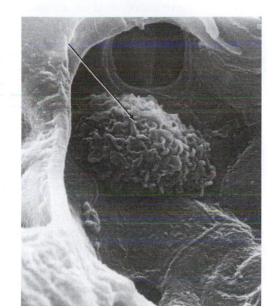

Figure 394
A scanning electron micrograph of a wandering tissue macrophage on the surface of an air sac located just below the entrance to the lung. Note the ruffled surface of the cell.

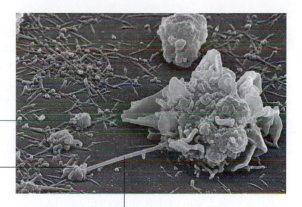

Figure 393
Phagocytosis in action. A phagocyte showing a long extension (pseudopodium) attached to a clump of bacteria.

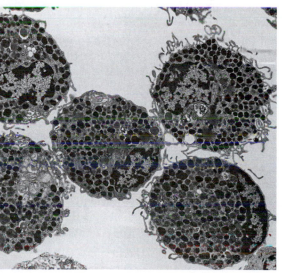

Figure 395
An ultrathin section through a cluster of mast cells. Note the large number of intracellular granules. These cells are important participants in certain allergic reactions.

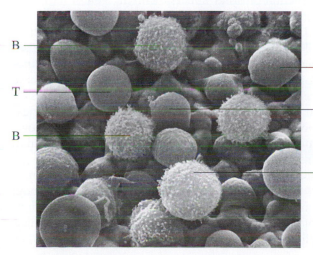

Figure 396
A scanning electron micrograph of the cells in a peripheral blood sample. T-(T) and B-(B) lymphocytes are scattered among a large number of red blood cells.

This area is entirely dedicated to immunoglobulin (antibody) production. Plasma cells are typically found in the various body organs such as the spleen, and the medulla of lymph nodes (Figure 398A), in mucosal associated lymphoid tissue, which includes certain areas of the gastrointestinal, respiratory, and urogenital systems, and in small numbers at inflammation sites.

The Thymus Gland and Other Selected Cellular Immune System Factors

Other components of the immune system include the thymus gland, where T lymphocytes develop and mature (Figure 398B); lymph nodes, where host immune responses are initiated; natural killer cells (NK), a type of lymphocyte that is capable of killing cancer and virus-infected cells (Figure 399), dendritic cells, a unique group of white blood cells that can capture and process antigens and present them to antigen-presenting cells (**APCs**). Lymph nodes, tonsils, and other secondary lymphoid tissues have specific areas known as *follicles* that serve as centers for T- and B-cell interaction (Figure 398A). In addition to several of the immune system components discussed earlier, other factors are also important to the responses of the immune system. These include the components of the **major histocompatibility complex** (MHC). This complex is made up of a cluster of genes important in immune recognition and signaling between cells of the immune system. The MHC encodes a large number of different protein molecules, including MHC Class 1 and Class 2 molecules. Class 1 consists of integral surface proteins of all nucleated cells. This class of proteins comprises the antigens involved in tissue graft rejection reactions. MHC Class 2 molecules are expressed on a number of body cells, including B lymphocytes, macrophages, monocytes, various antigen-presenting cells, and some T lymphocytes.

▲ A Brief View of Immune Responses

Most antigens must be picked up, processed, and carried by means of the body's lymphatic system to lymphoid organs such as lymph nodes and the spleen before they are presented to T lymphocytes in a form they can recognize (Figure 400). These functions are carried out by antigen-presenting cells such as **dendritic cells** in the medulla area of lymph nodes (Figure 398A) and the large phagocytic macrophages found in most tissues, the lungs, and in the linings of certain body cavities. The anatomy of a lymph node permits various types of cells, including lymphocytes and dendritic cells, to come into contact with antigens, or to communicate by means of cytokine molecules.

Both **humoral (antibody)** and **cell-mediated immune responses** develop in the lymphoid follicles of any secondary lymphoid tissue, such as lymph nodes, spleen, tonsils, and Peyer's patches in the intestine.

Following successful recognition of antigen on the surface of an antigen-presenting cell, such as a macrophage or a dendritic cell, a single Th cell responds by secreting the cytokine interleukin -2 or IL-2 (Figure 400). As indicated earlier, cytokines, in general, are protein messenger molecules which send signals between cells. IL-2 signals the Th cell which produced it to undergo cell division, thus giving rise to a number of descendants. The resulting Th cells are specific for the same antigens as the original parent cell. In addition, IL-2 can diffuse to nearby B or CD8+ T cells that have recognized an antigen, and deliver critical signals in the form of other cytokines which are necessary for such cells to respond to the antigen's presence. The presence of Th cells with the ability to secrete IL-2 in response to antigen is essential to initiate both humoral and cell-mediated responses. Following the initial secretion of IL-2 by Th cells in response to antigen, Th cells secrete additional cytokines to further the maturation of B and/or CD8+ T cells that have bound antigen. B cells fully develop into antibody-secreting plasma cells (PC), and CD8+ T cells become cytotoxic T lymphocytes (CTLs) capable of killing. This process of maturing and acquiring new functions is known as **differentiation.** In the case of B cells, Th cells provide help by establishing antigen- specific cell-to-cell contact, delivering both cell-surface signals and cytokines. In the case of CD8+ T cells, no physical contact is necessary, and Th cells provide help solely through the secretion of cytokines in the vicinity of a CD8+ T cell that has recognized an antigen on the surface of a cell. As a result of one or the other or both of these types of interactions, humoral and/or cellular immunity is/are generated in response to an antigen. This serves to eliminate or lead to the destruction or other effects on an antigen or pathogen (Figure 400).

In addition to dealing with the antigen at the time of first exposure, antigen-specific immune responses by Th, and B lymphocytes also give rise to immunologic T- and B-memory cells. The existence of these cells allows the immune system to mount a faster, and much more effective response on subsequent exposures to the same antigen(s). It is the immunologic memory in these cells which provides the protection from the same infectious disease after initial exposure and recovery, otherwise referred to as *immunity*. Both the cell-mediated and the humoral (antibody) immune responses are specifically acquired functions of the immune system and selectively recognize, eliminate, and remember individual antigens.

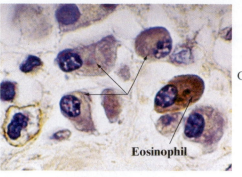
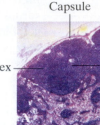

Eosinophil

Figure 397
Plasma cells (arrows) the antibody producing cells of the body. Note the presence of eosinophils (cells with red-staining granules).

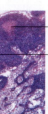
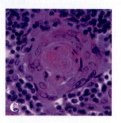
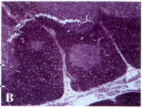

Capsule
Germinal center
Cortex
Medulla
Medulla
Cortex

Figure 398
Lymphatic tissue representatives. (A) A stained section through a lymph node showing some of its parts including the outer cortex which usually contains T lymphocytes and macrophages, a lighter germinal center a Hassal's corpuscle (HC), and the inner medulla region. (B) A view of the internal parts of the thymus gland and some of its parts including the sections known as lobules, the darker staining outer cortex, and a lighter staining medula. (C) A higher view of a thymus lobule showing a characteristic thymic or Hassal's corpuscle. The concentric cellular layers of this medulla component are clearly shown.

NK

Figure 399
A scanning electron micrograph showing an attack of a large cell by a much smaller natural killer cell (NK). The effects of the attack can be seen as blister-like formations (blebs) on the surfaces of the larger cell.

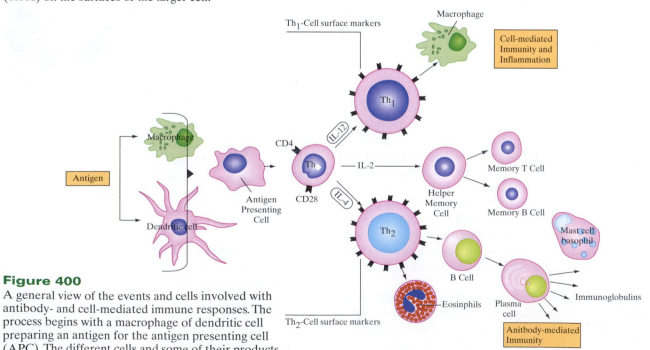

Figure 400
A general view of the events and cells involved with antibody- and cell-mediated immune responses. The process begins with a macrophage of dendritic cell preparing an antigen for the antigen presenting cell (APC). The different cells and some of their products, interleukins (ILs) also can be seen. (After Becton, Dickinson and Company, Hotline. 10:13 figure, 2006.)

Table 21 Immunologic Procedures Used in Diagnosis and/or Microbial Identification

Procedure	Principle Involved	Positive Test Result	Application(s)
Agglutination	Antibody clumps cells or other particulate antigen preparations (insoluble particles coated with antigens—e.g.,latex particles, *Staphylococcus* protein A, etc).	Aggregates (clumps) of antigens (Figure 402)	1. Diagnosis of typhus, Rocky Mountain spotted fever (Weil-Felix test), typhoid fever (Widal test), and infectious mononucleosis 2. Identification of disease agents including *Haemophilus influenzae* type b, *Neisseria meningitidis*, and *Streptococcus pneumoniae*.
Enzyme-linked immunoabsorbent assay (ELISA)	Antigen or antibody from specimens trapped by corresponding specific antibody or antigen coating a solid phase support combines with enzyme labeled specific antibody. The formed complex reacts with an added enzyme substrate in proportion to the amount of antigen or antibody first bound by the antibody or antigen coating.	Color changes occurring with the addition of an enzyme substrate are proportional to either antibody or antigen in specimens (Figure 407)	1. Detection of IgM to rubella and influenza A 2. Identification and/or detection of herpes simplex viruses types 1 and 2, cytomegalovirus, measles, hepatitis B, and AIDS viruses 3. Detection of antibodies to bacterial antigens
Hemagglutination	Homologous antibody causes (hemagglutinin) aggregates of red blood cells to form[a].	Aggregates of red blood cells (Figures 402 and 403)	Blood typing
Hemagglutination-inhibition (viral)	Antibody inhibits the agglutination of red blood cells by coating hemagglutinating virus.	Formation of a circle of unagglutinated cells (Figure 403B)	1. Determining the immune status toward rubella (German measles) 2. Virus identification
Immunodiffusion	Antibody and soluble antigen diffuse toward one another through an agar gel and react where homologous antibody is in proper proportion to homologous antigen	Lines of precipitate form within the agar (Figure 404)	Antigen and/or antibody identification
Immunofluorescent microscopy	Antibody (usually) or antigen is labeled with a fluorescent dye, which fluoresces on exposure to ultraviolet or blue light.	Glowing on exposure to UV light (Figure 409)	1. Detection of antigen or antibody 2. Identification of microbial pathogens of diseases such as rabies, syphilis, Legionnaires' disease, etc.
Precipitin	Antibody and soluble antigen react when they are in proper proportion to one another.	Lines of precipitate form (Figure 404)	1. Diagnosis of microbial diseases 2. Detection of antigens
Western blot	Proteins of antigen are separated by electrophoresis, transferred to and immobilized on nitrocellulose strips, and then exposed to serum specimens. Antigen-antibody reactions are detected by an added enzyme-linked antihuman immunoglobulin reagent.	Formation of a black precipitate in the regions where the enzyme-immunoglobulin reagent is bound (Figure 406)	1. Diagnosis of infectious diseases such as AIDS 2. Detection of antibody against different antigenic components

[a]Hemagglutination reactions caused by certain viruses and bacteria generally do not involve antibody.

Diagnostic Tests

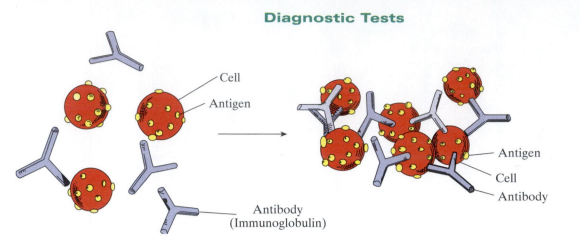

Figure 401

A diagrammatic view of agglutination. Agglutination occurs when antibodies bind the antigens on cells together to form large clumps. The reaction is useful in the laboratory for a variety of diagnostic tests.

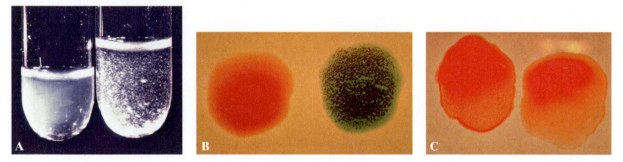

Figure 402

Agglutination reactions. (A) Bacterial agglutination (clumping) in a tube. (B) Blood typing. An A+ (clumping) reaction showing agglutination of the A blood factor (antigen) by antibodies against the factor. (C) A positive (clumping) reaction for the Rho, or D, factor.

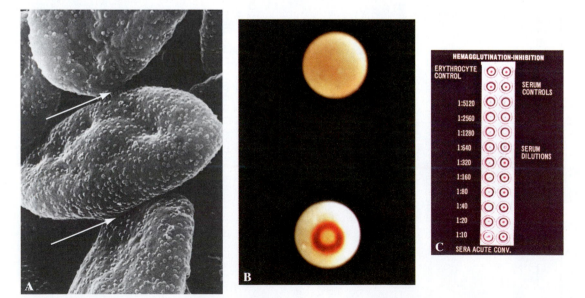

Figure 403

(A) A scanning micrograph showing the hemagglutinating action of viruses. Viruses forming a connection between red blood cells are indicated by arrows. (B) Hemagglutination reactions. The red doughnut reaction is a negative result. (C) Hemagglutination-inhibition (HAI). A pattern of reactions shown with a series of serum dilutions from 1:10 to 1:5,120. The dotlike results are negative reactions. Erythrocyte and serum controls also are shown.

The main difference between precipitin reactions and those involving agglutination is the state of the antigen used. In the precipitin reactions, the antigen molecules are soluble, so that solutions containing them are clear, whereas agglutination molecules in suspension form cloudy solutions.

Applications of immunoprecipitin reactions can be performed in agar gel and include the the single diffusion (Oudin) test and the double-diffusion (Ouchterlony test (Figure 404). Only the double-diffusion test will be described to show the features of precipitin reactions. In this procedure, antigen-containing solutions are added to certain circular or square holes or wells cut in an agar-plate or specially prepared slides with an agar surface. An antibody-containing preparation is placed in another well (Figure 404A). These preparations are incubated to allow the respective reactants to diffuse freely toward each other. Visible precipitin lines or bands develop where antigenic components react with homologous antibodies that have diffused from its well.

Three band patterns that represent possible reactions are shown in Figure 404. These are the *reaction of identity*, which occurs when pure antigen preparations are placed in two wells adjacent to a centrally located well containing homologous antibody (Figure 404A); the *reaction of nonidentity*, which results in the formation of two distinct bands that cross and occurs when two unrelated antigens are placed in two separate wells adjacent to a centrally located well containing antibody for each of the antigens (Figure 404B); and the *reaction of partial identity* or *cross-reaction*, which results when one antigen well contains a cross-reacting preparation and the other an antigen that is homologous for the antibody preparation (Figure 404C).

Gel Electrophoresis. Electrophoresis is an analytical technique in which charged molecules in solution, mainly proteins, protein-related compounds, and nucleic acids, migrate in response to an electrical field. It is a simple, rapid, and highly sensitive technique used to identify and study the properties of a single charged molecule and to separate a mixture of charged molecules. Identifying and determining the concentrations of normal blood serum components, such as the alpha, beta, gamma globulins and albumin and any unusual proteins in disease states, establish an immunoelectrophoretic pattern that is useful in diagnosis when compared to normal specimens (Figure 405A) The rate of molecule movement depends on several factors, including the net electrical charge, the size and shape of the molecules, the strength of the electrical field, and the thickness and temperature of the substance (medium) through which the molecules move. The procedure may be carried out in a supporting medium such as gels formed in tubes, slabs, or flat beds.

After the electrophoresis of a protein or nucleic acid sample is completed, the gel or other supporting matrix can be analyzed after staining by one or several methods. The most common staining techniques include Commassie blue or Ponceau S for proteins and ethidium bromide, a fluorescent dye that produces an orange glow when exposed to ultraviolet light, for nucleic acids (Figure 405B).

Gel electrophoresis is one of the highest-resolution methods for separating important biological molecules. One application involves the polymerase chain reaction (PCR) devised in the mid-1980s by Kary Mullis. With this technique, enormous numbers of copies of a single deoxyribonucleic acid molecule can be made in a relatively short time. This process of *in vitro* amplification of nucleic acid segments makes use of the heat-stable enzyme DNA polymerase, which copies DNA molecules. Heat is used to separate (denature) the DNA strands, each of which is copied by the enzyme. The products of the PCR reaction can be detected by gel electrophoresis (Figure 405B). Techniques that combine gel electrophoresis with those of immunodiffusion are called *immunoelectrophoresis*. Such techniques are the bases of certain diagnostic tests, including the Western blot procedure, an example of immunoblotting.

Immunoblotting. Immunoblotting is an immunodiagnostic procedure to detect protein-specific antibodies. The technique is often referred to as Western blotting. In a typical immunoblotting procedure, specific protein antigens such as those of the human immunodeficiency virus are separated by gel electrophoresis and then blotted onto a nitrocellulose or nylon membrane. The membrane blot is then used to detect the presence of specific antibodies in a specimen directed against the blotted antigens. Strips of the membrane are incubated with the sample, and antibodies that react with the membrane-bound antigens are revealed by reacting with a second labeled antibody (Figure 406). Such labels include horseradish peroxidase, and alkaline phosphatase.

Enzyme-Linked Immunoassay (EIA). Enzyme-linked immunoassays, an example of which is the enzyme-linked immunosorbent assay (ELISA), have become one of the more prominent procedures in immunodiagnosis. Two EIA methods are in use: the direct or double-antibody sandwich technique, which determines the level of antibody from an individual, and the indirect technique, which measures an individual's antibody level to a specific antigen. There are many materials that can be used for these procedures. One popular version of the techniques makes use of plastic plates with 96 wells. Such microliter plates are ideal for testing large numbers of specimens (Figure 407).

Diagnostic Tests *(Continued)*

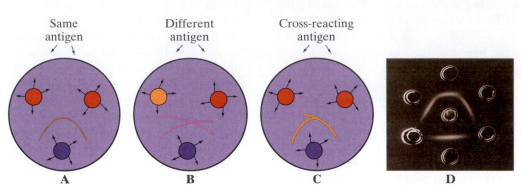

Same antigen   Different antigen   Cross-reacting antigen

A   B   C   D

Figure 404

Double–diffusion reactions. The small arrows around each well indicate the direction of diffusion. (A) Reaction of (antigen) identity. Both upper wells contain the same antigen, and the lower well contains homologous antibody. (B) Reaction of antigen nonidentity. Each of the upper wells contains a different antigen, and the lower well contains antibodies to both antigens. (C) Reaction of partial identity or cross reaction. The upper, left well contains a cross-reacting antigen, and the upper right well is filled with an antigen that is homologous for the antibody in the lower well. (D) A photograph of actual reactions of (antigen) identity.

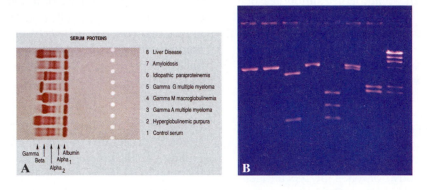

SERUM PROTEINS

8 Liver Disease
7 Amyloidosis
6 Idiopathic paraproteinemia
5 Gamma G multiple myeloma
4 Gamma M macroglobulinemia
3 Gamma A multiple myeloma
2 Hyperglobulinemic purpura
1 Control serum

Gamma Beta Alpha₂ Alpha₁ Albumin

A   B

Figure 405

Gel electrophoretic applications. (A) Immunoelectrophoretic patterns showing the concentration of serum proteins in normal and abnormal situations. The pattern of normal serum blood proteins. Alpha, beta, and gamma globulins, and albumin are compared with the same proteins in various disease states. Increased or lower concentrations of the different proteins in the disease patterns shown are indicated by the darker and larger sizes of the respective areas. (B) Patterns obtained with polymerase chain show the reaction (PCR) products. The violet background is due to the ultraviolet which is used to pattern during analysis. (B) Electrophoretic patterns obtained with polymerase chain reaction (PCR) products. The violet background is provided by ultraviolet light, which was used to show the pattern.

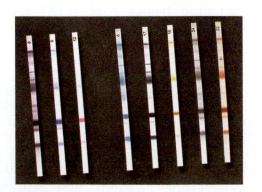

Figure 406

Results of the Western blot assay. This procedure is used to detect specific proteins.

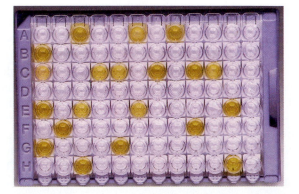

Figure 407

The dramatic difference between positive and negative enzyme immunoassay results is shown. In this case the test results are shown for an EIA used to detect a Shiga toxin-producing bacterium such as *Escherichia coli*. The formation of a yellow color is a positive result. (Courtesy of Remel Inc., Lenexa, KS.)

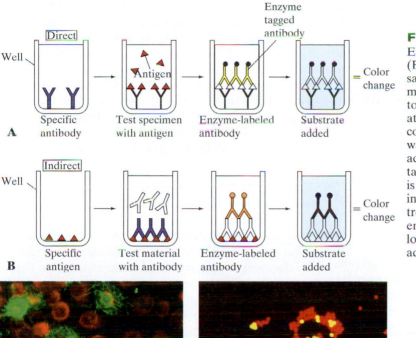

A Direct
Well — Specific antibody → Test specimen with antigen → Enzyme-labeled antibody → Substrate added = Color change

Enzyme tagged antibody

B Indirect
Well — Specific antigen → Test material with antibody → Enzyme-labeled antibody → Substrate added = Color change

Figure 408
Enzyme-linked immunosorbent assay (ELISA). Both the direct, or double sandwich, technique and the indirect method are shown. Antibody to the antigen to be detected (captured antibody) is attached to the well. A specimen possibly containing antigen (triangles) is added and washed. Antibody to the antigen is then added. In the direct method, this antibody is tagged with enzyme so that when substrate is added a color change occurs. In the indirect method, the captured antigen is first treated with heterologous antibody, then an enzymelabeled antibody to the hetero-logous antibody is added, before the addition of substrate.

Figure 409
Immunofluorescent microscopy. (A) The presence of human metapneumonia virus in infected nasopharyngeal cells as shown by the application of the direct fluorescent antibody technique. (From E. Percivalle, A. Sarasini, L. Visai. M. G. Revello, and G. Gerna. *Clin. Micro*. 43, (2005): 3443–3446 and courtesy of G. Gerna). (B) The use of specially prepared particles coated with fluorescent-tagged antibody particles to show the presence of cell surface antigens on leukemic cells. (Courtesy of Dr. Joseph Mirro, Jr.)

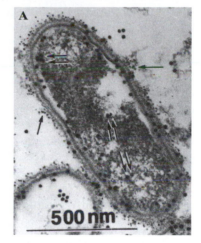

Figure 410
The results of an immunoelectron microscopy application. This trans-mission electron micrograph shows an ultrathin section of a bacterial cell labeled with gold probe spheres. The probes show the locations of both internal and external antigenic sites. (From K. E. Hechemy, *et al.*, *Clin. Micro*. 27, (1989): 377–384.)

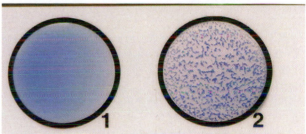

Figure 411
A commercial application of the enzyme immunoassay. This hand-held device is used to detect and differentiate among influenza groups. The test area (T) shows the presence of influenza A. The controls (C) are also shown. (Courtesy of Remel Inc., Lenexa, KS.)

Figure 412
The results of latex agglutination test for the detection of the bacterial pathogen *Listeria*. A positive result indicated by clumping particles is shown on the right. The control is on the left. (Courtesy of Becton, Dickinson and Company.)

In the direct procedure, a specimen containing the suspected antigen is added to polystyrene wells that have been coated with antibody specific for this antigen. This step is followed by the addition of a conjugate (combination) of enzyme-labeled specific antibody (Figure 408A). An enzyme substrate is the last reagent added to the test system. In a positive test, hydrolysis of the enzyme substrate occurs, which is proportional to the amount of antigen bound.

In the indirect method (Figure 408B), the polystyrene wells are coated with a specific test antigen. This step is followed by the addition of a test subject's serum, and then by the addition of an enzyme-labeled anti-human antibody. An enzyme substrate is added to the system. Again, as in the direct method, a positive test will result in the hydrolysis of the enzyme substrate. The extent of the reaction will be proportional to the antibody level in the test subject's serum.

Wash steps in both methods are used to separate and remove the unbound components. The use of an antibody or antigen is conjugated with an enzyme that, upon reacting with its substrate, forms a color change, or a fluorescent reaction product. Color changes can be monitored visually (Figure 407) or with the use of a spectrophotometer to determine the proportionality between the amount of color and the amount of antigen bound. Variations of enzyme immunoassays can be used to detect the presence of specific antigens in cancer cells as well as to demonstrate the presence of virus-infected cells.

Immunofluorescent Microscopy. Antibodies labeled (tagged) with fluorochromes (fluroescein dyes) have been used since 1941 for the detection of microbial and other antigens and antibodies in tissue and other types of samples. The degree of fluorescence observed depends on the ability of such dyes to absorb the energy of nonvisible ultraviolet light and short visible wavelengths, become excited, and release the energy in the form of longer visible wavelengths. One of the widely used dyes is fluorescein isothiocyanate.

Fluorescent-antibody techniques can be used either directly or indirectly. Antigens are usually detected by the direct test (Figure 409) with a number of different specimens, such as cells from lesions, body fluids, and nasal washings. Antibodies, on the other hand, are generally detected by indirect testing procedures. Fluorescent-tagged antibodies are applied to slides containing fixed microbial antigens.

▲ Electron Microscopy

Immunoelectron microscopy is a relatively new analytical approach used to identify viruses, to demonstrate differences among microbial structures, and to detect minute amounts of immunoglobulins and related substances. Probes made of gold particles combined with specific immunoglobulins can be used to detect antigens and chemical components of cells (Figure 410).

▲ Commercial Devices

A number of commercial devices are available to either directly or indirectly diagnose diseases. These include agglutination tests (Figure 412) and applications of enzyme immunoassays (Figure 411). One common EIA variation is the dot blot assay which makes use of nitrocellulose membranes. In this system, the antigen or antibody is coupled to the membrane, and specimens are applied to its surface. Usually a color reaction of some sort becomes visible shortly afterwards, thus providing a qualitative assay. There are several examples of application, including at-home testing kits for pregnancy and the detection of virus and bacterial infections (Figure 411).

APPENDIX

Appendix Figures

The following figures are included in the *Color Atlas* to provide some perspective as to the body regions involved with microbial and helminthic diseases. Note that different disease agents (e.g., bacteria, fungi, protozoa, and viruses) are shown in different colors. **Red:** viruses: **Green:** bacteria: **Blue:** fungi; **Orange:** protozoa: **Purple:** prions; **Black:** helminths.

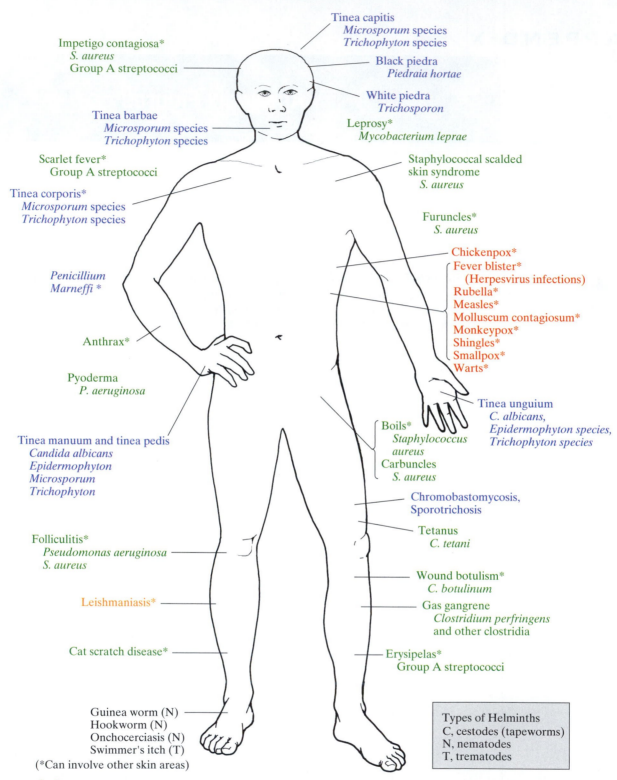

Tinea capitis
Microsporum species
Trichophyton species

Impetigo contagiosa*
S. aureus
Group A streptococci

Black piedra
Piedraia hortae

White piedra
Trichosporon

Tinea barbae
Microsporum species
Trichophyton species

Leprosy*
Mycobacterium leprae

Scarlet fever*
Group A streptococci

Staphylococcal scalded
skin syndrome
S. aureus

Tinea corporis*
Microsporum species
Trichophyton species

Furuncles*
S. aureus

Chickenpox*
Fever blister*
(Herpesvirus infections)
Rubella*
Measles*
Molluscum contagiosum*
Monkeypox*
Shingles*
Smallpox*
Warts*

Penicillium
Marneffi *

Anthrax*

Tinea unguium
C. albicans,
Epidermophyton species,
Trichophyton species

Pyoderma
P. aeruginosa

Boils*
Staphylococcus
aureus
Carbuncles
S. aureus

Tinea manuum and tinea pedis
Candida albicans
Epidermophyton
Microsporum
Trichophyton

Chromobastomycosis,
Sporotrichosis

Tetanus
C. tetani

Folliculitis*
Pseudomonas aeruginosa
S. aureus

Wound botulism*
C. botulinum

Leishmaniasis*

Gas gangrene
Clostridium perfringens
and other clostridia

Cat scratch disease*

Erysipelas*
Group A streptococci

Guinea worm (N)
Hookworm (N)
Onchocerciasis (N)
Swimmer's itch (T)
(*Can involve other skin areas)

Types of Helminths
C, cestodes (tapeworms)
N, nematodes
T, trematodes

Figure A–1
Infectious Diseases and Disease Agents Affecting the Skin, Nails, and Hair (Integumentary System).

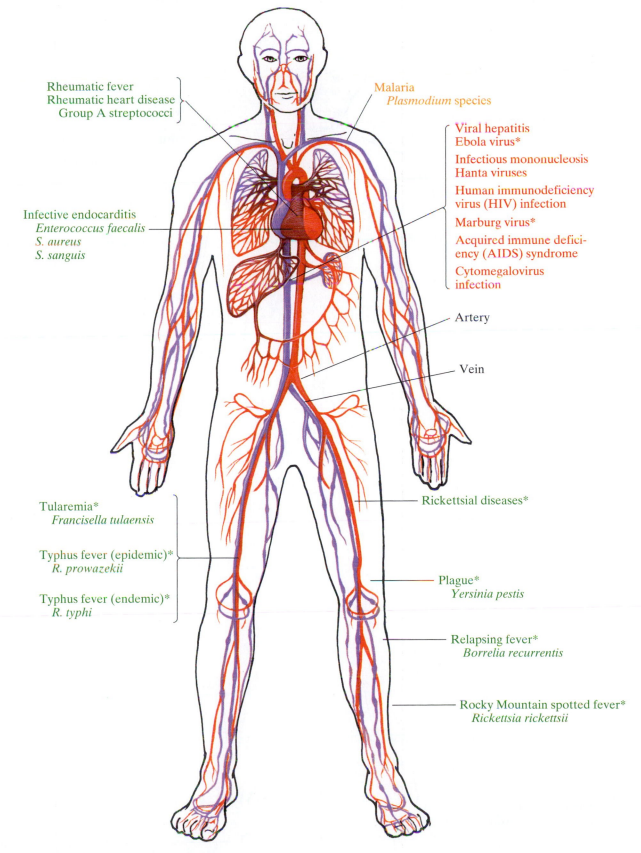

Rheumatic fever
Rheumatic heart disease
Group A streptococci

Malaria
Plasmodium species

Viral hepatitis
Ebola virus*

Infectious mononucleosis
Hanta viruses

Human immunodeficiency
virus (HIV) infection

Marburg virus*

Acquired immune defici-
ency (AIDS) syndrome

Cytomegalovirus
infection

Infective endocarditis
Enterococcus faecalis
S. aureus
S. sanguis

Artery

Vein

Tularemia*
Francisella tulaensis

Typhus fever (epidemic)*
R. prowazekii

Typhus fever (endemic)*
R. typhi

Rickettsial diseases*

Plague*
Yersinia pestis

Relapsing fever*
Borrelia recurrentis

Rocky Mountain spotted fever*
Rickettsia rickettsii

*Involves the system in general

Figure A–2
Infectious Diseases and Disease Agents Affecting the Cardiovascular System.

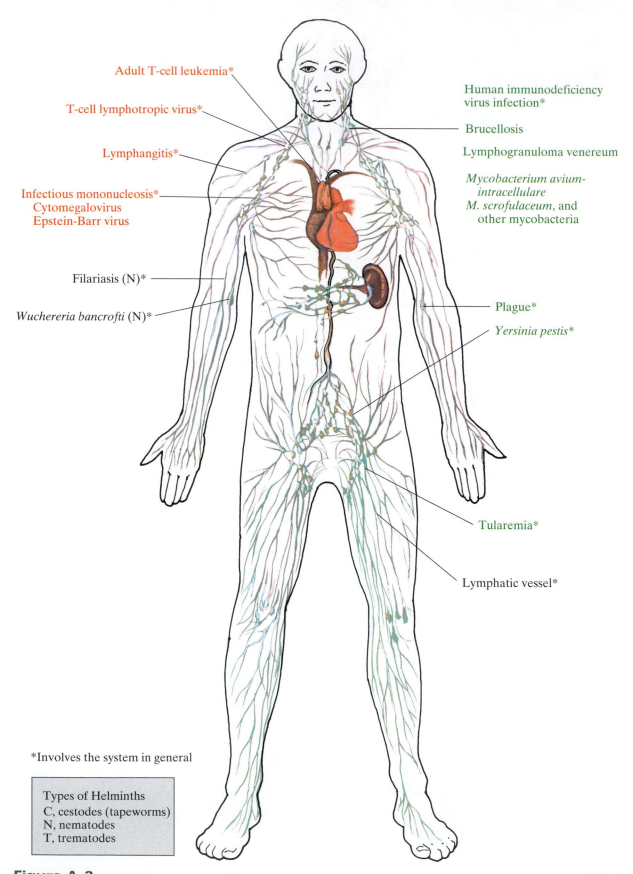

Adult T-cell leukemia*

T-cell lymphotropic virus*

Lymphangitis*

Infectious mononucleosis*
Cytomegalovirus
Epstein-Barr virus

Filariasis (N)*

Wuchereria bancrofti (N)*

Human immunodeficiency
virus infection*

Brucellosis

Lymphogranuloma venereum

*Mycobacterium avium-
intracellulare*
M. scrofulaceum, and
other mycobacteria

Plague*

Yersinia pestis

Tularemia*

Lymphatic vessel*

*Involves the system in general

Types of Helminths
C, cestodes (tapeworms)
N, nematodes
T, trematodes

Figure A–3
Infectious Diseases and Disease Agents Affecting the Lymphatic System.

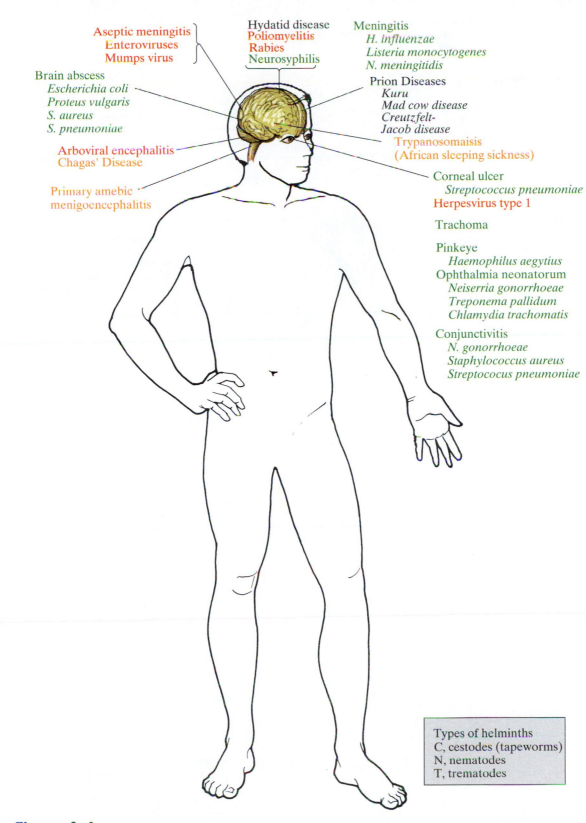

Aseptic meningitis
Enteroviruses
Mumps virus

Brain abscess
Escherichia coli
Proteus vulgaris
S. aureus
S. pneumoniae

Arboviral encephalitis
Chagas' Disease

Primary amebic
menigoencephalitis

Hydatid disease
Poliomyelitis
Rabies
Neurosyphilis

Meningitis
H. influenzae
Listeria monocytogenes
N. meningitidis

Prion Diseases
Kuru
Mad cow disease
Creutzfelt-
Jacob disease

Trypanosomaisis
(African sleeping sickness)

Corneal ulcer
Streptococcus pneumoniae
Herpesvirus type 1

Trachoma

Pinkeye
Haemophilus aegytius
Ophthalmia neonatorum
Neiserria gonorrhoeae
Treponema pallidum
Chlamydia trachomatis

Conjunctivitis
N. gonorrhoeae
Staphylococcus aureus
Streptococus pneumoniae

Types of helminths
C, cestodes (tapeworms)
N, nematodes
T, trematodes

Figure A–4
Infectious Diseases and Disease Agents Affecting the Nervous System.

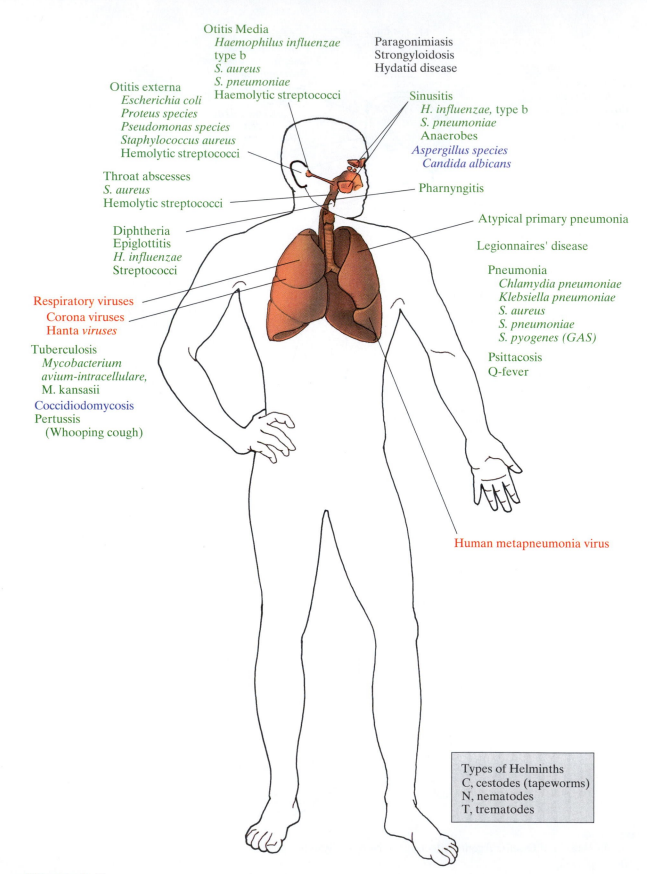

Otitis Media
Haemophilus influenzae
type b
S. aureus
S. pneumoniae
Haemolytic streptococci

Paragonimiasis
Strongyloidosis
Hydatid disease

Otitis externa
Escherichia coli
Proteus species
Pseudomonas species
Staphylococcus aureus
Hemolytic streptococci

Sinusitis
H. influenzae, type b
S. pneumoniae
Anaerobes
Aspergillus species
Candida albicans

Throat abscesses
S. aureus
Hemolytic streptococci

Pharyngitis

Diphtheria
Epiglottitis
H. influenzae
Streptococci

Atypical primary pneumonia

Legionnaires' disease

Respiratory viruses
Corona viruses
Hanta *viruses*

Pneumonia
Chlamydia pneumoniae
Klebsiella pneumoniae
S. aureus
S. pneumoniae
S. pyogenes (GAS)

Tuberculosis
Mycobacterium
avium-intracellulare,
M. kansasii
Coccidiodomycosis
Pertussis
(Whooping cough)

Psittacosis
Q-fever

Human metapneumonia virus

Types of Helminths
C, cestodes (tapeworms)
N, nematodes
T, trematodes

Figure A–5
Infectious Diseases and Disease Agents Affecting the Respiratory System.

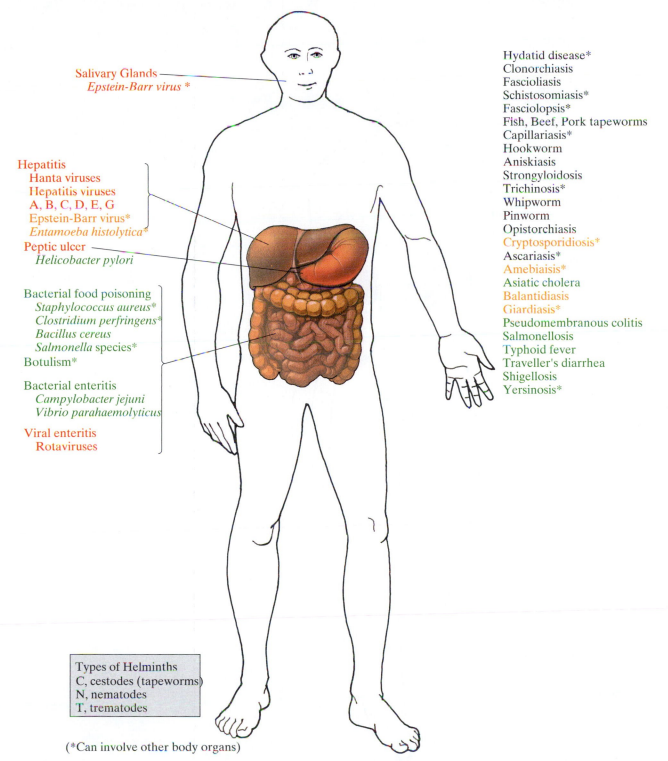

Salivary Glands
  Epstein-Barr virus *

Hepatitis
  Hanta viruses
  Hepatitis viruses
  A, B, C, D, E, G
  Epstein-Barr virus*
  Entamoeba histolytica
Peptic ulcer
  Helicobacter pylori

Bacterial food poisoning
  *Staphylococcus aureus**
  *Clostridium perfringens**
  Bacillus cereus
  Salmonella species*
Botulism*

Bacterial enteritis
  Campylobacter jejuni
  Vibrio parahaemolyticus

Viral enteritis
  Rotaviruses

Hydatid disease*
Clonorchiasis
Fascioliasis
Schistosomiasis*
Fasciolopsis*
Fish, Beef, Pork tapeworms
Capillariasis*
Hookworm
Aniskiasis
Strongyloidosis
Trichinosis*
Whipworm
Pinworm
Opistorchiasis
Cryptosporidiosis*
Ascariasis*
Amebiasis*
Asiatic cholera
Balantidiasis
Giardiasis*
Pseudomembranous colitis
Salmonellosis
Typhoid fever
Traveller's diarrhea
Shigellosis
Yersinosis*

Types of Helminths
C, cestodes (tapeworms)
N, nematodes
T, trematodes

(*Can involve other body organs)

Figure A–6
Infectious Diseases and Disease Agents Affecting the Digestive System.

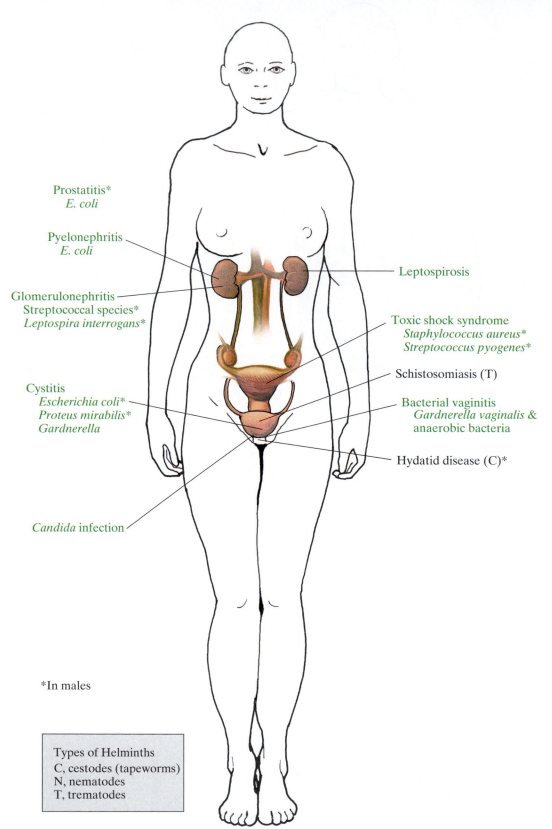

Prostatitis*
E. coli

Pyelonephritis
E. coli

Glomerulonephritis
Streptococcal species*
Leptospira interrogans*

Leptospirosis

Toxic shock syndrome
Staphylococcus aureus*
Streptococcus pyogenes*

Schistosomiasis (T)

Cystitis
Escherichia coli*
Proteus mirabilis*
Gardnerella

Bacterial vaginitis
Gardnerella vaginalis &
anaerobic bacteria

Hydatid disease (C)*

Candida infection

*In males

Types of Helminths
C, cestodes (tapeworms)
N, nematodes
T, trematodes

*Can involve other body organs.

Figure A–7
Infectious Diseases and Disease Agents Affecting the Urinary System.

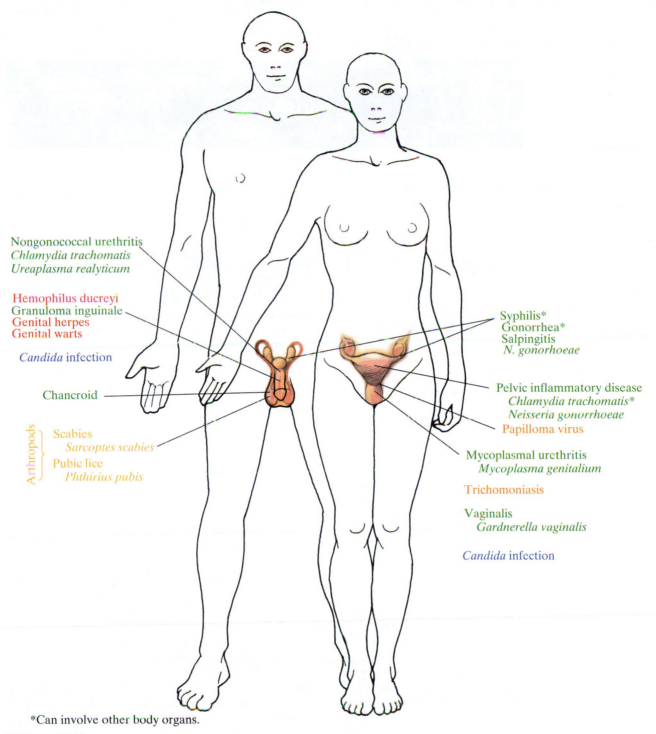

Nongonococcal urethritis
Chlamydia trachomatis
Ureaplasma realyticum

Hemophilus ducreyi
Granuloma inguinale
Genital herpes
Genital warts

Candida infection

Chancroid

Arthropods {
Scabies
   Sarcoptes scabies
Pubic lice
   Phthirius pubis
}

Syphilis*
Gonorrhea*
Salpingitis
   N. gonorhoeae

Pelvic inflammatory disease
   *Chlamydia trachomatis**
   Neisseria gonorrhoeae
Papilloma virus

Mycoplasmal urethritis
   Mycoplasma genitalium

Trichomoniasis

Vaginalis
   Gardnerella vaginalis

Candida infection

*Can involve other body organs.

Figure A–8
Infectious Diseases and Disease Agents Affecting the Reproductive System.

Glossary

abscess (AB-ses): an accumulation of pus in a cavity hollowed out by tissue damage.

aerobe: a cell that is able to use oxygen as a terminal electron acceptor.

acid-fast reaction: a staining reaction in which organisms resist decolorization with acid-alcohol and retain the primary dye; acid-fast bacteria contain large amounts of long-chain fatty acids in their cell walls.

aerial mycelium (er-$\overline{\text{E}}$-al): the portion of a mycelium that develops on surfaces.

aerosol ($\overline{\text{Er}}$-$\overline{\text{o}}$-sol): atomized particles or droplets suspended in air.

aerotolerant anaerobes: bacteria that can survive in the presence of oxygen but do not use it in their metabolism

agar ($\overline{\text{A}}$-gar): a dried polysaccharide extract of red algae used as a solidifying agent in microbiological media; agar generally melts at 99°–100°C, is solid at 42°C, and is attacked enzymatically by very few microbial species.

agglutination(a-gloo-ti-N$\overline{\text{A}}$-shun): the clumping of cells.

agranulocyte(a-GRAN-yoo-lō-sit): a leukocyte lacking cytoplasmic granules and having rounded nuclei.

AIDS: acquired immune deficiency syndrome; a disease caused by human immunodeficiency viruses (HIVs), resulting in the destruction of T_4 (CD_4) or T helper lymphocytes and other components of the immune system.

alga (pl, algae): any member of a group of eukaryotic, photosynthetic, and unicellular or multicellular forms of life; members belong to either the kingdom Plantae or Protista.

alkaline (AL-ka-lin): the condition caused by an abundance of hydroxyl ions (OH⁻), resulting in a pH greater than 7.0.

alpha-hemolysis: partial or incomplete breakdown of hemoglobin by bacteria or other microorganisms growing on a blood agar medium; results in green zones surrounding bacterial colonies.

amastigote: the intracellular form of *Leishmania* species and *Trypanosoma cruzi* found in the human and reservoir hosts; the flagellum does not extend beyond the margin of the protozoan parasite.

amphitrichous (am-F$\overline{\text{E}}$-tri-us): describes a cell that has a single flagellum at each end.

amylase (AM-i-lās): an enzyme that hydrolyzes starch.

anaerobe (an-ER-ob): an organism that cannot grow in an environment containing oxygen (air); a compound other than oxygen is the terminal electron acceptor; O_2 is toxic for the organism.

annulus(AN-ū-lus): a ring-shaped or collar structure found on the stak of some mushrooms.

antibiotic (an-ti-bi-OT-ik): a chemical produced by microorganisms and/or synthesized commercially that can inhibit the growth of or kill other microbes.

antibiotic resistance: the genetic or acquired capability of a microorganism to grow and perform its essential activities in the presence of an antibiotic to which the microorganism is usually sensitive.

antibody (immunoglobulin): a protein produced in response to an antigen that is capable of binding specifically to that antigen.

antigen (AN-ti-jen) (immunogen): a substance that stimulates the immune system to produce specific proteins known as antibodies (immunoglobulins) that react with the antigen and/or activate specific cells of the immune system.

Apicomplexa: intracellular protozoan parasites that have their organelles organized into an apical complex; such organelles are used to penetrate host cells.

Archeae: one of the two domains of prokaryotes which includes the methogens, halophiles (most extreme salt-loving microorganisms), and hyperthermophiles.

arthropod (AR-thrō-pod): an invertebrate with a segmented body and jointed legs, such as a mosquito, tick, or related forms.

arthrospore (AR-thrō-spōr): an asexual spore formed by the fragmentation of specific fungal mycelial filaments; also called an arthroconidium.

ascospore (AS-kō-spōr): a sexual spore, characteristic of the Ascomycetes (Ascomycotina), formed in a saclike structure (*ascus*).

asexual spores: spores formed by a cell, without the fusion of nuclei of two different cells (sexual reproduction).

autotrophy: an organism able to use CO_2 as a sole source of carbon.

B lymphocyte: a cell of the immune system that differentiates into a B memory cell or into the immunoglobulin-producing cell known as the plasma cell.

bacterial colony: a visible accumulation of bacteria on a solid culture medium.

bactericidal: refers to a factor capable of killing bacteria.

bacteriophage (bak-TĒ-rē-ō-fāj): a bacterial virus; also referred to as *phage*.

bacteriostatic: refers to a factor capable of inhibiting bacterial growth without killing.

bacteriuria (bak-tē-rē-yoo-rē-a): the presence of bacteria in urine.

Basidiomycotina (ba-sid-ē-ō-mī-KŌ-tē-na): filamentous fungi that form spores externally on a bulb-shaped cell, the *basidium* (ba-sid-Ē-um).

basidiospore (ba-sid-ē-Ō-spor): a specialized asexual reproduction unit found with filamentous fungi in the Basidiomycotina.

basophil (BĀ-sō-fil): a class of leukocyte that is characterized by being stained with basic dyes producing blue-colored granules and lobed nuclei.

beta (β) hemolysis: a complete breakdown of hemoglobin, resulting in sharply defined clar zones surrounding colonies on blood agar.

beta-lactam: refers to an antibiotics such as penicillin that contain the four-numbered heterocyclic beta-lactam ring.

binomial system of nomenclature: the scientific method of classifying and naming organisms, using a *Genus* and *species* designation.

biological warfare: the use of animals, plants, or microorganisms and/or their products as agents to incapacitate, injure, or kill humans, lower animals, or plants.

blackwater fever: a complication of falciparum malaria consisting of massive hemolysis resulting in bloody urine.

blood agar: a differential and enriched medium used to identify microorganisms that cause hemolysis.

blood group antigens: genetically determined antigens (immunogens) found in the membranes of red blood cells and related tissues.

bloom: a heavy growth of algae or cyanobacteria on a water surface.

bradyzoite: the stage of *Toxoplasma gondii* that develops within tissue cysts.

brood capsule: a cystlike structure that originates from the germinal membrane of hydatid cysts associated with the worm *Echinococcus* species; the brood capsule contains primitive worm scolexes.

bubo (BOO-bō):a swollen lymph node, especially in the groin and armpit; a characteristic lesion in bubonic plague, chancroid, and certain other infectious diseases.

budding: a form of asexual reproduction in which a new cell is formed as an outgrowth from a parent cell; typical of several yeast species.

capsid: the protein shell of a virus particle.

capsule: a polysaccharide or protein-containing structure surrounding and external to the cell walls of certain microorganisms.

catalase: an enzyme that catalyzes the breakdown of hydrogen perioxide (H_2O_2) to water and oxygen.

CD_4 cell: specific lymphocytes also known as helper cells that serve as the cornerstone of the immune response; these cells are among the primary targets for the human immunodeficiency virus (HIV)

cell wall: a structure of unique chemical composition that lies close to and external to the plasma membrane; it confers rigidity and shape to bacteria, algae, and fungi.

cercaria (ser-KA-rē-a):the larval stage of trematodes (flukes) that, after its development, leaves the snail intermediate host and invades a final host.

chancre (SHANG-ker): the initial lesion in syphilis; a hard, painless, nondischarging lesion.

chemically defined media: media for the exact chemical composition is known.

chlamydospore (kla-MID-ō-spōr):a thick-walled, fungal resistant spore formed by the direct differentiation of mycelia cells; also referred to as a chlamydoconidium (kla-MID-ō-kōn-id-ē-um).

chromatodial body: deeply staining bundles of crystalline RNA found in young cysts of the genus *Entamoeba*.

coagulase (kō-AG-yoo-lās): an enzyme produced by pathogenic staphylococci that coagulates human blood plasma.

cocco-bacillus: an oval bacterium that is intermediate between the coccus and bacillus shapes.

coenocytic (SĒ-nō-sit-ik): a condition in which crosswalls are absent, thus allowing cellular material to flow uninterrupted in certain hyphae.

colony: a visible accumulation of microbial growth on the surface of a solid culture medium.

complement fixation: the binding (fixing) of complement to an antigen-antibody complex so that the complement is unavailable for a subsequent reaction.

conidiophore (kon-ID-ē-ō-for): a hypha that holds (bears) condiospores.

conidium (kō-NID-ē-um): an asexual fungal spore that may be one- or multi-celled and of various sizes and shapes; also called conidiospore.

crutose (krus-TŌS): a flat, crustlike growth of lichens.

cyanobacteria(sin-an'-ō-bak-TĒ-rē-a): photosynthesizing prokaryotes that contain chlorophyll and phycocyainin.

cyst (sist): with bacteria, these are spherical, thick-walled resting cells; with protozoa, thick-walled environmentally resistant structures formed for reproduction purposes.

cysticercus (sis-ti-SER-kus): the larval encysted stage of the tapeworms *Taenia solum* and *T. saginata* containing fluid and a single head, or scolex.

cytopathic effect: the visible destructive effects frequently seen with virus-infected tissue culture cells.

dark-field microscopy: a form of microscopy in which organisms appear white against a dark background; spirochetes that stain poorly or not at all with the usual dyes are best observed in the living state by this method.

defined medium: a preparation that contains known specific kinds and concentrations of chemical substances.

definitive (de-FIN-i-tiv) **host**: the final host.

denitrification: the process by which nitrates (NO$_3$) are reduced to nitrous oxide (NO) or nitrogen gas (N$_2$).

dermatophytes (der-ma'-tō-FĪTS): fungi that invade the skin, nails, and hair; ringworm.

Deuteromycotina: a group of fungi in which no sexual stage occurs; also known as the Fungi Imperfecti.

diatoms (DĪ-a-toms): plantlike protists that are characterized by their glasslike outer shell.

differential medium: a culture medium capable of distinguishing bacterial species from one another by differences in colonial appearances or by changes produced in the medium.

differential stains: a procedure that uses more than one stain to distinguish parts of a cell from one another.

dimorphic: having two morphological forms, one displayed in the tissue of a host, the other in nature and commonly in cultures in the laboratory.

Domain: the highest biological classification level. The three domains currently recognized are the **Bacteria**, **Archeae**, and **Eukarya**.

ectoparasite: a form of life that lives on the surface of another organism.

ectothrix: a sheath of fungal spores on the outside of a hair shaft.

electrophoresis (e-lek-trō-fōr-Ē-sis): a laboratory technique used to separate proteins and certain other large molecules by passing an electrical current through a specimen on a gel.

elephantiasis: an enlargement of the legs and/or scrotum in chronic lymphatic filariasis, a nematode infection.

ELISA (enzyme-linked immunoabsorbent assay): a highly specific immunologic diagnostic procedure that involves the linking of soluble antigens or antibodies to an insoluble solid surface to retain the reactivity of the antigen or antibody.

endothrix: a cluster of arthroconidia (arthrospores) within a hair shaft.

euglenoid (yoo-glen-oid): an alga containing chlorophyll *a* and *b* and belonging to the taxonomic division Euglenophyta.

Eukarya: the phylogenetic domain consisting of all eukaryotic organisms, animals, plants, protozoa, algae, and fungi.

eukaryote: a cell with a well-defined nucleus surrounded by a nuclear membranous envelope and having other membrane-bound organelles such as mitochondria.

exospore: a spore resistant to heat and drying that is formed external to the vegetative cell by budding.

facultative: a term used to indicate an organism's ability to grow in either the presence or absence of an environmental factor; an adjustable organism.

facultative anaerobe: an organism that grows well under both aerobic and anaerobic conditions and for which oxygen is not toxic.

fermentation: the metabolic process in which the final electron acceptor is an organic compound.

fluorescence (floo-RES-ents) **microscopy**: a type of microscopy in which cells, their parts, or related structures are stained with a fluorescent dye and when exposed to ultraviolet light appear as glowing objects.

fluke (trematode): with the exception of the schistosomes, flukes are flat, leaf-shaped worms having both sexes (hermaphroditic) and oral and ventral suckers.

foliose: leaflike.

fruticose (FRU-ti-cōs): shrublike.

fruiting body: a macroscopic reproductive structure formed by some bacteria and fungi.

gametocyte: either the male or female form of the malaria and related parasites (sex cell).

GC (G+C) ratio: refers to the percentage of total nucleic acid that consists of the nitrogenous base guanine plus the base cytosine in an organism's DNA or RNA.

gamma hemolysis: the absence of an enzymatic breakdown of hemoglobin; red blood cells in a blood agar medium remain intact.

gelatinase: the exoenzyme (extracellular) that degrades the protein gelatin.

Gram stain: the differential staining procedure developed by C. Gram to distinguish two groups of bacterial cells from one another, gram-positive and gram-negative.

granulocytes: one group of white blood cells having differently staining granules within their respective cytoplasms; neutrophils, eosinophils, and basophils.

gumma: granulomatous lesion found in the skin, bone, and liver during the tertiary stage of syphilis.

helminth: a worm.

hemolysis (hē-MOL-i-sis): disruption of red blood cells.

hermaphroditism (her-MAF-rō-dit-izm): the presence of both female and male organs in the same individual.

hydatid cyst: the larval stage of the sheep tapeworm *Echinococcus granulosus* consisting of a fluid-filled cystic (bladderlike) structure.

hypha (HĪ-fa): the structural vegetative unit of a fungal mycelium.

icosahedron: a geometrical figure consisting of 20 triangular surfaces.

indicator system: one of the components of the complement fixation test; consists of hemolysin (antisheep red blood cells) and sheep red blood cells.

immunogen (im-MYOO-nō-jen): a substance that stimulates the formation of immunoglobulins (antibodies).

immunoglobulin (im-mu-nō-GLOB-yoo-lin): refers to protein molecules produced in response to immunogens (antigens).

intermediate (in-ter-MĒ-dē-at): a form of life used for development of larval stages of helminths.

in vitro($V\bar{E}$-tr$\bar{o}$): refers to procedures and/or tests performed outside of the body, usually in test tubes, microtiter plates, etc.

larva (lar-VAH): developing stage of an insect or a worm.

lesion: an area of injury or a circumscribed pathological tissue change.

lichen: a plantlike form of life consisting of a symbiotic relationship between a fungus and a photosynthetic alga or cyanobacterium.

litmus: plant extract used as a pH indicator; also an oxidation-reduction indicator; turns blue when alkaline and red when acid in reaction.

lymphocyte: a type of leukocyte arising in lymphoid tissues and the most important cell in specific immunity.

lysis: disruption, or breaking apart, of cells.

macroconidium: the larger of two types of conidia produced by a single species of fungus.

macrogametocyte: the female form of the malaria and related parasites usually found within red blood cells.

magnetosomes: small iron-containing particles present in cells that exhibit movement directed by such particles toward a magnetic field.

medium: a nutrient preparation used to grow microorganisms; it may be a liquid or a solid.

merozoite: the form of the malaria parasite that results from asexual multiplication within the RBC; the form that is released from the mature schizont to infect other RBCs.

metabolism: the total chemical activities of an organism; consists of anabolism and catabolism.

metacercaria: an encysted larval trematode stage; it is found in second intermediate hosts and is infectious for the final (definitive) host.

metachromatic (met-a-kr$\bar{o}$-MAT-ik) **granules:** a reservoir of inorganic phosphate within a bacterium that is stainable by basic dyes.

microaerophilic (m$\bar{i}$-kr$\bar{o}$-AR-$\bar{o}$-fil-ik): refers to aerobes that require environments with small amounts of oxygen or less than atmospheric oxygen levels for growth.

microconidium: the smaller of two types of conidia produced by a single species of fungus.

microfilaria: the embryonic roundworm produced by adult female filarial parasites, including *Onchocerca volvulus*.

microgametocyte: the male form of malaria and related parasites within the RBC.

micrometer: unit of measurement for microorganisms and other forms of life; 1 micrometer (μm) equals 0.001 mm or 1/25,400 in. (*micrometer* replaces the older term *micron*.)

microsporidia: protozoa that are characterized by the production of spores containing coiled polar tubes.

miracidium (meh-ra-SID-$\bar{e}$-um): a ciliated free-swimming development form of a fluke that hatches from a helminth egg.

mitosis: nuclear division that follows duplication of chromosomes and results in daughter nuclei with chromosomes identical to the parent cell.

molds: filamentous fungi.

monoclonal antibodies: antibodies of a specific type produced by cells arising from a single clone of antibody-producing cells.

monocyte: actively phagocytic, monoculear white cell found in the bloodstream; as monocytes mature, they emigrate into tissues and differentiate into macrophages.

monotrichous: having a single flagellum at one end of the cell.

mordant: a substance that fixes (precipitates) a stain; e.g., iodine in the Gram stain.

negative stain: a staining procedure in which the background surrounding a specimen is stained, but the specimen is not.

nematode (NEM-a-t$\bar{o}$d): a roundworm.

nitrate reduction: reduction of nitrates to nitrites or ammonia.

nitrification (n$\bar{i}$-tri-fi-K$\bar{A}$-shun): the conversion of ammonia to nitrate.

nonseptate (non-SEP-t$\bar{a}$t): having no dividing walls (septa) in hyphae (filaments).

nosocomial (n$\bar{o}$s-$\bar{o}$-K$\bar{O}$-m$\bar{e}$-al) **infection:** an infection acquired in a hospital or other health care facility.

nuceloid: the aggregated mass of deoxyribonucleic acid that comprises the chromosome of prokaryotic cells.

nucleic acid probe: a procedure that uses a nucleic acid strand that can be labeled (tagged) and used to hybridize (combine) to a complementary molecule from a mixture of other nucleic acid strands; the technique can be used to identify and/or to show relationships among microbes.

nuceloside: a molecule consisting of a purine or pyridine and a 5-carbon sugar such as ribose or deoxyribose.

nucleotide: a basic unit of nucleic acids consisting of a purine or pyridine, a 5-carbon sugar such as ribose or deoxyribose, and a phosphate unit.

obligate aerobes: organisms that must use oxygen as their final electron acceptor.

obligate anaerobes: organisms unable to grow in the presence of oxygen.

oncosphere: the embryo in the eggs of tapeworms; it has six hooks.

oocyst: a stage of development in the life cycle of apicomplexans. The mature oocyst contains infectious sporozoites; it is the diagnostic form found in feces in infections with *Cryptosporidium, Cyclospora,* and *Isospora.*

ookinete: a motile zygote formed by sporozoans such as the malarial parasite (*Plasmodium* species).

operculum: the lidlike portion of the eggshell of most trematodes and the fish tapeworm *Diphyllobothrium latum;* the larva hatches from the egg through the opening made by detachment of the operculum.

opportunistic infection: infectious disease caused by a microorganism without major virulence factors that takes advantage of a host's immunosuppression.

ovum ($\bar{O}$-vum): the female reproductive cell; egg.

oxidase test: a test to distinguish colonies of *Neisseria* from other bacteria; does not differentiate among *Neisseria* species.

oxidation: a process by which a compound gives up electrons (or hydrogen atoms) and thereby becomes oxidized.

pilus: a submicroscopic structure found mainly with prokaryotes; several types are known. They include the sex pili used for the transfer of genetic material from one prokaryote to another, and others which are used for attachment.

plaque (viral): a clear area in the confluent growth of a bacterial or cell culture due to lysis by a phage or virus.

pock: a local blisterlike formation caused by virus infection of the chorioallantoic membrane.

pour (PŌR) **plate**: a basic technique used in the culturing and/or isolation of bacteria in which melted, yet sufficiently cooled, medium is inoculated with a bacterial culture, introduced into a sterile Petri dish, and allowed to harden; the individual bacteria trapped within the medium grow and eventually form colonies.

precipitin (prē-SIP-ē-tin) **reaction**: the basis of an immunologic reaction in which antigens and antibodies diffuse toward one another, resulting in a precipitate in which these reactants are in equivalent proportions.

prion: currently the smallest infectious agent, consisting of only protein.

proglottid: an individual segment of a tapeworm; a mature proglottid contains both male and female sex organs, and a gravid proglottid contains eggs (ova).

prokaryote: a cell lacking a membrane-bound nucleus and membrane-bound organelles.

promastigote: the form of *Leishmania* species found in the sand fly vector and in culture; it is elongate and has a prominent anterior flagellum.

Rh (factor) blood group: a blood group agglutinogen discovered on the red blood cells of Rhesus monkeys and given the designation *Rh*; the factors are found to a variable degree in human blood cells.

rhizoids: rootlike structures characteristic of some molds.

schizont: a stage in the prerythrocytic reproductive cycle of the malaria parasite.

scolex (SKŌ-leks): the head of a tapeworm.

selective and differential medium: a preparation that incorporates the features of both selective and differential media.

serology: systematic study of blood serum; e.g., reactions between antibodies and antigens.

spore print: a technique used to determine the properties of spores and lamellae.

sporocyst (SPŌR-ō-sist): a developmental stage of a fluke formed from a miracidium containing reproductive cells and found in snail (intermediate host) tissues.

sporozoite: form of malaria parasite injected into the bloodstream by an infected mosquito.

streak plate: basic technique used in the culturing and/or isolation of bacteria in which an inoculum is spread over the surface of a medium by means of an inoculating loop or needle; the isolated bacteria in the inoculum grow and form colonies on the surface of the medium.

substrate: substance acted on by an enzyme.

T cell: a specific type of lymphocyte responsible for specific immunological reactions.

tachyzoite: the rapidly dividing, crescent-shaped form of *Toxoplasma gondii* and related parasites.

tapeworm: a ribbonlike helminth (cestode) that has an attachment organ called a scolex and developing segments called proglottids.

tinea: the medical term for the ringworms.

trophozoite: the motile, feeding form of protozoa.

trypomastigote: also called trypanosome; this form of the genus *Trypanosoma* is found in the human host; the organism is elongate and has an undulating membrane that runs the length of the body and extends as a flagellum from the anterior end.

vector: any agent that carries a disease from one host to another; may be animate, as insects, or may be inanimate (nonliving), as soil.

Voges-Proskauer reaction: test for acetylmethylcarbinol (also called acetoin) production; used to differentiate *Escherichia* and *Enterobacter* of the coliform group.

Ziehl-Neelsen (zēl-NĒL-sen) **method**: a differential staining procedure used to distinguish species of *Mycobacteria* and *Nocardia* and certain pathogenic protozoa; these microorganisms are acid-fast.

zoonosis (zō-NŌ-sis): a disease communicable from lower animals to humans under natural conditions.

zygospore: sexual spore that results from the fusion of like gametes; characteristic of Zygomycetes (Zygomycotina).

Index

Streptococcus, 86, 88*f,* 90
  group B, 90
  identification tests, 89*f*
Streptococcus pneumoniae, 78, 88*f,* 90
  group B, 79
Streptomyces azureus, 158*f*
Strongyloides stercoralis, 186, 189, 190*f*
Subcutaneous mycoses, 124, 128
  agents of systemic, 124
  Blastomyces dermatitidis, 124
  Chromoblastomycosis, 124
  Coccidiodes immitis, 124, 128
  Histoplasma capsulatum, 128
  sporotrichosis, 124
Substrates, 30
Succinic acid excretion, 36
Suckers, 184
Sulfanilic acid, 38
Sulfide, indole, motility medium, 44
Superficial mycoses, 120, 124
Superoxide dismutase (SD), 40
Suppressor (Ts) cells, 200
Surface replicas, 3*t*
Surface replica technique, 3*t*
Symbiotic nitrogen fixation, 49, 50*f*
Symbiotic relationship, 144
Synthetic media, 23
Syphilis
  early stages of, 104*f*
  latent, 103
  primary, 101
  secondary, 101
  tertiary, 103

T

Tachyzoites, 155
Taenia saginata, 184
Taenia solium, 184, 185*f*
Tapeworm
  fish, 184
  pork, 184
Taxa, 3
Tegument, 175
Temporary wet mount, 115–16, 117*f*
Tentacle, 144*t*
Teritiary stage syphilis, 103
Tetrads, 16, 17*f*
Thalassionema, 141*f*
Thallassiosira, 141*f*
Theileria parva, 156*f*
Thin sectioning technique, 3*t,* 159
Thrush, 132
Thymus gland, 202
Ticks, 196*f*
Tinea barbae, 128*t*
Tinea capitis, 128*t*
Tinea coporis, 123*f*
Tinea corporis, 128*t*
Tinea cruris, 128*t*

Tinea pedis, 128*t*
Tinea unguium, 128*t*
Titer, 204
T lymphocytes, 200, 201
  Th1, 200
  Th2, 200
Toxoplasma gondii, 152, 154*f,* 155
Toxoplasmosis, 154*f*
Transmission electron microscope (TEM),
    1, 5*t,* 7*f,* 10*f,* 181*f*
  bacterial specialized parts, 23*f*
  techniques, 4*t*
T-regulatory cells, 200
Trematodes, 189, 191, 194
  Clonorchis (Opisthorchis) sinensis, 189
  Fasciola hepatica, 191
  Fasciolopsis buski, 191
  features of, 182*t*
  life cycle of, 191*f*
  Paragonimus westermani, 191, 194
  Schistosoma, 194
Treponema pallidum, 101, 103, 104*f*
Trichinella spiralis, 189, 190*f*
Trichocyst, 144*t*
Trichome, 106
Trichomonas vaginalis, 149, 150*f*
Trichophyton rubrum, 126*f,* 127*f*
Trichophyton tonorans, 126*f*
Trichuris trichiura, 189, 190*f*
2,3,5-triphenyltetrazolium chloride (TTC),
    40
Triple sugar iron agar (TSIA), 43*f,* 44*t*
Trophozoites, 145
Trypanosoma brucei gambiense, 149, 150*f*
Trypanosoma brucei rhodesiense, 149
Trypanosoma cruzi, 149, 151*f*
Tuberculate macroconidium, 128
Turbidity, 27
Typhus fever, 51*t*
Tzanck test, 176*f*

U

Ulcer, 79*f*
Ultrasonography, 11
Ulva lactuca, 141*f*
Undulating membrane, 144*t*
United States Public Health System, 165
Ureaplasma urealyticum, 90–91
Urease, 37*t,* 38*f*
Urinary system, infectious diseases of, 218*f*

V

Vaccinia virus, 180
Valves, 140
Varicella-zoster virus, 175, 178*f*
Variola major, 177
Variola minor, 177
Veils, mushroom, 116

Vesicle, 165
Vibrio cholerae, 73, 74*f,* 75
Vibrios, 18*f*
Virions, 157, 168, 169, 181*f*
Viroids, 157
  effects of, 160*f*
  properties of, 157*t*
Virology, 1, 157–81
  classification, 165
  cultivation, 162, 165
  diagnosis, 165
  DNA viruses, 172, 174–75, 177, 180
  electron microscopy, history of, 159
  infections with rash, features of, 165, 167
  RNA viruses, 168–72
  viruses and effects, 157–59
  virus structure, 159, 162
Viruses, 9
  as biological weapons, 165, 167
  diagnosis, 165
  DNA, 172, 174–75, 177, 180
  effects of, 157–59
  infections with rash, features of, 165,
    167
  influenza, 161*f*
  plant, 158*f*
  properties of, 157*t*
  replication, 163*f*
  RNA, 168–72
  structure of, 159, 162
Voges-Proskauer test, 36, 37*f,* 37*t,* 41, 75
Volvox, 141*f*
Vorticella, 151*f*
Vulvovaginitis, 132

W

Wandering tissue macrophage, 201
Wangiella dermatitidis, 127*f*
Western blot, 205*t,* 208*f*
Whey, 42
White piedra, 120
Whooping cough, 55
Worm burden, 186
Worms, 183*f*
  hermaphroditic, 184
Wuchereria bancrofti, 189, 190*f*

Y

Yaws, nasal destruction in, 104*f*
Yeasts, 110, 120
Yersinia enterocolitica, 75, 76*f*
Yersinia pestis, 75, 76*f*
Yersinia pseudotuberculosis, 75

Z

Zygomycota, 114*t*
Zygote, 114